Springer Complexity

Springer Complexity is an interdisciplinary program publishing the best research and academic-level teaching on both fundamental and applied aspects of complex systems—cutting across all traditional disciplines of the natural and life sciences, engineering, economics, medicine, neuroscience, social and computer science.

Complex Systems are systems that comprise many interacting parts with the ability to generate a new quality of macroscopic collective behavior the manifestations of which are the spontaneous formation of distinctive temporal, spatial or functional structures. Models of such systems can be successfully mapped onto quite diverse "real-life" situations like the climate, the coherent emission of light from lasers, chemical reaction-diffusion systems, biological cellular networks, the dynamics of stock markets and of the internet, earthquake statistics and prediction, freeway traffic, the human brain, or the formation of opinions in social systems, to name just some of the popular applications.

Although their scope and methodologies overlap somewhat, one can distinguish the following main concepts and tools: self-organization, nonlinear dynamics, synergetics, turbulence, dynamical systems, catastrophes, instabilities, stochastic processes, chaos, graphs and networks, cellular automata, adaptive systems, genetic algorithms and computational intelligence.

The three major book publication platforms of the Springer Complexity program are the monograph series "Understanding Complex Systems" focusing on the various applications of complexity, the "Springer Series in Synergetics", which is devoted to the quantitative theoretical and methodological foundations, and the "Springer Briefs in Complexity" which are concise and topical working reports, case studies, surveys, essays and lecture notes of relevance to the field. In addition to the books in these two core series, the program also incorporates individual titles ranging from textbooks to major reference works.

Understanding Complex Systems

Founding Editor: S. Kelso

Future scientific and technological developments in many fields will necessarily depend upon coming to grips with complex systems. Such systems are complex in both their composition–typically many different kinds of components interacting simultaneously and nonlinearly with each other and their environments on multiple levels–and in the rich diversity of behavior of which they are capable.

The Springer Series in Understanding Complex Systems series (UCS) promotes new strategies and paradigms for understanding and realizing applications of complex systems research in a wide variety of fields and endeavors. UCS is explicitly transdisciplinary. It has three main goals: First, to elaborate the concepts, methods and tools of complex systems at all levels of description and in all scientific fields, especially newly emerging areas within the life, social, behavioral, economic, neuro- and cognitive sciences (and derivatives thereof); second, to encourage novel applications of these ideas in various fields of engineering and computation such as robotics, nano-technology, and informatics; third, to provide a single forum within which commonalities and differences in the workings of complex systems may be discerned, hence leading to deeper insight and understanding.

UCS will publish monographs, lecture notes, and selected edited contributions aimed at communicating new findings to a large multidisciplinary audience.

More information about this series at http://www.springer.com/series/5394

M. Reza Rahimi Tabar

Analysis and Data-Based Reconstruction of Complex Nonlinear Dynamical Systems

Using the Methods of Stochastic Processes

Springer

M. Reza Rahimi Tabar
Department of Physics
Sharif University of Technology
Tehran, Iran

ISSN 1860-0832 ISSN 1860-0840 (electronic)
Understanding Complex Systems
ISBN 978-3-030-18471-1 ISBN 978-3-030-18472-8 (eBook)
https://doi.org/10.1007/978-3-030-18472-8

This Springer imprint is published by the registered company Springer Nature Switzerland AG
The registered company address is: Gewerbestrasse 11, 6330 Cham, Switzerland

In memory of Professor Rudolf Friedrich (1956–2012) whose original and innovative thinking with Professor Joachim Peinke on the methods that we describe in this book motivated a large body of work by others on the subject.

Preface

The data analysis of physical observables has a long tradition in the field of nonlinear dynamics and complex systems. Much effort has been devoted to answering the question of how to extract a "deterministic" dynamical system from a suitable analysis of experimental data, given that an appropriate analysis can yield important information on dynamical properties of the system under consideration. Fluctuations in these time series are usually considered as a purely random or uncorrelated variable, which is additively superimposed on a trajectory generated by a deterministic dynamical system. The problem of dynamical noise, i.e., fluctuations that interfere with the dynamical evolution, has not been addressed in much detail, although it is of utmost importance for the analysis of fluctuating time series.

This book focuses on a central question in the field of complex systems: *Given a fluctuating (in time or space), uni- or multi-variant sequentially measured set of experimental data (even noisy data), how should one analyse non-parametrically the data, assess underlying trends, uncover characteristics of the fluctuations (including diffusion and jump contributions), and construct a stochastic evolution equation?* Here, the term "non-parametrically" exemplifies that all the functions and parameters of the constructed stochastic evolution equation can be determined directly from the measured data.

In recent years, significant progress has been made when addressing this question to the classes of continuous stochastic processes and of processes with jump discontinuities. These can be modeled by nonlinear generalized Langevin equations that include additive as well as multiplicative diffusive and even jump parts. An important building block for the analysis approach presented in this book is a Markovian property, which can be detected in real systems above a certain time or length scale. This scale is referred to as the *Markov–Einstein* scale, and has turned out to be an important characteristic of complex time series. The *Markov–Einstein* time scale is the minimum scale above which the data can be considered as a Markov process, and one can estimate it directly from observations. The main advantage of the analysis approach is that is *completely data-based* and thus allows one to find all functions and parameters of the modeling directly from measured data. Due to its feasibility and simplicity, it has been successfully applied to fluctuating time series and spatially disordered structures of complex systems studied in scientific fields such as physics, astrophysics, meteorology, earth science, engineering, finance, medicine, and the neurosciences, and has led to many important results.

This book provides an overview of methods that have been developed for the analysis of fluctuating time series and of spatially disordered structures.

The book also offers numerical and analytical approaches to the analysis of complex time series that are most common in the physical and natural sciences. It is self-contained and readily accessible to students, scientists, and researchers who are familiar with traditional methods of mathematics, such as ordinary, and partial differential equations. Codes for analysing *continuous* time series are available in an R package developed under the supervision of Joachim Peinke by the research group Turbulence, Wind energy, and Stochastics (TWiSt) at the Carl von Ossietzky University of Oldenburg. This package allows one to extract the (stochastic) evolution equation underlying a set of data or measurements.

The book is divided into three main parts: I (Chaps. 1–9), II (Chaps. 10–21), and III (Chaps 22–23).

Chapter 1 provides an introduction and an overview of topics covered in this book. Chapter 2 reviews essentials of stochastic processes, namely the statistical description of stochastic processes, stationary processes, classification of stochastic processes, the Chapman–Kolmogorov equation as a necessary condition for Markov processes, statistical continuous processes, as well as stochastic processes in the presence of jump discontinuities. In Chap. 3, we present details of the Kramers–Moyal expansion, the Pawula theorem, the Fokker–Planck equation and its short-term propagator, and derive the master equation from the Chapman–Kolmogorov equation. In Chap. 4, we provide Lindeberg's condition for the continuity of stochastic trajectories. It is shown that the Fokker–Planck equation describes a continuous stochastic process. We derive the stationary solutions of the Fokker–Planck equation and define a potential function for dynamics. In Chap. 5, we introduce the Langevin equation and Wiener processes along with their statistical properties. Chapter 6 reviews the Itô and the Stratonovich calculus. We prove Itô's lemma and describe the Itô calculus for multiplicative noise. In Chap. 7, we show the equivalence between the Langevin approach and the Fokker–Planck equation and derive equations for statistical moments of a process whose dynamics is given by the Langevin equation. In Chap. 8, we provide examples for stochastic calculus using the Kubo–Anderson process, the Ornstein–Uhlenbeck process, and the Black–Scholes process (or geometric Brownian motion). Chapter 9 covers the following topics: Langevin dynamics in higher dimension, the Fokker–Planck equation in higher dimension, finite-time propagators of a d-dimensional Fokker–Planck equation, as well as discrete time evolution and discrete time approximation of stochastic evolution equations. Chapters 1–9 can be skipped by readers who are familiar with the standard notions of stochastic processes, or they may be useful for examples and applications.

In Chap. 10, we introduce the Lévy noise-driven Langevin equation and the fractional Fokker–Planck equations, derive the short-time propagator of Lévy noise-driven processes, and provide limit theorems for Wiener and Lévy processes. Finally, a non-parametric determination of Lévy noise-driven Langevin dynamics from time series will be described. In Chap. 11, we study stochastic processes with jump discontinuities and discuss the meaning of nonvanishing higher-order Kramers–Moyal coefficients. We address in detail the physical meaning of non-vanishing fourth-order Kramers–Moyal coefficients, stochastic processes with jumps, as well as stochastic properties and statistical moments of Poisson jump processes. In Chap. 12, we introduce the jump-diffusion processes with Gaussian and mixed-Gaussian jumps. In Chap. 13, we introduce bi-variant jump-diffusion equations and in Chap. 14, we describe different numerical schemes for the integration of Langevin and jump-diffusion stochastic differential equations, such as the Euler–Maruyama scheme, the Milstein scheme, and Runge-Kutta-like methods. This chapter closes with an introduction of packages in R and Python for the numerical integration of stochastic differential equations. In Chap. 15, we discuss the analysis of spatially disordered structures and provide a physical picture of the fluctuation cascade from large to small scales. Moreover, this section introduces the multipliers in cascade processes, and we derive a scale-dependent solution of the Fokker–Planck equation and present the Castaing equation.

An answer to the question of how to set up stochastic equations for real-world processes is presented in Chaps. 16–21. In Chap 16, the reader is familiarized with the methods for estimating the Kramers–Moyal coefficients, and we introduce the Markov–Einstein time (length) scale of a data set. This chapter also contains important technical aspects of the method for estimating drift and diffusion coefficients as well as higher-order Kramers–Moyal coefficients from time series. In Chap. 17, we explain how to derive the Kramers–Moyal coefficients from non-stationary time series using the Nadaraya–Watson estimator and we investigate Kramers–Moyal coefficients in the presence of microstructure (measurement) noise. In Chap. 18, we study the influence of a finite time step on the estimation of the Kramers–Moyal coefficients from diffusive and jumpy data. In Chap. 19, we analytically derive a criterion (as a necessary condition) that allows one to check whether for a given, even noisy time series the underlying process has a continuous (diffusive) or a discontinuous (jumpy) trajectory. In Chap. 20, the steps of deriving a Langevin equation from diffusive experimental time series are given, and we finish the chapter with an introduction of an R Package for the modeling of one- and two-dimensional continuous stochastic processes. In addition, the steps for deriving a jump-diffusion stochastic equation from experimental time series with jumps are presented. Also, two other methods for a reconstruction of time series will be reviewed shortly. In Chap. 21 we reconstruct, as examples, some stochastic dynamical equations from various synthetic continuous time series, from time series with jump discontinuities and from time series generated by Lévy noise-driven Langevin dynamics.

Chapter 22 briefly reviews applications of the presented method (Chaps. 16–21) to the analysis of real-world time series and ends with an outlook. As an example of the analysis methods to real-world time series, we present in Chap. 23 results derived from analyses of electroencephalographic time series.

I would like to thank N. Abedpour, M. Anvari, A. Barhraminasab, D. Bastine, F. Böttcher, J. Davoudi, F. Ghasemi, J. Gottschall, Z. Fakhraai, S. M. Fazeli, J. Friedrich, T. Jadidi, G. R. Jafari, A. Hadjihosseini, N. Hamedai-Raja, A. M. Hashtroud, H. Heiβelmann, J. Heysel, M. Hölling, C. Honisch, O. Kamps, D. Kleinhans, M. Kohandel, P. G. Lind, G. Lohmann, St. Lück, P. Manshour, P. Milan, E. Mirzahossein, S. Moghimi, S. M. Mousavi, S. M. S. Movahed, I. Neunaber, M. D. Niry, J. Puczylowski, N. Reinke, Ch. Renner, P. Rinn, V. Rezvani, F. Shahbazi, A. Sheikhan, M. Siefert, S. Siegert, F. T. Shahri, S. M. Vaez Allaei, M. Wächter, L. Zabawa, and F. Zarei for useful discussions and whose Ph.D. theses have contributed to this book. I am also thankful to Daniel Nickelsen and Adrian Baule for sharing their ideas presented in Sects. 22.2.3 and 22.2.5.

Special acknowledgments should be given to Uriel Frisch (Observatoire de la Côte d'Azur, Nice), Joachim Peinke (Carl-von-Ossietzky University Oldenburg), Muhammad Sahimi (University of Southern California), and Holger Kantz (Max–Planck Institute for Physics of Complex Systems, Dresden). I would also like to thank the Alexander von Humboldt Foundation for financial support and Klaus Lehnertz (Department of Epileptology, University of Bonn) for many detailed discussions, the kind hospitality, and proofreading of Chap. 23.

It is greatly appreciated if readers could forward any errors, misprints, or suggested improvements to: tabar@uni-oldenburg.de or rahimitabar@sharif.edu.

Oldenburg, Bonn M. Reza Rahimi Tabar
2019

The original version of the book was revised: Belated corrections have been incorporated. The corrections to the book is available at https://doi.org/10.1007/978-3-030-18472-8_24

Contents

Chapter 1
Introduction

1.1 Time Series of Complex Systems

Complex systems are composed of a large number of subsystems that may interact with each other. The typically nonlinear and multiscale interactions often lead to large-scale behaviors, which are not easily predicted from the knowledge of only the behavior of individual subsystems. Such large-scale collective emergent behaviors may be desired or undesired. Examples of undesired emergent behaviors include short- and long-term climate changes, hurricanes, cascading failures, and epileptic seizures. Among the desired ones are evolution, adaptation, learning, and intelligence, to name just a few [1–11].

In complex systems the fluctuations stemming from the microscopic degrees of freedom play a fundamental role in introducing temporal variations on a fast time scale that, quite often, can be treated as short-time correlated source of fluctuations. In such systems, self-organized behaviour arises due to the emergence of collective variables, or order parameters that, compared to the time or length scales of the microscopic subsystems, vary on slower temporal and larger spatial scales. The interaction of the order parameters are typically nonlinear, with the microscopic degrees of freedom showing up in the fluctuations that participate in the order parameter dynamics and results in complex time series; see Fig. 1.1. Thus, the analysis of the behaviour of complex systems must be based on the assessment of the nonlinear mutual interactions, as well as the determination of the characteristics and strength of the fluctuating forces. This immediately leads to the problem of retrieving a stochastic dynamical system from data; see Fig. 1.2 for typical stochastic time series.

Analysis of time series has a long history in the field of nonlinear dynamics [12–16]. The problem of dynamical noise, i.e., fluctuations that interfere with the dynamical evolution, has not been addressed in much details, although it is of utmost importance for the analysis of strongly fluctuating time series [17, 18]. In this book, we provide detailed description and discussion of a non-parametric method, known as the *reconstruction method*, which has been developed for analyzing *continuous* stochastic processes and stochastic data with jumps in time and/or length scales.

M. R. Rahimi Tabar, *Analysis and Data-Based Reconstruction of Complex Nonlinear Dynamical Systems*, Understanding Complex Systems,
https://doi.org/10.1007/978-3-030-18472-8_1

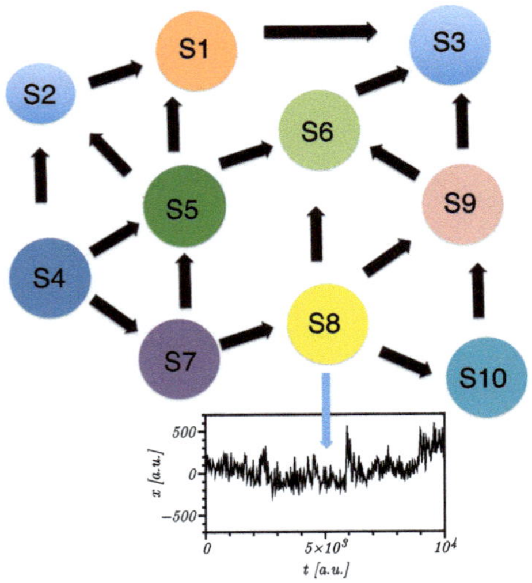

Fig. 1.1 Complex systems are composed of a large number of subsystems behaving in a "collective manner". In such systems, which are far from equilibrium, "collectivity" arises due to self-organization. It results in the formation of temporal, spatial, spatio-temporal and functional structures. The state of the subsystems change over time and results in stochastic dynamics (as shown for subsystem 8). The dynamics of order parameters in complex systems are generally non-stationary, and the subsystems interact with each other in nonlinear manner. The arrows indicate causal directions of influence

Fig. 1.2 Segments of intracranial electroencephalographic (iEEG) time series, recorded during a seizure-free interval from within the epileptic focus (red) and from a distant brain region (black). *Source* from [21]

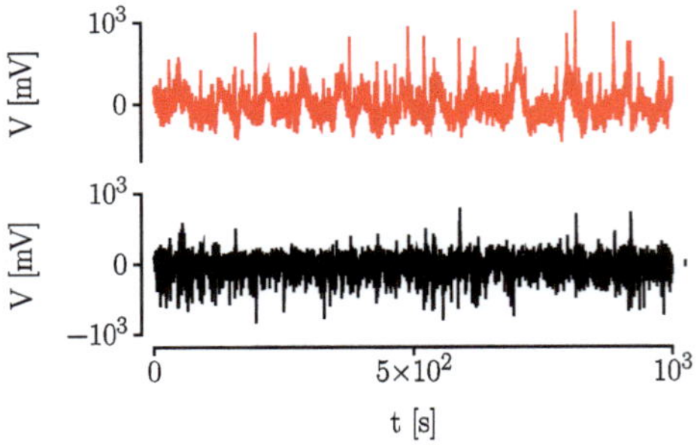

The development of such methods has been stimulated by research on turbulent flows and neuroscience [19–21], which has demonstrated the necessity of treating the fluctuations as dynamical variables that interfere with the deterministic dynamics.

1.2 Stochastic Continuous Time Series

Systems under the influence of random forcing, or in the presence of nonlinear interactions with other systems, can behave in a very complex stochastic manner [17, 18, 22, 23]. The corresponding time series of such systems generally have continuous trajectories, or may possess jump discontinuities. To decipher the problem of retrieving a stochastic dynamical system from data, the main assumption is that the measured time series represents a *Markov process*. A stochastic process with a degree of stochasticity may have a finite Markov-Einstein (ME) time scale t_M—the minimum time interval over which the data can be considered as a Markov process. Therefore one should first estimate t_M for the measured time series, after which one can apply the method described in this book to reconstruct the corresponding dynamical stochastic equation. We will introduce various methods to estimate t_M in Chap. 16.

A process $x(t)$ has continuous trajectories if the following relations for conditional averaging, known as the Kramers–Moyal (KM) conditional moments, hold for infinitesimal dt:

$$\begin{aligned}
\langle (x(t+dt) - x(t))^1 |_{x(t)=x} \rangle &= a(x,t)\, dt \\
\langle (x(t+dt) - x(t))^2 |_{x(t)=x} \rangle &= b^2(x,t)\, dt \\
\langle (x(t+dt) - x(t))^{2+s} |_{x(t)=x} \rangle &= \mathcal{O}(dt^{1+\chi})
\end{aligned} \tag{1.1}$$

with $s > 0$ and $\chi > 0$. The KM coefficients are defined as $D^{(j)}(x) = \lim_{dt\to 0} \frac{1}{j!}\frac{1}{dt}$ $K^{(j)}(x)$, where the conditional moments $K^{(j)}(x) = \langle [x(t+dt) - x(t)]^j |_{x(t)=x} \rangle$, and can be determined non-parametrically, i.e., directly form the measured time series [17]. Throughout this book we will use two definitions for the KM coefficients, $D^{(j)}(x)$ and $M^{(j)}(x)$, which are related to each other by, $D^{(j)}(x) = 1/j!\, M^{(j)}(x)$.

The dynamics of continuous stochastic processes is governed by the Langevin dynamics that has the following expression [24–26],

$$dx(t) = a(x,t)\, dt + b(x,t)\, dW(t), \tag{1.2}$$

where $\{W(t), t \geq 0\}$ is a scalar Wiener (Brownian) motion, and the functions $a(x,t)$ and $b^2(x,t)/2$ are known as drift and diffusion coefficients, respectively. In terms of the conditional probability distribution, a continuous Markov process $x(t)$ satisfies the following continuity condition, given some $\delta > 0$ [25],

$$C = \lim_{dt\to 0} \frac{Prob[\ |\Delta x(t)| > \delta \quad |_{x(t)=x}]}{dt} = 0 \tag{1.3}$$

where $\Delta x(t) = x(t+dt) - x(t)$. This means that jumps in the process are unlikely. This condition is called Lindeberg's continuity condition.

Generalization of Langevin dynamics for d-dimensional time series is straightforward. For d-dimensional continuous time series, a necessary ingredient of the system under consideration is that its dynamical behavior should be described by a set of macroscopic order parameters $\mathbf{x}(t)$ that are governed by the nonlinear Langevin equations [17],

$$\frac{d}{dt}\mathbf{x} = \mathbf{a}(\mathbf{x}) + \mathbf{b}(\mathbf{x})\,\Gamma(t) \tag{1.4}$$

where $\mathbf{x}(t)$ denotes the d-dimensional state vector, $\mathbf{a}(\mathbf{x})$ the drift vector, and the matrix $\mathbf{b}(\mathbf{x})$ is related to the diffusion matrix (second-order KM matrix) according to, $D_{ij}^{(2)}(\mathbf{x}) = \sum_{k=1}^{d} b_{ik}(\mathbf{x})\, b_{jk}(\mathbf{x})$. The noise terms $\Gamma_l(t)$, lumped together in the vector $\Gamma(t)$, are fluctuating forces with Gaussian statistics and rapidly decaying temporal correlations, such that delta-correlation in time can be assumed, i.e., $\langle\Gamma(t)\rangle = 0$, and, $\langle\Gamma_l(t)\Gamma_k(t')\rangle = \delta_{lk}\delta(t - t')$.

1.3 Time Series with Jump Discontinuity

In Eq. (1.2) for smooth functions $a(x, t)$ and $b(x, t)$, the process $x(t)$ is diffusion process and has *continuous* trajectory. In general, non-vanishing C in Eq. (1.3) or non-vanishing higher order KM coefficients indicate that a measured time series does not belong to the class of continuous diffusion processes; see Chap. 12. Accordingly, an improvement of the Langevin-type modeling, i.e. Eq. (1.2), is needed. An important generalization of the Langevin-type modeling is to include jump processes, also known as the jump-diffusion processes, with properties that can also be determined from measured time series. The jump-diffusion processes are given by dynamical stochastic equation:

$$dx(t) = a(x, t)\, dt + b(x, t)\, dW(t) + \xi\, dJ(t), \tag{1.5}$$

where $\{W(t), t \geq 0\}$ is again a scalar Wiener process, $a(x, t)$ and $b(x, t)$ are the drift and multiplicative diffusion functions, respectively, and $J(t)$ is a time-homogeneous Poisson jump process. The jump has state-dependent rate $\lambda(x)$ and jump size ξ according to some distribution with variance σ_ξ^2. We assume that ξ has Gaussian distribution, or follows any symmetric distribution with finite moments. This represents the minimal modeling of a measured time series that contains jumps. In general, one may assume any distributions for the jump events. In this book, also we will describe how one models the jumps characteristics—its rate and variance σ_ξ^2—with mixed Gaussian jumps, where the variance is distributed according to a Gamma distribution. We describe in Chap. 12 how all the unknown functions and coefficients in Eq. (1.5) are determined/computed directly from measured time series. Two typical trajectories corresponding to continuous (Brownian type) process and stochastic process with jumps are shown in Fig. 1.3. Jumps with amplitudes ξ_1 and ξ_2 are marked with arrows.

In Chap. 18, we demonstrate that sampling at discrete times gives rise to nonvanishing higher-order conditional moments $K^{(j)}(x)$ with $j > 2$, even if the underlying path is continuous. In Chap. 19 we will derive analytically a criterion that allows one to check whether for a given, even noisy time series the underlying process has a continuous (diffusive) or jumpy trajectories.

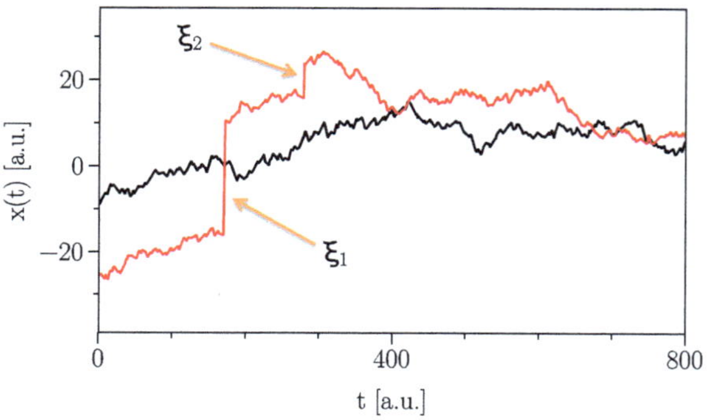

Fig. 1.3 Exemples of trajectories of a continuous process (Brownian type, black line) and a synthetic discontinuous stochastic process (red line). For the latter, jumps with amplitudes ξ_1 and ξ_2 are marked with arrows

1.4 Microstructural (Measurement) Noise

A measured time series may also contain some other noise, which is not assimilated by the stochastic process. In this case, the time series is written as $y(t) = x(t) + \epsilon(t)$, where $x(t)$ denotes the pure stochastic variable, and $\epsilon(t)$ is an additional noise that is assumed to be a short-range correlated noise and statistically independent of $x(t)$. In general, such noise may have its origin in intrinsic components of the complex dynamics, or can be caused by an external disturbance, e.g., that added to the time series by the measurement process.

In the literature, such spoiling noise is called differently, either as observational or measurement noise, or as microstructural noise; an example is in the financial sciences. The method described in this book is also able to separate and determine the stochastic behavior of the pure stochastic variable $x(t)$ and the statistical parameters of the noise $\epsilon(t)$, such as its variance and higher order statistical moments, etc. Therefore, we will be able to estimate precisely the "noise" contributions in a given time series. The microstructural noise is closely linked to the so-called Epps effect of financial data. Epps observed that there is a dramatic drop of the absolute value of correlations among stocks, when the sampling frequency increases. In such short time scales the noise $\epsilon(t)$ contributes much to the dynamics of $y(t)$ and, due to its strong fluctuations over short time scales, the predictability of $y(t)$ is decreased. We note that generally, an original series $x(t)$ has slower dynamics.

1.5 Intermittency

It turns out that the non-parametric method of analysis can also be successfully applied to not only fluctuating time series, but also to the analysis of spatio-temporal disordered systems, such as fluid turbulence [18, 19], or characterization of rough surfaces [27], and the porosity distributions in large-scale porous media. Such structures can be analyzed as scale-dependent stochastic processes. Experimental observ-

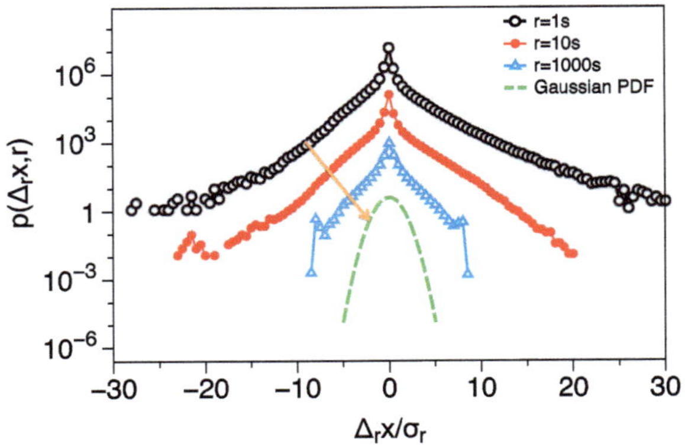

Fig. 1.4 Probability distribution functions (PDF) of the increment statistics, $p(\Delta_r x, r)$ for wind power fluctuations. Continuous deformation of the increments' PDFs for time lags $r = 1, 10, 1000$ sec in log-linear scale are shown. For better clarity the PDFs have been shifted in the vertical direction, and $\Delta_r x$'s are measured in units of their standard deviation σ_r. Extreme events up to about $20\sigma_{r=1}$ are recorded. A Gaussian PDF with unit variance is plotted for comparison. The method described in Chap. 15 provides an evolution equation for the change of the shape or deformation of the PDF of $p(\Delta_r x, r)$ with respect to scale r. *Source* from [28]

ables include the field increments, i.e., the difference in the field between two points separated by a distance or lag r, $\Delta_r x(t) = x(t + r) - x(t)$. The change of the increments as a function of the scale r can then be viewed as a stochastic process over a length scale. The method described in this book provides an evolution equation for the change of shape or deformation of the probability distribution function (PDF) of $p(\Delta_r x, r)$ with respect to scale r; see Fig. 1.4.

To study the fractal and multifractal behavior of given time series, the approach can be viewed as an extension of the multifractal description of stochastic processes [17]. The multifractal description focusses on the scaling behavior of the moments of quantities of interest, such as velocity or temperature increments, as a function of the scale. The complete information on an increment, $\Delta_r x(t)$, is contained in the probability distribution function (PDF) $p(\Delta_r x, r)$ for a certain scale r. For a self-similar process, it is assumed that the increments exhibit scaling behavior, $\Delta_r x \approx r^\xi$, which means $p(\Delta_r x, r) = \tilde{p}(\Delta_r x / r^\xi)/r^\xi$, where $\int_{-\infty}^{\infty} \tilde{p}(u) du = 1$. With locally-varying scaling exponent ξ the PDF is constructed as a superposition of the probability distributions

$$p(\Delta_r x, r) = \int d\xi \, \frac{1}{r^\xi} \, \tilde{p}(\Delta_r x / r^\xi) \, f(\xi, r) \,, \tag{1.6}$$

where the measure $f(\xi, r)$ characterizes the distribution of the regions with different scaling behavior. Knowledge of the deformation equation for $p(\Delta_r x, r)$ with respect to the scale r enables one to study the scaling exponents of increments for given time series; see Chap. 15.

In addition, the approach that we describe, which is based on the characterization of fluctuating fields by a scale-dependent stochastic process, can describe the joint statistics of the chosen stochastic variable, i.e. increments, on many different scales. This is achieved by the knowledge of the joint PDF, $p(\Delta_{r_1}x, r_1; \ldots; \Delta_{r_n}x, r_n)$. Using such joint PDFs, the correlations between the scales are also worked out, demonstrating how the complexity is linked between the scales. If the statistics of the scale-dependent measure can be regarded as a Markov process evolving in r, knowledge of two-scale conditional PDF is sufficient for a complete description of multiscale joint PDF; see Chap. 15 for more details.

Another consequence of the method that we describe in this book is that it enables us to understand the cascade nature of scale-dependent processes, as well as the intermittency of the time series. In the intermittent time series the structures arise in a time series and exhibit themselves as correlated high peaks at random times. The intervals between them are characterized by a low intensity and a large size. Rare peaks are the hallmarks of non-gaussian tails of the PDFs. In Fig. 1.4, as an example, the PDF of the increment statistics, $p(\Delta_r x, r)$, for wind power fluctuations are shown [28]. Continuous deformation of the increment PDFs for time lags $r = 1, 10, 1000$ sec in log-linear scale are shown. Here, $\Delta_r x$ is the wind power increments in lag r and measured in units of their standard deviation σ_r. Extreme events up to about 20 $\sigma_{r=1}$ are recorded that deviates strongly from Gaussian PDF.

References

1. H. Haken, *Synergetics, An Introduction* (Springer, Berlin, 1983)
2. H. Haken, *Advanced Synergetics* (Springer, Berlin, 1987)
3. H. Haken, *Information and Self-Organization: A Macroscopic Approach to Complex Systems* (Springer, Berlin, 2000)
4. H. Haken, *Synergetics: Introduction and Advanced Topics* (Springer, Berlin, 2004)
5. G. Nicolis, I. Prigogine, *Exploring Complexity* (W. H. Freeman & Co., San Francisco, 1989)
6. P. Bak, *How Nature Works: The Science of Self-Organized Criticality* (Oxford University Press, Oxford, 1999)
7. L. Schimansky-Geier, T. Poeschel, *Stochastic Dynamics* (Springer, Berlin, 1997)
8. F. Schweitzer, *Self-Organization of Complex Structures: From Individual to Collective Dynamics* (Gordon and Breach, London, 1997)
9. R.N. Mantegna, H.E. Stanley, *An Introduction to Econophysics: Correlations and Complexity in Finance* (Cambridge University Press, New York, 2000)
10. P.L. Gentili et al., *Untangling Complex Systems: A Grand Challenge for Science* vol. 36, Issue 8, (Rowman & Littlefield Publishers, 2018)
11. D. Sornette, *Critical Phenomena in Natural Sciences*, 2nd edn. (Springer, Heidelberg, 2003)
12. P. Grassberger, T. Schreiber, C. Schaffrath, Nonlinear time sequence analysis. Int. J. Bifurc. Chaos **1**, 521 (1991)
13. J.D. Hamilton, *Time Series Analysis* (Princeton University Press, Princeton, 1994)
14. R. Hegger, H. Kantz, T. Schreiber, Practical implementation of nonlinear time series methods: the TISEAN package. Chaos **9**, 413 (1999)
15. H. Kantz, T. Schreiber, *Nonlinear Time Series Analysis* (Cambridge University Press, Cambridge, 2003)

16. J. Argyris, G. Faust, M. Haase, R. Friedrich, *An Exploration of Dynamical Systems and Chaos* (Springer, New York, 2015)
17. R. Friedrich, J. Peinke, M. Sahimi, M.R. Rahimi Tabar, Phys. Rep. **506**, 87 (2011)
18. J. Peinke, M.R. Rahimi Tabar, M. Wächter, Annu. Rev. Condens. Matter Phys. **10** (2019)
19. R. Friedrich, J. Peinke, Phys. Rev. Lett. **78**, 863 (1997)
20. J. Davoudi, M.R. Rahimi Tabar, Phys. Rev. Lett. **82**, 1680 (1999)
21. M. Anvari, K. Lehnertz, M.R. Rahimi Tabar, J. Peinke, Scientific Reports **6**, 35435 (2016)
22. P.T. Clemson, A. Stefanovska, Phy. Rep. **542**, 297 (2014)
23. W.-X. Wang, Y.-C. Lai, C. Grebogie, Phys. Rep. **644**, 1 (2016)
24. H. Risken, *The Fokker-Planck Equation* (Springer, Berlin, 1989)
25. C.W. Gardiner, *Handbook of Stochastic Methods* (Springer, Berlin, 1983)
26. Z. Schuss, *Theory and Applications of Stochastic Processes: An Analytical Approach* (Springer, Berlin, 2010)
27. G.R. Jafari, S.M. Fazeli, F. Ghasemi, S.M. Vaez Allaei, M.R. Rahimi Tabar, A. Iraji Zad, G. Kavei, Phys. Rev. Lett. **91**, 226101 (2003)
28. M. Anvari, G. Lohmann, M. Wächter, P. Milan, E. Lorenz, D. Heinemann, M.R. Rahimi Tabar, J. Peinke, New J. Phys. **18**, 063027 (2016)

Chapter 2
Introduction to Stochastic Processes

In this chapter we provide mathematical tools to study the stochastic process from the physical point of view.

2.1 Introduction

The term *stochastic process* is intuitively associated with a trajectory that randomly fluctuates and, therefore, it requires a probabilistic description.[1]

---

[1] To define a stochastic process, let us at first provide the definition of probability space. A probability space associated with a random experiment is a triple $(\Omega,\mathcal{F},P)$, where, (i) Ω is a nonempty set, whose elements are known as outcomes or states, and is called the sample state; (ii) $\mathcal{F}$ is a family of subsets of Ω, which has the structure of a Borel σ-field, this means that:
(a) $\emptyset \in \mathcal{F}$
(b) If $A \in \mathcal{F}$, then its complement A^c also belongs to $\mathcal{F}$
(c) If $A_1, A_2, \ldots \in \mathcal{F}$ then $\bigcup_{i=1}^{\infty} A_i \in \mathcal{F}$,
(iii) p is a function which associated a number $p(A)$ to each set $A \in \mathcal{F}$ with the following properties:
(a) $0 \leq p(A) \leq 1$
(b) $p(\Omega) = 1$
(c) If $A_1, A_2, \ldots$ are pairwise disjoints set in $\mathcal{F}$ (that is $A_i \cap A_j = \emptyset$, whenever $i \neq j$), then $p(\bigcup_{i=1}^{\infty} A_i) = \sum_{i=1}^{\infty} p(A_i)$.

The elements of the σ-field $\mathcal{F}$ are called events and the mapping P is called a probability measure.

For one flip of a coin, $\Omega = \{Head = H, Tail = T\}$. The events $\mathcal{F}$ along with the corresponding $\mathcal{F} = \Pi(\Omega)$ contains all subsets of Ω, i.e. $\mathcal{F} = \{\{\emptyset\}, \{H\}, \{T\}, \{H, T\}\}$ and $p(H) = p(T) = \frac{1}{2}$. $\{\emptyset\}$ neither heads nor tails and $\{H, T\}$ to having simultaneously H and T, with probabilities 0.

Definition: Let $(\Omega,\mathcal{F},P)$ be a probability space and let T be an arbitrary set (called the index set). Any collection of random variables $x = \{x_t : t \in T\}$ defined on $(\Omega,\mathcal{F},P)$ is called a *stochastic process* with index set T.

If $x_{t_1}, x_{t_2}, \ldots, x_{t_n}$ are random variables defined on some common probability space, then $\mathbf{x}_t = (x_{t_1}, x_{t_2}, \ldots, x_{t_n})$ defines an $\mathbb{R}^n$ valued random variable, also called a random vector. *Stochastic processes* are also often called random processes.

M. R. Rahimi Tabar, *Analysis and Data-Based Reconstruction of Complex Nonlinear Dynamical Systems*, Understanding Complex Systems,
https://doi.org/10.1007/978-3-030-18472-8_2

Let us start with one dimensional array of random variables $\{\zeta_1, \zeta_2, \ldots, \zeta_N\}$, where $N > 1$. The possible values of ζ_n is known as its *state space*. The states can be either continues or discrete, depending on the nature of ζ_n. The position of Brownian particle and possible appearing of numbers in dice are examples of continues and discrete random variables. We can also imagine a random variable ζ_t the varies with the parameter t, where t is continuous parameter, in this case we deal with a stochastic process with continuous time. In what follows we provide some statistical tools to study the *continuous-state* processes, where ζ_n can have any real value.

2.2 Statistical Description of Time Series

We summarize the corresponding statistical description of a given time series. Such a description is achieved by introducing suitable statistical averages. We shall denote the averages by the brackets $\langle \cdots \rangle$. For stationary processes the averages can be viewed as time averages. For nonstationary processes the averages are defined as ensemble averages, i.e., averages over an ensemble of experimental (or numerical) realizations of the stochastic process. Here, we provide a review of the statistical description of stochastic processes, following the exposition of Risken [1].

Let us start with a one-dimensional time series ζ_n whose N realizations are:

$$\zeta_1, \ldots, \zeta_N. \tag{2.1}$$

We can also define ensembles of identical systems and make simultaneous experiments with different initial conditions.[2] Although the values of $\zeta_1, \ldots, \zeta_N$ cannot be predicted, some averaging in the limit $N \to \infty$ may be predictable, which then yield the same value for identical systems. For instance, the mean value of ζ_ns is defined as,

$$\langle \zeta \rangle = \lim_{N\to\infty} \frac{1}{N}\{\zeta_1 + \zeta_2 + \cdots + \zeta_N\}. \tag{2.2}$$

For some arbitrary function of ζ, i.e. $f(\zeta)$, its mean is defined as,

$$\langle f(\zeta) \rangle = \lim_{N\to\infty} \frac{1}{N}\{f(\zeta_1) + f(\zeta_2) + \cdots + f(\zeta_N)\}. \tag{2.3}$$

---

[2]This is true for an ergodic process. A stochastic process is said to be ergodic if its statistical properties can be deduced from a single, sufficiently long, random sample of the process.

2.2.1 The Probability Density

Let us assume that $f(\zeta)$ to be,

$$f(\zeta) = \theta(x - \zeta) \tag{2.4}$$

where $\theta(x)$ is the step function, defined as,

$$\theta(x) = \begin{cases} 1 & x > 0 \\ \frac{1}{2} & x = 0 \\ 0 & x < 0. \end{cases} \tag{2.5}$$

Using the Eqs. (2.3) and (2.5) we write,

$$\langle \theta(x - \zeta) \rangle = \lim_{N \to \infty} \frac{1}{N} \left\{ n(x > \zeta) + \frac{1}{2} n(x = \zeta) + 0 \right\} \tag{2.6}$$

where $n(x > \zeta)$ and $n(x = \zeta)$ are the numbers of events for ζ with $x > \zeta$ and ζ to be equal x, respectively. In terms of probability it can be written as,

$$\langle \theta(x - \zeta) \rangle = P(x > \zeta) + \frac{1}{2} P(x = \zeta). \tag{2.7}$$

For the processes with continuous states the probability of having an exact value x for process ζ is $P(x = \zeta) = 0$. Therefore, we find $\langle \theta(x - \zeta) \rangle = P(x > \zeta)$. The probability density $p_\zeta(x)$ of random variable ζ is defied as,

$$p_\zeta(x) = \frac{d}{dx} P(\zeta \le x) = \frac{d}{dx} \langle \theta(x - \zeta) \rangle = \langle \frac{d}{dx} \theta(x - \zeta) \rangle = \langle \delta(x - \zeta) \rangle \tag{2.8}$$

where $\delta(x - \zeta)$ is the Dirac δ-function. Assuming the differentiability of P, we find,

$$P(\zeta \le x + dx) - P(\zeta \le x) = \frac{d}{dx} P(\zeta \le x)\, dx = p_\zeta(x)\, dx. \tag{2.9}$$

Here $dP = p_\zeta(x) dx$ is the probability that the stochastic process ζ belongs to the interval $x \le \zeta \le x + dx$.

Using the following identity for δ-function,

$$f(\zeta) = \int f(x)\, \delta(x - \zeta)\, dx \tag{2.10}$$

one can define the average of $f(\zeta)$ over the distribution $p_\zeta(x)$ as,

$$\langle f(\zeta)\rangle = \langle \int f(x)\delta(x-\zeta)dx\rangle = \int f(x)\langle\delta(x-\zeta)\rangle dx = \int f(x)p_\zeta(x)dx \tag{2.11}$$

with normalisation condition, $\int p_\zeta(x)dx = 1$ (with $f(\zeta) = 1$). In the following, we omit the subscript ζ in distribution functions to enhance readability.

2.2.2 *Joint and Conditional Probability Distribution Functions*

Let us define $p_r(x_1, \ldots, x_r)dx_1dx_2\ldots dx_r$ to be the probability that the random variables $\zeta_1, \ldots, \zeta_r$ with $r \leq N$ (here N is the number of data points) belong to the intervals $x_1 \leq \zeta_1 \leq x_1 + dx_1, \ldots, x_r \leq \zeta_r \leq x_r + dx_r$. The probability density $p_r(x_1, \ldots, x_r)$ is the $r - point$ joint probability density function (PDF). We define the conditional PDF $p(x_1|x_2, \ldots, x_r)$ in terms of joint PDF as,

$$p(x_1|x_2, \ldots, x_r) = \frac{p_r(x_1, \ldots, x_r)}{p_{r-1}(x_2, \ldots, x_r)} \tag{2.12}$$

where is rewritten also as,

$$p(x_1|x_2, \ldots, x_r) = \frac{p_r(x_1, \ldots, x_r)}{\int p_r(x_1, \ldots, x_r)dx_1}. \tag{2.13}$$

By extension of definition (2.8) for one-point PDF to the $r-$point joint PDF, we write $p_r(x_1, \ldots, x_r)$ as

$$p_r(x_1, \ldots, x_r) = \langle\delta(x_1 - \zeta(t_1)) \cdots \delta(x_r - \zeta(t_r))\rangle \tag{2.14}$$

where $\zeta(t_i) = \zeta_i$. We call a stochastic process to be *solvable*, if for given r, one finds $r-$point joint probability distribution function $p_r(x_1, \ldots, x_r)$. Having these distributions enables us to calculate the $r-$point correlation functions $\langle x_1 \ldots x_r\rangle$, at given times $t_1, \ldots, t_r$, where $x_i = x(t_i)$.

2.3 Classification of Stochastic Processes

The conditional PDF $p_{1|r-1} = p(x_r, t_r|x_{r-1}, t_{r-1}; x_{r-2}, t_{r-2}; \ldots; x_1, t_1)$ is the conditional 1-point PDF, where it is conditioned on the values of x on succeeding time spots $t_{r-1}, \ldots, t_1$, and is defined as,

$$\begin{aligned} &p(x_r, t_r|x_{r-1}, t_{r-1}; x_{r-2}, t_{r-2}; \ldots; x_1, t_1) \\ &= \langle\delta(x_r - \zeta(t_r))\rangle|_{\zeta(t_{r-1})=x_{r-1},\ldots,\zeta(t_1)=x_1}. \end{aligned} \tag{2.15}$$

In terms of the $r-point$ joint PDF, it can be written as,

$$p(x_r, t_r|x_{r-1}, t_{r-1}; x_{r-2}, t_{r-2}; \ldots; x_1, t_1)$$

$$= \frac{p_r(x_r, t_r; \ldots; x_1, t_1)}{p_{r-1}(x_{r-1}, t_{r-1}; \ldots; x_1, t_1)} = \frac{p_r(x_r, t_r; \ldots; x_1, t_1)}{\int p_r(x_r, t_r; \ldots; x_1, t_1)dx_r}. \tag{2.16}$$

Distinct dependence of the conditional PDF, $p(x_r, t_r|x_{r-1}, t_{r-1}; \ldots; x_1, t_1)$, on its succeeding values $x_{r-1}, \ldots, x_1$ classifies the stochastic process x. Following Wand and Uhlenbeck [2], we classify the stochastic processes as [1]:

2.3.1 *Purely Random Processes*

For purely random process the conditional PDF is given by:

$$p(x_r, t_r|x_{r-1}, t_{r-1}; x_{r-2}, t_{r-2}; \ldots; x_1, t_1) = p(x_r, t_r). \tag{2.17}$$

which means that it forgets all of values at earlier time. For this process, the $r-$point joint PDF is written as,

$$p_r(x_r, t_r; \ldots; x_1, t_1) = p(x_r, t_r)\; p_{r-1}(x_{r-1}, t_{r-1}; \ldots; x_1, t_1). \tag{2.18}$$

Applying same argument to p_{r-1}, we obtain factorized PDF:

$$p_r(x_r, t_r; \ldots; x_1, t_1) = p(x_r, t_r) \cdots p(x_1, t_1) \tag{2.19}$$

which means that the complete information of the processes is contained in one-point PDF, also known as the marginal PDF.

2.3.2 *Markov Processes*

Markov processes are defined with the condition that,

$$p(x_r, t_r|x_{r-1}, t_{r-1}; x_{r-2}, t_{r-2}; \ldots; x_1, t_1) = p(x_r, t_r|x_{r-1}, t_{r-1}) \tag{2.20}$$

which means that conditional PDF depends only on the value $x(t_{r-1}) = x_{r-1}$ at the next earlier time. For a Markov process we can write,

$$p_r(x_r, t_r; \ldots; x_1, t_1)$$

$$= p(x_r, t_r|x_{r-1}, t_{r-1})\; p_{r-1}(x_{r-1}, t_{r-1}; \ldots; x_1, t_1). \tag{2.21}$$

Now, using the same argument as (2.20) for p_{r-1}, we find the following relation for the $r-$point joint PDF of Markov processes,

$$p_r(x_r, t_r; \ldots; x_1, t_1) = p(x_r, t_r|x_{r-1}, t_{r-1}) \cdots p(x_2, t_2|x_1, t_1)\ p(x_1, t_1). \quad (2.22)$$

Therefore, the marginal PDF $p(x_1, t_1)$ and the conditional PDF $p(x_k, t_k|x_{k-1}, t_{k-1})$ are sufficient to describe Markov processes. They have the property that if one knows that the process x is in the state x_{r-1} at the current time t_{r-1}, the probability of finding the realization x_r in the future time t_r does not change, if we add further information about the past of the process before time t_{r-1}. This means that the future of the process does not depend on the entire past, but only on the present state of the process.

2.3.3 Higher Order Processes

For higher order processes, one needs conditional PDF

$$p(x_r, t_r|x_{r-1}, t_{r-1}; \ldots; x_m, t_m)$$

for $m \in \{1, \ldots, r-2\}$ to describe these processes.

2.4 Stationary Processes

If for a stochastic process the r-point joint probability distribution function is invariant under constant time shift, $t_i \to t_i + \tau$, i.e.,

$$p_r(x_r, t_r; \ldots; x_2, t_2; x_1, t_1) = p_r(x_r, t_r + \tau; \ldots; x_2, t_2 + \tau; x_1, t_1 + \tau)$$

then we say it is a stationary stochastic process. The immediate consequences are that the mean value of the process $\langle x \rangle = \bar{x}$ is constant, and the correlation function $\langle (x(t_1) - \bar{x})(x(t_2) - \bar{x}) \rangle$ is function of $t_2 - t_1$.

In general, an experimental data set is either stationary or nonstationary. Establishing or rejecting the stationarity of the set can be done by several statistical tests. Non-stationary behaviors can be trends, cycles, level shifts, regime shifts, time-dependent stochastic behavior, etc. In principle, for non-stationary processes one should study the ensemble averages. In real world data, however, such as stock market index, or pattern of cosmic background radiation, there is a unique trajectory. We refer the reader to Sect. 16.4 for technical details, where we discuss how one verifies the stationarity or nonstationarity of given time series with *unique trajectory*.

2.5 The Chapman–Kolmogorov Equation and the Necessary Condition for a Process to Be Markov

Next, we extend the definitions of previous sections for d-dimensional stochastic processes $\boldsymbol{\zeta}(t)$ and derive the necessary condition that a process to be Markov process. Let us consider the joint probability density

$$p_r(\mathbf{x}_r, t_r; \ldots; \mathbf{x}_1, t_1) \tag{2.23}$$

which is related to the probability finding the system at times t_i in the volume ΔV_i in the phase (state) space, which can be written as:

$$p_r(\mathbf{x}_r, t_r; \ldots; \mathbf{x}_2, t_2; \mathbf{x}_1, t_1) = \langle \delta(\mathbf{x}_r - \boldsymbol{\zeta}(t_r)) \cdots \delta(\mathbf{x}_1 - \boldsymbol{\zeta}(t_1)) \rangle \tag{2.24}$$

where the brackets indicate the statistical average, which may be a time average for stationary processes, or an ensemble average. As we defined earlier, an important subclass of stochastic processes are Markov processes. For such processes the r-point joint probability distribution can be constructed from the knowledge of the conditional probability distributions (also known as the transition probability)

$$p(\mathbf{x}_{i+1}, t_{i+1} | \mathbf{x}_i, t_i) = \frac{p_2(\mathbf{x}_{i+1}, t_{i+1}; \mathbf{x}_i, t_i)}{p(\mathbf{x}_i, t_i)} \tag{2.25}$$

according to

$$\begin{aligned} & p_r(\mathbf{x}_r, t_r; \ldots; \mathbf{x}_2, t_2; \mathbf{x}_1, t_1) \\ &= p(\mathbf{x}_r, t_r | \mathbf{x}_{r-1}, t_{r-1}) \cdots p(\mathbf{x}_2, t_2 | \mathbf{x}_1, t_1) p(\mathbf{x}_1, t_1). \end{aligned} \tag{2.26}$$

Here, the Markov property of a process for the multiply conditioned probabilities was used. Again similar in one-dimensional Markov processes, knowledge of the transition probabilities, together with the marginal probability distribution $p(\mathbf{x}_1, t_1)$, suffice to define the r-point probability distribution.

It is straightforward to prove the Chapman–Kolmogorov equation using the three-point probability distribution for Markov processes[3]:

$$p(\mathbf{x}_j, t_j | \mathbf{x}_i, t_i) = \int d\mathbf{x}_k \; p(\mathbf{x}_j, t_j | \mathbf{x}_k, t_k) \; p(\mathbf{x}_k, t_k | \mathbf{x}_i, t_i). \tag{2.27}$$

---

[3] All stochastic processes satisfy the relation $p(\mathbf{x}_3, t_3) = \int d\mathbf{x}_2 p_2(\mathbf{x}_3, t_3; \mathbf{x}_2, t_2) = \int d\mathbf{x}_2 p(\mathbf{x}_3, t_3 | \mathbf{x}_2, t_2) p(\mathbf{x}_2, t_2)$. Moreover, the conditional PDF can be written as $p(\mathbf{x}_3, t_3 | \mathbf{x}_1, t_1) = \int d\mathbf{x}_2 p(\mathbf{x}_3, t_3; \mathbf{x}_2, t_2 | \mathbf{x}_1, t_1) = \int d\mathbf{x}_2 p(\mathbf{x}_3, t_3 | \mathbf{x}_2, t_2; \mathbf{x}_1, t_1) p(\mathbf{x}_2, t_2 | x_1, t_1)$. Taking into account the Markov assumption, if $t_3 > t_2 > t_1$, we can ignore the $\mathbf{x}_1$ dependence. Therefore, we find, $p(\mathbf{x}_3, t_3 | \mathbf{x}_1, t_1) = \int d\mathbf{x}_2 p(\mathbf{x}_3, t_3 | \mathbf{x}_2, t_2) p(\mathbf{x}_2, t_2 | \mathbf{x}_1, t_1)$, which is the Chapman–Kolmogorov equation.

This relation is valid for all times $t_i < t_k < t_j$, and is the necessary condition (because we used only the three-point distribution function in its proof) that a process belongs to the class of Markov processes.

2.6 Continuous Stochastic Markov Processes

Now let us define *continuous stochastic Markov processes*. In terms of the conditional probability distribution a continuous Markov process x satisfies the following continuity condition, given some $\delta > 0$ [3],

$$\begin{aligned} C(t) \triangleq C(t, x, \delta) &= \lim_{\tau \to 0^+} \frac{Prob[\ |\Delta x(t)| > \delta \ \ |_{x(t)=x}]}{\tau} \\ &= \lim_{\tau \to 0^+} \frac{\int_{|\Delta x(t)|=|x'-x|>\delta} p(x', t+\tau|x, t)\, dx'}{\tau} = 0 \end{aligned}$$

holds for all $\delta > 0$, uniformly for all x and t. Here, $\Delta x = x(t+\tau) - x(t)$ and $p(x', t+\tau|x, t)$ is the conditional probability distribution. Noting the condition $\delta > 0$, this means that jumps in the process are unlikely. This is called Lindeberg's condition for continuity of stochastic trajectories [4], and states that the trajectories of a Markov process are continuous if $C = 0$ holds for all $\delta > 0$, uniformly for all x and t. This condition means that the probability of moving more than δ in a short time τ decays faster than linearly with τ. Therefore, a vanishing or non vanishing C indicates that the processes are continuous or that they may have jump discontinuity, respectively. See Chaps. 4 and 11 for examples of continuous Markov processes and those with jumps.

Problems

2.1 Statistical moment-generating function

(a) Let $\mathbf{x} = (x_1, \ldots, x_n)^T$ be a random vector, and $\mathbf{u} = (u_1, \ldots, u_n)^T \in \mathbb{R}^n$, where $(\cdots)^T$ be the transpose of vector $(\cdots)$. The statistical moment-generating function is defined by

$$Z_{\mathbf{x}}(\mathbf{u}) = \langle \mathbf{e}^{\mathbf{u}^{\mathbf{T}}\mathbf{x}} \rangle$$

for all $\mathbf{u}$ for which the average exists (is finite). Show that the statistical moments of order k can be determined using the following relation:

$$\frac{\partial^k}{\partial u_1^{k_1} \cdots \partial u_n^{k_n}} Z_{\mathbf{x}}(\mathbf{u})|_{\mathbf{u}=\mathbf{0}} = \langle x_1^{k_1} \cdots x_n^{k_n} \rangle$$

where $k = k_1 + \cdots + k_n$.

(b) The density function of the univariate normal distribution is given by

$$f(x) = \frac{1}{\sqrt{2\pi}\sigma} \exp\left\{-\frac{1}{2}\left(\frac{x-\mu}{\sigma}\right)^2\right\}$$

for $-\infty < x < \infty$, where μ is the mean and $\sigma^2 > 0$ is the variance. Show that

$$Z_x(u) = \exp\{\mu u + \frac{\sigma^2 u^2}{2}\}$$

and prove that, $\langle (x-\mu)^{2n} \rangle = \frac{2n!}{2^n n!} \langle (x-\mu)^2 \rangle^n$ and $\langle (x-\mu)^{2n+1} \rangle = 0$, where $n = 1, 2, \ldots$.

2.2 Bivariate normal distribution

The density function of the bivariate normal distribution is given by

$$p(x, y) = \frac{\exp\left\{-\frac{1}{2(1-\rho^2)}\left[\left(\frac{x-\mu_x}{\sigma_x}\right)^2 - 2\rho\left(\frac{x-\mu_x}{\sigma_x}\right)\left(\frac{y-\mu_y}{\sigma_y}\right) + \left(\frac{y-\mu_y}{\sigma_y}\right)^2\right]\right\}}{2\pi\sigma_x\sigma_y\sqrt{1-\rho^2}}$$

where (μ_x, μ_y) is the mean vector and the variance-covariance matrix is

$$\begin{pmatrix} Var(x) & Cov(x, y) \\ Cov(x, y) & Var(y) \end{pmatrix} = \begin{pmatrix} \sigma_x^2 & \rho\sigma_x\sigma_y \\ \rho\sigma_x\sigma_y & \sigma_y^2 \end{pmatrix}.$$

The constraints are $\sigma_x^2 > 0$, $\sigma_y^2 > 0$ and $-1 < \rho < 1$, where ρ is the correlation coefficient, $\rho = Cov(x, y)/\sigma_x\sigma_y$ and $Cov(x, y) = \langle (x-\mu_x)(y-\mu_y) \rangle$.
Derive the following conditional averaging and show that,
(a) $\langle y|x \rangle = \mu_y + \rho \, \frac{\sigma_y}{\sigma_x} \, (x - \mu_x)$
(b) $\sigma_{y|x}^2 = \sigma_y^2 \, (1 - \rho^2)$.

2.3 p-variate normal distribution

The density function of the p-variate normal distribution is given by

$$f(\mathbf{x}) = \frac{1}{(2\pi)^{p/2} \, |\mathbf{g}|^{1/2}} \exp\left\{-\frac{1}{2}\,(\mathbf{x} - \boldsymbol{\mu})^T \, \mathbf{g}^{-1} \, (\mathbf{x} - \boldsymbol{\mu})\right\}$$

where $\mathbf{x}^T = (x_1, \ldots, x_p)$, $\boldsymbol{\mu}^T = (\mu_1, \ldots, \mu_p)$ and $\mathbf{g}$ is a full rank variance-covariance matrix, i.e.,

$$\mathbf{g}_{ij} = Cov(x_i, x_j) = \langle (x_i - \mu_i)(x_j - \mu_j) \rangle$$

where $\mathbf{g}^{-1}$ and $|\mathbf{g}|$ are the inverse and the determinant of $\mathbf{g}$.

(a) Derive the statistical moment-generating function for p-variate normal distribution.

(b) Prove the following relation for the fourth-order correlation function (Wick's Theorem):

$$\langle (x_i - \mu_i)(x_j - \mu_j)(x_k - \mu_k)(x_l - \mu_l) \rangle = \mathbf{g}_{ij}\mathbf{g}_{kl} + \mathbf{g}_{ik}\mathbf{g}_{jl} + \mathbf{g}_{il}\mathbf{g}_{jk}.$$

2.4 Chapman–Kolmogorov equation

Show that the following conditional density functions, (a) Brownian motion and (b) Cauchy process satisfy the Chapman–Kolmogorov equation

$$(a) \quad p(x_2, t_2 | x_1, t_1) = \frac{1}{\sqrt{2\pi(t_2 - t_1)}} \exp\left\{ -\frac{(x_2 - x_1)^2}{2(t_2 - t_1)} \right\},$$

$$(b) \quad p(x_2, t_2 | x_1, t_1) = \frac{1}{\pi} \frac{t_2 - t_1}{(t_2 - t_1)^2 + (x_2 - x_1)^2}$$

where $t_2 > t_1$.

References

1. H. Risken, *The Fokker-Planck Equation* (Springer, Berlin, 1989)
2. M.C. Wand, G.E. Uhlenbeck, Rev. Mod. Phys. **17**, 323 (1945)
3. C.W. Gardiner, *Handbook of Stochastic Methods* (Springer, Berlin, 1983)
4. I.V. Girsanov, On transforming a certain class of stochastic processes by absolutely continuous substitution of measures. Theory Probab. Appl. **5**, 285 (1960)

Chapter 3
Kramers–Moyal Expansion and Fokker–Planck Equation

In this chapter, we present the details of Kramers–Moyal (KM) expansion and prove the Pawula theorem. The Fokker–Planck equation is then introduced and its short-term propagator is presented. Finally, we derive the master equation from the Chapman–Kolmogorov equation.

3.1 Kramers–Moyal Expansion

The probability distributions of Markov processes satisfy a first-order partial differential equation in time and order infinity in the state variable. The governing equation is known as the Kramers–Moyal equation. In this section we provide a proof for the Kramers–Moyal expansion [1]. The probability density $p(x, t + \tau)$ at time $t + \tau$ and $p(x', t)$ at time t are related to each others via:

$$p(x, t + \tau) = \int p(x, t + \tau|x', t)\ p(x', t)\ dx' \tag{3.1}$$

where $p(x, t + \tau|x', t)$ is the conditional probability distribution function. We assume that all of the conditional moments $K^{(n)}(x', t, \tau)$, which are defined as follows, exist

$$\begin{aligned} K^{(n)}(x', t, \tau) &= \langle [x(t + \tau) - x(t)]^n \rangle|_{x(t)=x'} \\ &= \int (x - x')^n\ p(x, t + \tau|x', t)\ dx. \end{aligned} \tag{3.2}$$

To derive the Kramers–Moyal equation we note that the conditional PDF $p(x, t + \tau|x', t)$ can be written as:

M. R. Rahimi Tabar, *Analysis and Data-Based Reconstruction of Complex Nonlinear Dynamical Systems*, Understanding Complex Systems,
https://doi.org/10.1007/978-3-030-18472-8_3

$$p(x, t+\tau|x', t) = \int \delta(y-x)\; p(y, t+\tau|x', t)\, dy. \tag{3.3}$$

Having a expansion of δ-function, one finds:

$$\begin{aligned}\delta(y-x) = \delta(x'-x+y-x') &= \sum_{n=0}^{\infty} \frac{(y-x')^n}{n!} \left(\frac{\partial}{\partial x'}\right)^n \delta(x'-x) \\ &= \sum_{n=0}^{\infty} \frac{(y-x')^n}{n!} \left(-\frac{\partial}{\partial x}\right)^n \delta(x'-x).\end{aligned} \tag{3.4}$$

Here, we used the relation $\frac{\partial}{\partial x'} f(x-x') = -\frac{\partial}{\partial x} f(x-x')$. Using Eq. (3.3) the conditional PDF $p(x, t+\tau|x', t)$ will have the following expression:

$$\begin{aligned}p(x, t+\tau|x', t) &= \sum_{n=0}^{\infty} \frac{1}{n!} \left(-\frac{\partial}{\partial x}\right)^n \int dy\; (y-x')^n\; p(y, t+\tau|x', t)\; \delta(x'-x) \\ &= \left[1 + \sum_{n=1}^{\infty} \frac{1}{n!} \left(-\frac{\partial}{\partial x}\right)^n K^{(n)}(x', t, \tau)\right] \delta(x'-x).\end{aligned} \tag{3.5}$$

By substituting Eq. (3.5) in Eq. (3.1) we find,

$$\begin{aligned}p(x, t+\tau) - p(x, t) &= \frac{\partial p(x,t)}{\partial t} \tau + \mathcal{O}(\tau^2) \\ &= \sum_{n=1}^{\infty} \frac{1}{n!} \left(-\frac{\partial}{\partial x}\right)^n \int \delta(x'-x)\; K^{(n)}(x', t, \tau)\; p(x', t)\, dx' \\ &= \sum_{n=1}^{\infty} \left(-\frac{\partial}{\partial x}\right)^n \left[\frac{K^{(n)}(x, t, \tau)}{n!}\right] p(x, t).\end{aligned} \tag{3.6}$$

Dividing both sides of Eq. (3.6) by τ and taking the limit $\tau \to 0$, we obtain the Kramers–Moyal equation for the probability density function $p(x, t)$

$$\begin{aligned}\frac{\partial p(x,t)}{\partial t} &= \sum_{n=1}^{\infty} \left(-\frac{\partial}{\partial x}\right)^n \left[D^{(n)}(x, t)\; p(x, t)\right] \\ &\equiv \mathcal{L}_{KM}\; p(x, t)\end{aligned} \tag{3.7}$$

where Kramers–Moyal operator $\mathcal{L}_{KM}$ is given by

$$\mathcal{L}_{KM} = \sum_{n=1}^{\infty} \left(-\frac{\partial}{\partial x}\right)^n D^{(n)}(x, t).$$

Here $D^{(n)}(x,t)$ are the Kramers–Moyal coefficients and are given in terms of conditional moments $K^{(n)}(x,t,\tau)$ as,

$$D^{(n)}(x,t) = \frac{1}{n!}\lim_{\tau\to 0}\frac{1}{\tau}K^{(n)}(x,t,\tau) = \frac{1}{n!}\lim_{\tau\to 0}\frac{1}{\tau}\langle[x(t+\tau)-x(t)]^n\rangle|_{x(t)=x}. \quad (3.8)$$

We note that the Kramers–Moyal coefficients $D^{(n)}(x,t)$ can be determined directly from time series under study. However, to estimate these coefficients one needs to take care of some technical aspects that we will describe in details in Chaps. 16–21.

3.2 Pawula Theorem and Fokker–Planck Equation

Pawula theorem states that there are only three possible cases in the KM expansion: (i) The Kramers–Moyal expansion is truncated at $n = 1$, meaning that the process is deterministic; (ii) the KM expansion stops at $n = 2$, with the resulting equation being the Fokker–Planck equation, and describes diffusion processes and, finally, (iii) The KM expansion contains all the term up to $n = \infty$. Any truncation of expansion at a finite order $n > 2$ would produce non-positive probability density $p(x,t)$ [1, 2].

To prove the Pawula theorem, we consider the Schwartz inequality that holds for non-negative function $p(x)$ and arbitrary functions $f(x)$ and $g(x)$ as:

$$\left\{\int f(x)g(x)p(x)dx\right\}^2 \le \int f^2(x)p(x)dx \int g^2(x)p(x)dx. \quad (3.9)$$

Choosing the functions in (3.9) as $f(x) = (x-x')^n$, $g(x) = (x-x')^{n+m}$ and $p(x) = p(x,t+\tau|x',t)$, we obtain the conditional moments inequality as (in the following, we omit the x- and t-dependence of conditional moments and KM coefficients):

$$\left(K^{(2n+m)}\right)^2 \le K^{(2n)}\ K^{(2n+2m)}. \quad (3.10)$$

By dividing both sides of (3.10) by τ^{4n+2m} and assuming the finiteness of conditional moments $K^{(n)}$, one finds that the KM coefficients satisfy the following inequality,

$$\left[(2n+m)!D^{(2n+m)}\right]^2 \le (2n)!\ (2n+2m)!\ D^{(2n)}\ D^{(2n+2m)}. \quad (3.11)$$

For $n = 2$ we find,

$$\left[(4+m)!D^{(4+m)}\right]^2 \le 4!(4+2m)!\ D^{(4)}\ D^{(4+2m)} \quad (3.12)$$

which shows that for vanishing fourth order KM coefficients $D^{(4)} = 0$, we have

$$\left[(4+m)!D^{(4+m)}\right]^2 \leq 0. \tag{3.13}$$

We note that in order to have finite coefficients $D^{(4+2m)}$ in right side of (3.12), the possible values for m will be $m = -1, 0, 1, 2, \ldots$ (noting the fact that $D^{(0)}$ diverges with $1/\tau$ as $\tau \to 0$, which excludes the case $m = -2$). Using the allowed values for m and from (3.13), we find $D^{(4+m)} = 0$ which means that

$$D^{(4)} = 0 \to D^{(3)} = D^{(5)} = D^{(6)} = \cdots = 0. \tag{3.14}$$

Moreover, if one assumes in (3.11) that $D^{(2n)} = 0$, it will results in $D^{(2n+1)} = D^{(2n+2)} = \cdots = 0$ with $n \geq 1$. This means that for $n = 1$ a vanishing $D^{(2)} = 0$ results in $D^{(3)} = D^{(4)} = D^{(5)} = \cdots = 0$, and only the first KM coefficient $D^{(1)}$ can be nonzero. Truncation of KM expansion at any finite order l implies, $D^{(3)} = D^{(4)} = D^{(5)} = \cdots = 0$, which is in contradiction with non-zero values of KM coefficients with $k < l$.

For the case in which $D^{(4)} = 0$, the KM expansion reduces to the Fokker–Planck equation:

$$\frac{\partial p(x,t)}{\partial t} = -\frac{\partial}{\partial x}\left(D^{(1)}(x,t)p(x,t)\right) + \frac{\partial^2}{\partial x^2}\left(D^{(2)}(x,t)p(x,t)\right)$$
$$\frac{\partial p(x,t)}{\partial t} \equiv \mathcal{L}_{FP}\ p(x,t). \tag{3.15}$$

Here, we defined the Fokker–Planck operator $\mathcal{L}_{FP}$. For a general stochastic process, all the KM coefficients may be nonzero. Provided, however, that $D^{(4)}$ vanishes, truncation of the KM expansion after the second term is meaningful in the statistical sense, and the probability density of the processes can be described using the Fokker–Planck equation. In the Chap. 11 we will discuss the implications of $D^{(4)} \neq 0$. Also in Chap. 18 we introduce a criterion to validate numerically the Pawula theorem, to judge whether the $D^{(4)}$ tends to zero.

3.3 Short-Time Propagator of Fokker–Planck Equation in One Dimension

As long as the KM coefficients $D^{(1)}(x,t)$ and $D^{(2)}(x,t)$ depend explicitly on time, the formal solution of the Fokker–Planck (FP) equation is given by a time-ordered Dyson series [1]. In this section we derive the short-term propagator of the FP equation and assume that coefficients $D^{(1)}(x,t)$ and $D^{(2)}(x,t)$ are smooth and not changing dramatically over a short time interval.

For a process whose PDF satisfies the Fokker–Planck (FP) equation, the conditional PDF $p(x, t+\tau|x', t)$ also obeys the same equation. This can be seen by simply writing, $p(x,t) = \int p(x,t|x',t')\, p(x',t')dx'$, and by inserting in the FP equation we find,

$$\frac{\partial p(x,t|x',t')}{\partial t} = \mathcal{L}_{FP}\, p(x,t|x',t') \tag{3.16}$$

with the initial condition $p(x,t|x',t) = \delta(x-x')$. Define $p(x,t+\tau|x',t)$ for small τ, as the short-time propagator. Therefore, it can be written as,

$$p(x,t+\tau|x',t) = \exp\left(\tau\mathcal{L}_{FP}\right)\, p(x,t|x',t) = \exp\left(\tau\mathcal{L}_{FP}\right)\,\delta(x-x'). \tag{3.17}$$

Expansion of $\exp\left(\tau\mathcal{L}_{FP}\right)$ in powers of τ gives,

$$\begin{aligned} p(x,t+\tau|x',t) &= \left\{1 + \mathcal{L}_{FP}\tau + \mathcal{O}(\tau^2) + \cdots\right\}\delta(x-x') \\ &= \left\{1 - \tau\frac{\partial}{\partial x}D^{(1)}(x,t) + \tau\frac{\partial^2}{\partial x^2}D^{(2)}(x,t) + \mathcal{O}(\tau^2) + \cdots\right\}\delta(x-x') \\ &= \left\{1 - \tau\frac{\partial}{\partial x}D^{(1)}(x',t) + \tau\frac{\partial^2}{\partial x^2}D^{(2)}(x',t) + \mathcal{O}(\tau^2) + \cdots\right\}\delta(x-x') \end{aligned}$$

where in the last step we replaced x by x' in the coefficients $D^{(1)}$ and $D^{(2)}$. Writing the Dirac δ-function as $\delta(x-x') = 1/2\pi\int_{-\infty}^{\infty}\exp(iu(x-x'))du$ and replacing $\frac{\partial}{\partial x} \to iu$ one gets,

$$p(x,t+\tau|x',t) = \frac{1}{2\pi}\int_{-\infty}^{\infty}\exp\left\{iu(x-x') - iu\tau D^{(1)}(x',t) - u^2\tau D^{(2)}(x',t)\right\}du$$

and, finally, integration of the Gaussian integral yields,

$$p(x,t+\tau|x',t) = \frac{1}{2\sqrt{\pi\tau D^{(2)}(x',t)}}\exp\left\{-\frac{(x-x'-\tau D^{(1)}(x',t))^2}{4\tau D^{(2)}(x',t)}\right\}. \tag{3.18}$$

It appears, according to Eq. (3.18), that the short-term propagator $p(x,t+\tau|x',t)$ for Fokker–Planck equation is a Gaussian distribution with mean value $(x'+\tau D^{(1)}(x',t))$ and variance $2D^{(2)}(x',t)\tau$. This enables us to understand intuitively the KM coefficients $D^{(1)}(x,t)$ and $D^{(2)}(x,t)$. Indeed, $D^{(1)}(x,t)$ is the drift that moves the mean from x' to $(x'+\tau D^{(1)}(x',t))$, and determines where the process will go to on average. The second coefficient $D^{(2)}(x,t)$ is the diffusion coefficient and determines the strength of the stochastic influence and, therefore, the uncertainty about what the process will evolve in the short-term τ [3]. In the Chap. 9 we will drive the short-time propagator of the FP equation in higher dimensions.

3.4 Master Equation

The most general equation specifying a Markov process for a continuous-time process $x(t)$ is the master equation. Here, we derive the master equation from the Chapman–Kolmogorov equation, i.e.

$$p(x_3, t_3|x_1, t_1) = \int dx_2 \, p(x_3, t_3|x_2, t_2) \, p(x_2, t_2|x_1, t_1). \tag{3.19}$$

By assuming stationarity of the process x, we write $p(x_2, t_2|x_1, t_1) \equiv T_{t_2-t_1}(x_2|x_1)$. If we define $\tau' = t_2 - t_1$ and $\tau = t_3 - t_2$, then with $\tau + \tau' = t_3 - t_1$ we find,

$$T_{\tau+\tau'}(x_3|x_1) = \int dx_2 \, T_\tau(x_3|x_2) \, T_{\tau'}(x_2|x_1). \tag{3.20}$$

Expanding the transition probability $T_\tau(x_3|x_2)$ for a small time difference τ and then taking the limit $\tau \to 0$, we derive an equation for its time evolution. For small $\tau << 1$, up to $\mathcal{O}(\tau)$ we have,

$$T_\tau(x_3|x_1) = \delta(x_3 - x_1) + \tau \partial_\tau T_\tau(x_3|x_1)|_{\tau=0} + \mathcal{O}(\tau^2). \tag{3.21}$$

The delta function expresses the probability of staying at the same state after time zero equals one, whereas the probability to change the state after time zero will vanish. Define $\omega(x_3|x_1) = \partial_\tau T_\tau(x_3|x_1)|_{\tau=0}$, which is the time derivative of the transition probability at $\tau = 0$, which we call the transition probability per unit time. The expression (3.21) must satisfy the normalization condition. Therefore, the integral over x_3 must equal unity. To have normalisation condition, Eq. (3.21) must be corrected in the following way:

$$T_\tau(x_3|x_2) = (1 - \beta(x_2)\,\tau)\,\delta(x_3 - x_2) + \tau\,\omega(x_3|x_2) + \mathcal{O}(\tau^2). \tag{3.22}$$

Here, the delta function has been corrected by the coefficient $(1 - \beta(x_2)\tau)$, which corresponds to the probability for no transition to have taken place at time duration τ. Normalisation condition gives us,

$$\beta(x_2) = \int \omega(x_3|x_2)\, dx_3. \tag{3.23}$$

Using Eq. (3.22) in the Chapman–Kolmogorov equation gives,

$$T_{\tau+\tau'}(x_3|x_1) = \int dx_2 T_\tau(x_3|x_2)\, T_{\tau'}(x_2|x_1)$$

$$= \int \{[1 - \beta(x_2)\tau]\,\delta(x_3 - x_2) + \tau\omega(x_3|x_2)\}\, T_{\tau'}(x_2|x_1) dx_2 + \mathcal{O}(\tau^2)$$

$$= T_{\tau'}(x_3|x_1) - \beta(x_3)\tau T_{\tau'}(x_3|x_1) + \int \tau\omega(x_3|x_2)T_{\tau'}(x_2|x_1)dx_2 + \mathcal{O}(\tau^2).$$

By rearrangement the terms and dividing both sides by τ we find,

$$\frac{T_{\tau+\tau'}(x_3|x_1) - T_{\tau'}(x_3|x_1)}{\tau}$$
$$= -\beta(x_3)T_{\tau'}(x_3|x_1) + \int \omega(x_3|x_2)T_{\tau'}(x_2|x_1)dx_2 + \mathcal{O}(\tau). \quad (3.24)$$

Define $t = \tau'$ and in the limit $\tau \to 0$ we find

$$\frac{\partial}{\partial t}T_t(x_3|x_1) = -\beta(x_3)T_t(x_3|x_1) + \int \omega(x_3|x_2)T_t(x_2|x_1)dx_2. \quad (3.25)$$

Using the expression, $\beta(x_3) = \int \omega(x_2|x_3)dx_2$, one finally finds the so-called master equation:

$$\frac{\partial}{\partial t}T_t(x_3|x_1) = \int \Big[\omega(x_3|x_2)T_t(x_2|x_1) - \omega(x_2|x_3)T_t(x_3|x_1)\Big]dx_2. \quad (3.26)$$

It is useful to write the master equation in terms of marginal PDF $p(x_3, t)$. By multiplying both side of Eq. (3.26) by $p(x_1, t)dx_1$ and integrating over the variable x_1, we rewrite the master equation as:

$$\frac{\partial}{\partial t}p(x_3, t) = \int \Big[\omega(x_3|x_2)\ p(x_2, t) - \omega(x_2|x_3)\ p(x_3, t)\Big]\ dx_2. \quad (3.27)$$

Now let us consider the case that x belongs to a discrete set of states, labeled with n. Then Eq. (3.27) reduces to:

$$\frac{\partial}{\partial t}p_n(t) = \sum_{n'} \Big[w_{n,n'}\ p_{n'}(t) - w_{n',n}\ p_n(t)\Big]. \quad (3.28)$$

At the steady state, the left side of the master equation equals zero and, therefore, one finds [4]

$$\sum_{n'} w_{n,n'}\ p_{n'}(t) = \sum_{n'} w_{n',n}\ p_n(t).$$

This relation expresses the fact that in the steady state, the sum of all transitions per unit time into any state n must be balanced by the sum of all transitions from n into other states n'. We can, however, define the detailed balance condition, which is a stronger assertion that for each pair n and n' separately, the transitions must balance, i.e. $w_{n,n'}\, p_{n'}(t) = w_{n',n}\, p_n(t)$. This means that there is no net current, whereas in a stationary state without detailed balance a net probability current is possible in the form of a closed-loop current.

The discrete version of master equation can be visualized as a weighted bidirectional graph. Every possible discrete state n is represented by a node of the graph, while $w_{n,n'}$ represents the directed, weighted edge connecting the node n to the node n'. Some properties of the physical system follow from the topology of the graph, while others depend on the exact values of the weights.

The main difference between the Chapman–Kolmogorov equation and the master equation is that the former is a nonlinear equation (in the transition probabilities) that expresses the Markov character of the process, but contains no information about the local dynamics of any particular Markov process. In the master equation, however, one considers the transition probability at short times, $\omega(x_j|x_i)$, as a given function determined by the specific physical system, and the resulting equation is linear in the conditional probability density that determines the state of that system [1].

Problems

3.1 Equation of statistical moments on the Kramers–Moyal equation

From the more general Kramers–Moyal expansion (3.7) for the probability density $p(x,t)$, derive the following differential equations for the nth-order statistical moments (assuming that they exist) of x,

$$\frac{\partial}{\partial t}\left\langle x^n\right\rangle = \sum_{k=1}^{n} \frac{n!}{(n-k)!}\left\langle x^{n-k}\, D^{(k)}(x,t)\right\rangle .$$

3.2 Path integral solution of the Fokker–Planck equation

The short-term propagator (or transition probability) (3.18) is needed for path integral solution of the Fokker–Planck equation. Dividing time difference $t - t_0$ into N small intervals of length $\tau = (t-t_0)/N$, defining $t_n = t_0 + n\tau$, and by repeatedly applying the Chapman–Kolmogorov equation (3.19) show that in the limit $N \to \infty$ (or $\tau \to 0$)

(a) the probability distribution function $p(x,t)$ can be written in terms of initial probability density $p(x_0,t_0)$ as:

$$p(x,t) = \lim_{N\to\infty} \int dx_{N-1} \cdots \int dx_0\; p(x,t|x_{N-1},t_{N-1}) \cdots p(x_1,t_1|x_0,t_0)\; p(x_0,t_0).$$

(b) Using Eq. (3.18), derive the path integral solution of the Fokker–Planck equation as,

$$p(x,t) = \lim_{N\to\infty} \underbrace{\int \cdots \int}_{\text{N times}} \left\{ \Pi_{i=0}^{N-1} \left(4\pi D^{(2)}(x_i,t_i)\tau\right)^{-1/2} dx_i \right\}$$
$$\times \exp\left(- \sum_{i=0}^{N-1} \frac{\left[x_{i+1} - x_i - D^{(1)}(x_i,t_i)\tau\right]^2}{4D^{(2)}(x_i,t_i)\tau} \right) p(x_0,t_0)$$

where

$$\lim_{N\to\infty} \sum_{i=0}^{N-1} \frac{\left[x_{i+1} - x_i - D^{(1)}(x_i,t_i)\tau\right]^2}{4D^{(2)}(x_i,t_i)\tau} = \int_{t_0}^{t} \frac{\left[\dot{x}(t') - D^{(1)}(x(t'),t')\right]^2}{4D^{(2)}(x(t'),t')} dt'$$

which means that for small diffusion coefficient $D^{(2)}(x(t),t)$, only the paths near the deterministic solution $\dot{x}(t) = D^{(1)}(x(t),t)$, contribute to $p(x,t)$.

(c) Use the result of part (b) and argue that the distribution function $p(x,t)$ must remain positive, if one starts with a positive distribution $p(x_0,t_0)$.

3.3 Backward Kramers–Moyal equation

Starting from the following Chapman–Kolmogorov equation

$$p(x,t|x',t') = \int p(x,t|x'',t'+\tau)\; p(x'',t'+\tau|x',t')dx''$$

with $t \geq t' + \tau \geq t'$,

(a) Show that $p(x,t|x',t')$ obey the following backward Kramers–Moyal equation

$$\frac{\partial p(x,t|x',t')}{\partial t'} = -\sum_{n=1}^{\infty} D^{(n)}(x',t')\left(\frac{\partial}{\partial x'}\right)^n p(x,t|x',t').$$

(b) Show that the operator

$$\mathcal{L}_{KM}^{\dagger} = \sum_{n=1}^{\infty} D^{(n)}(x',t')\left(\frac{\partial}{\partial x'}\right)^n$$

is the adjoint operator of[1]

$$\mathcal{L}_{KM} = \sum_{n=1}^{\infty}\left(-\frac{\partial}{\partial x'}\right)^n D^{(n)}(x',t').$$

3.4 The master equation: Random walk and diffusion equation

Let $p(i,N)$ denote the probability that a random walker is at site i after N steps. Assume that walkers has an equal probability to walk one step left and right.

[1]The L^2-adjoint of $\mathcal{L}_{KM}$ is defined as $\int h\mathcal{L}_{KM} f\, dx = \int f\mathcal{L}_{KM}^{\dagger} h\, dx$, assuming that f and h decay sufficiently fast at infinity.

(a) Use the master equation and show that

$$p(i, N) = \frac{1}{2} p(i+1, N-1) + \frac{1}{2} p(i-1, N-1).$$

(b) To obtain the continuum limit of this equation, define $t = N\tau$ and $x = ia$, by assuming that $D = a^2/2\tau$ is finite in the limit $\tau \to 0$ and $a \to 0$, show that $p(x, t)$ satisfies the diffusion equation,

$$\frac{\partial p(x,t)}{\partial t} = D \frac{\partial^2 p(x,t)}{\partial x^2}.$$

where D is the diffusion constant.

(c) Show that the solution of diffusion equation is given by:

$$p(x,t) = \frac{1}{\sqrt{4\pi D t}} \exp\left\{-\frac{x^2}{4Dt}\right\}.$$

(d) Show that the conditional probability distribution of the diffusion equation with initial condition $p(x', t|x, t) = \delta(x' - x)$ is given by:

$$p(x', t+\tau|x, t) = \frac{1}{\sqrt{4\pi D\tau}} \exp\left\{-\frac{(x-x')^2}{4D\tau}\right\}.$$

(e) Show that second statistical moment of x is given by

$$\langle x^2(t)\rangle = 2D\, t.$$

3.5 The master equation: Poisson process

Poisson process is defined by the jump rate, $w_{n,n'} = \lambda\, \delta_{n,n'+1}$ for $n = 0, 1, 2, \ldots$ and positive constant $\lambda > 0$.

(a) Show that the Master equation (3.28) reduces to

$$\dot{p}_n = \lambda p_{n-1} - \lambda p_n \quad \forall\, n.$$

(b) Show that the solution for p_n is,

$$p_n = \frac{(\lambda t)^n}{n!} \exp\{-\lambda t\}.$$

(c) Compute the first and second statistical moments $\langle n\rangle$ and $\langle n^2\rangle$.

References

1. H. Risken, *The Fokker-Planck Equation* (Springer, Berlin, 1989)
2. R.F. Pawula, Phys. Rev. **162**, 186 (1967)
3. C. Honisch, *Analysis of Complex Systems: From Stochastic Time Series to Pattern Formation in Microscopic Fluidic Films* (Dissertation, University of Münster) (Westfalen, 2014)
4. J. Honerkamp, *Stochastic Dynamical Systems: Concepts, Numerical Methods, Data Analysis* (Wiley-VCH, New York, 1993)

Chapter 4
Continuous Stochastic Processes

In this chapter we define notions of stochastic continuity and differentiability and describe Lindeberg's condition for continuity of stochastic Markovian trajectories. We also show that the Fokker–Planck equation describes a continuous stochastic process. Finally, we derive the stationary solutions of the Fokker–Planck equation and define potential function of dynamics.

4.1 Stochastic Continuity

In this chapter, we modify the standard definitions of continuity and differentiability to include stochastic trajectories and show that the idea of convergence in the mean-square is essential. We show that wherever mean-squared convergence appears, the definition of continuity and differentiability can be rewritten in terms of the correlation functions [1].

4.1.1 Stochastic Mean-Square Continuity

We begin with the definition of mean-squared continuity [1, 2].

Definition: A process $x(t)$ is *mean-square continuous* at t, if and only if the limit

$$\lim_{h \to 0} \langle (x(t) - x(t-h))^2 \rangle = 0 \tag{4.1}$$

exists. The formal definition can be rewritten in a more useful form (for zero-mean processes) involving the correlation function $f(t_1, t_2) = \langle x(t_1)x(t_2) \rangle$. The aforementioned limit exists if and only if $f(t_1, t_2)$ is continuous in t_1 and t_2 at the point

M. R. Rahimi Tabar, *Analysis and Data-Based Reconstruction of Complex Nonlinear Dynamical Systems*, Understanding Complex Systems,
https://doi.org/10.1007/978-3-030-18472-8_4

$t_1 = t_2 = t$; that is, the limit

$$\lim_{h_1, h_2 \to 0} \left[f(t - h_1, t - h_2) - f(t, t) \right] = 0 \tag{4.2}$$

must exist. If $x(t)$ is stationary, then $f(t, t - \tau) = f(\tau)$ is a function of the time difference only and the condition for continuity is satisfied if and only if $f(\tau)$ is continuous at $\tau = 0$; that is, the limit

$$\lim_{h \to 0} [f(h) - f(0)] = 0 \tag{4.3}$$

must exist. It is then clear that if a stationary process is continuous at time t, it will also be continuous for all times.

4.1.2 Lindeberg's Continuity Condition for Markov Processes

Lindeberg's condition for continuity of stochastic trajectories asserts that the trajectories of a Markov process are continuous if

$$\begin{aligned} C(t) \triangleq C(x, t, \delta) &= \lim_{\tau \to 0^+} \frac{Prob[\ |\Delta x(t)| > \delta \quad |_{x(t)=x}]}{\tau} \\ &= \lim_{\tau \to 0^+} \frac{\int_{|\Delta x(t)|=|x'-x|>\delta} p(x', t+\tau|x, t)\, dx'}{\tau} = 0 \end{aligned} \tag{4.4}$$

holds for all $\delta > 0$, uniformly for all x and t. Here, $\Delta x = x(t+\tau) - x(t)$ and $p(x', t + \tau|x, t)$ is the conditional probability distribution. The Lindeberg's condition (4.4) means that the probability of moving more than δ in a short time τ decays faster than linearly with τ, and that finite jumps of arbitrarily small sizes become very improbable for sufficiently short time intervals.

Example 4.1 (**A process with continues trajectory**) The conditional probability distribution function (PDF) $p(x', t + \tau|x, t)$ for a random walk—Brownian motion-with a diffusion coefficient D is given by (see problem 2.4)

$$p(x', t+\tau|x, t) = \frac{1}{\sqrt{4\pi D\tau}} \exp - \left[(x - x')^2 / 4D\tau \right].$$

Here we show that it satisfies the Lindeberg's continuity condition. From (4.4) we have,

$$C(x, t, \delta) = \lim_{\tau \to 0^+} \frac{1}{\tau} \int_{|x'-x|>\delta} \frac{1}{\sqrt{4\pi D\tau}} \exp\left[-(x - x')^2 / 4D\tau \right] dx'$$

$$= \lim_{\tau\to 0^+} \frac{1}{\tau} \int_{|z|>\delta} \frac{1}{\sqrt{4\pi D\tau}} \exp\left[-z^2/4D\tau\right] dz$$
$$= \lim_{\tau\to 0^+} \frac{2}{\tau} \int_{\delta}^{\infty} \frac{1}{\sqrt{4\pi D\tau}} \exp\left[-z^2/4D\tau\right] dz$$
$$= \lim_{\tau\to 0^+} \frac{2}{\tau} \int_{\delta/\sqrt{2D\tau}}^{\infty} \frac{1}{\sqrt{4\pi}} \exp\left[-u^2/2\right] du$$
$$= \lim_{\tau\to 0^+} \frac{\delta}{\sqrt{4\pi D\tau^3}} \exp\left[-\delta^2/4D\tau\right] = 0. \tag{4.5}$$

The L'hopital rule was used in the last line. It follows that the conditional probability of diffusion equation ensures the continuity of the trajectories.

Example 4.2 (**A process with discontinues trajectory**) The conditional PDF $p(x', t+\tau|x, t)$ for Cauchy process is given by, (see problem 2.4)

$$p(x', t+\tau|x, t) = \frac{1}{\pi} \frac{\tau}{\tau^2 + (x-x')^2}.$$

Here, we examine the Lindeberg's continuity condition for trajectories of the Cauchy process. From Eq. (4.4) we have,

$$C(x, t, \delta) = \lim_{\tau\to 0^+} \frac{1}{\pi\tau} \int_{|x-x'|>\delta} \frac{\tau dx'}{(x-x')^2 + \tau^2} = \lim_{\tau\to 0^+} \frac{2}{\pi\tau} \int_{\delta/\tau}^{\infty} \frac{\tau^2}{\tau^2(z^2+1)} dz$$
$$\lim_{\tau\to 0^+} \frac{2}{\pi\tau} \left\{\arctan(\infty) - \arctan\frac{\delta}{\tau}\right\} = \frac{2}{\delta} \neq 0. \tag{4.6}$$

The L'hopital rule was invoked in the last line. It then follows that the continuity condition is not satisfied, so that the trajectories of the Cauchy process cannot be expected to be continuous. We note that Brownian and Cauchy processes both satisfy the Chapman-Kolmogorov equation.

4.2 Stochastic Differentiability

4.2.1 Mean-Square Differentiability of Stochastic Processes

The derivative of a stochastic process is defined in the following way.

Definition: A process $x(t)$ is mean-square differentiable at t, if and only if there exists a random variable $x'(t)$ such that the limit [1, 2]

$$\lim_{h\to 0} \left\langle \left(\frac{x(t+h) - x(t)}{h} - x'(t) \right)^2 \right\rangle = 0 \tag{4.7}$$

exists. This definition is difficult to implement in practice. As a more useful outcome, one can show that Eq. (4.7) is satisfied, if and only if the correlation function $f(t_1, t_2) = \langle x(t_1)x(t_2)\rangle$ is differentiable at point $t = t_1 = t_2$, i.e., the limit

$$\lim_{h_1, h_2 \to 0} \frac{1}{h_1 h_2}\Big[f(t-h_1, t-h_2) - f(t, t-h_2) - f(t-h_1, t) + f(t,t)\Big]$$

must exist. If the process $x(t)$ is stationary, then $f(t, t-\tau) = f(\tau)$ is a function of the time difference only, and the condition for differentiability is simplified to the existence of the limit

$$\lim_{h \to 0} \frac{1}{h^2}\left[f(h) - 2f(0) + f(-h)\right]. \tag{4.8}$$

4.2.2 *General Condition for Non-differentiability of Stochastic Processes*

A stochastic process $x(t)$ is not differentiable at t, if there exists finite probability

$$Prob\left\{\lim_{h\to 0^+} \left|\Big[x(t+h) - x(t)\Big]/h\right| > k\right\}$$

for any fixed finite k and $h > 0$.

4.3 Description of a Continuous Stochastic Process by a Fokker–Planck Equation

Assume that the first and second Kramers-Moyal (KM) coefficients, $D^{(1)}(x,t)$ and $D^{(2)}(x,t)$, are smooth and do not change dramatically over a short time interval τ (and $D^{(n)}(x,t) = 0$ for $n > 2$), by substituting $p(x', t+\tau|x,t)$, Eq. (3.18) (note the arguments of p), in (4.4) one obtains [3]

$$\begin{aligned} C(x,t,\delta) = &\lim_{\tau\to 0^+} \frac{1}{2\tau\sqrt{\pi D^{(2)}(x,t)\tau}}\left\{\int_{-\infty}^{-\delta+x} \exp\left(-\frac{(x'-x-D^{(1)}(x,t)\tau)^2}{4D^{(2)}(x,t)\tau}\right) dx'\right. \\ &+ \left.\int_{x+\delta}^{\infty} \exp\left(-\frac{(x'-x-D^{(1)}(x,t)\tau)^2}{4D^{(2)}(x,t)\tau}\right) dx'\right\} \\ \equiv &\lim_{\tau\to 0^+} \frac{1}{2\tau\sqrt{\pi D^{(2)}(x,t)\tau}} (I + II)\,. \end{aligned} \tag{4.9}$$

The first term in Eq. (4.9) is written as,

$$I = \int_{-\infty}^{-\delta+x} \exp\left(-\frac{(x' - x - D^{(1)}(x,t)\tau)^2}{4D^{(2)}(x,t)\tau}\right) dx'$$
$$= \sqrt{4D^{(2)}(x,t)\tau} \int_{-\infty}^{(-\delta - D^{(1)}(x,t)\tau)/\sqrt{4D^{(2)}(x,t)\tau}} \exp(-u^2)\, du$$
$$= \sqrt{4D^{(2)}(x,t)\tau}\, \frac{\sqrt{\pi}}{2}\, erfc\left[\frac{\delta + D^{(1)}(x,t)\tau}{4D^{(2)}(x,t)\tau}\right] \tag{4.10}$$

where $u = (x' - x - D^{(1)}(x,t)\tau)^2/\sqrt{4D^{(2)}(x,t)\tau}$ and erfc$[x]$ can be written in terms of the error function as erfc$[x]$ =1 – erf$[x]$, where erf$(x) = 2/\sqrt{\pi}\int_0^x \exp(-u^2)du$. Expanding the expression for I in terms of τ gives,

$$I = 2D^{(2)}(x,t)\tau \frac{\exp\left\{-(\delta + D^{(1)}(x,t)\tau)^2/4D^{(2)}(x,t)\tau\right\}}{\delta + D^{(1)}(x,t)\tau} \times$$
$$\left\{1 - \frac{2D^{(2)}(x,t)\tau}{(\delta + D^{(1)}(x,t)\tau)^2} + \frac{12(D^{(2)}(x,t)\tau)^2}{(\delta + D^{(1)}(x,t)\tau)^4} + \cdots\right\}. \tag{4.11}$$

In the limit $\tau \to 0^+$, we find that $\lim_{\tau\to 0^+} \frac{1}{\tau\sqrt{4\pi D^{(2)}(x,t)\tau}} I$ approaches zero as $\lim_{\tau\to 0^+} (1/\tau)^{1/2}\exp(-\delta^2/4D^{(2)}(x,t)\tau) \to 0$. Using similar analysis shows also that the second term in Eq. (4.9) approaches zero in the limit $\tau \to 0^+$. Therefore, we conclude that the short-time propagator of Fokker–Planck equation satisfies the continuity condition, Eq. (4.4).

In the Chap. 11 we will prove that non-vanishing higher order KM coefficients $D^{(n)}(x,t)$ with $n > 2$ provide an upper limit for the continuity condition (4.4).

4.4 Stationary Solution of the Fokker–Planck Equation and the Potential Function

The probability density functions of the processes with non-vanishing drift $D^{(1)}(x,t)$ and diffusion coefficients $D^{(2)}(x,t)$ satisfy the Fokker–Planck (FP) equation. Here, we are interested in to study the long time limit solution, i.e., the stationary solution of the equation. This is defined as the probability density $p_{st}(x) = \lim_{t\to\infty} p(x,t)$ that satisfies

$$\frac{\partial p_{st}(x,t)}{\partial t} = 0. \tag{4.12}$$

Let us consider time-homogeneous continues processes, which means that the drift $D^{(1)}(x,t)$ and diffusion coefficient $D^{(2)}(x,t)$ are independent of time. In this case the FP operator $\mathcal{L}_{FP}$ is written as,

$$\mathcal{L}_{FP} p_{st}(x) = -\frac{\partial}{\partial x}\left(D^{(1)}(x)\, p_{st}(x)\right) + \frac{\partial^2}{\partial x^2}\left(D^{(2)}(x)\, p_{st}(x)\right)$$
$$= \frac{d\, j(x)}{dx} = 0 \tag{4.13}$$

where the current $j(x)$ is given by;

$$j(x) = \left(-D^{(1)}(x) + \frac{d}{dx} D^{(2)}(x)\right) p_{st}(x). \tag{4.14}$$

The solution of Eq. (4.13) is a constant current, $j(x) = c$. If the process is limited to some interval, say $x(t) \in [x_1, x_2]$, then $j(x_1) = j(x_2) = c$. We assume a reflecting boundary condition in one of the boundaries, for instance at $x = x_1$, which means that $j(x_1) = 0$, hence fixing the constant c to be zero. Equation (4.14) with $j(x) = 0$ and for $g(x) = D^{(2)}(x)\, p_{st}(x)$ is then written as,

$$\frac{dg}{dx} = \left\{\frac{D^{(1)}(x)}{D^{(2)}(x)}\right\} g(x). \tag{4.15}$$

By solving the first-order differential Eq. (4.15), we determine the stationary solution of the FP equation to be

$$p_{st}(x) = \frac{A}{D^{(2)}(x)} \exp\left\{\int^{x} \frac{D^{(1)}(x')}{D^{(2)}(x')}\, dx'\right\} \tag{4.16}$$

where A is a normalization constant.

We can define a potential corresponding to the process in terms of stationary solution of FP equation. By writing $p_{st}(x)$ as $A \exp(-V(x))$, one defines the potential of dynamics as,

$$V(x) = \ln(D^{(2)}(x)) - \int^{x} \frac{D^{(1)}(x')}{D^{(2)}(x')}\, dx' \tag{4.17}$$

where for a constant diffusion coefficient the potential will be (ignoring a constant shift and multiplication factor)

$$V(x) = -\int^{x} D^{(1)}(x')\, dx'. \tag{4.18}$$

The valleys in the potential $V(x)$ represent stable attractors of the dynamics. In general, knowing the number of states in a stochastic process is crucial for understanding its underlying dynamics, as well as for nonparametric modeling of its behavior. In addition, for many systems we are interested in studying the changes in the number of states that represent bifurcations of the systems. Estimating the drift and diffusion coefficients from time series (see Chaps. 16–21) enables us to study the structure of the potential of the system and, henceforth, the number of stable attractors.

Problems

4.1 Second statistical moment from Fokker–Planck equation

Suppose that the drift and diffusion coefficients in the Fokker–Planck equation are given by: $D^{(1)}(x) = -\gamma x$ and $D^{(2)}(x) = \alpha + \beta x^2$, where $\alpha, \beta, \gamma > 0$.

(a) Derive the stationary solution of Fokker–Planck equation with vanishing current.

(b) Show that second statistical moment $\langle x^2 \rangle$ diverges for $\beta \geq \gamma$.

4.2 The Ornstein-Uhlenbeck process

The Ornstein-Uhlenbeck process is stationary at steady-state, and its correlation function is given by,

$$f(\tau) = \frac{1}{2\tau_c} \exp(-|\tau|/\tau_c)$$

where $\tau_c > 0$ is a constant (correlation time scale).

Using (4.3) and (4.8), check the mean-square continuity and differentiability of the Ornstein-Uhlenbeck process for finite and vanishing (limit) τ_c.

References

1. M. Scott, *Applied Stochastic Processes in Science and Engineering* (Springer, Berlin, 2013)
2. Z. Schuss, *Theory and Applications of Stochastic Processes: An Analytical Approach* (Springer, Berlin, 2010)
3. M. Anvari, K. Lehnertz, M.R. Rahimi Tabar, J. Peinke, Sci. Rep. **6**, 35435 (2016)

Chapter 5
The Langevin Equation and Wiener Process

In this chapter we introduce the Langevin equation and Wiener process. All the statistical properties of Wiener process will be presented and discussed.

5.1 The Langevin Equation

Suppose that the stochastic variable $x(t)$ satisfies the following dynamical equation,

$$\frac{dx}{dt} = a(x, t) + b(x, t)\ \Gamma(t) \tag{5.1}$$

where the functions $a(x, t)$ and $b(x, t)$ are arbitrary smooth functions and $\Gamma(t)$ is Gaussian zero mean, i.e., $\langle \Gamma(t) \rangle = 0$, and white-noise, which means that its correlation function is given by,

$$\langle \Gamma(t)\ \Gamma(t') \rangle = \delta(t - t')\ . \tag{5.2}$$

Here, $\delta(t - t')$ is the Dirac-delta function and, by convention, the intensity of noise is 1. $\Gamma(t)$ is known as white-noise because the Fourier transform of its correlation function has constant amplitude for every frequencies. The stochastic Eq. (5.1) is known as the white noise-driven *Langevin* equation. Depending on the functions $a(x, t)$ and $b(x, t)$, one obtains various stochastic processes. For constant $b(x, t) = b_0$ the process is known as one with additive noise, while for x-dependent (state-dependent) $b(x, t)$ the process $x(t)$ is one with multiplicative noise.

The Wiener process is defined by $a(x, t) = 0$ and $b(x, t) = 1$. Therefore, we have

$$\frac{dx(t)}{dt} = \Gamma(t). \tag{5.3}$$

In what follows, we denote the Wiener process by $W(t)$ ($x(t) \equiv W(t)$). It should be pointed out that the Wiener process is another name for a Brownian motion. It has

M. R. Rahimi Tabar, *Analysis and Data-Based Reconstruction of Complex Nonlinear Dynamical Systems*, Understanding Complex Systems,
https://doi.org/10.1007/978-3-030-18472-8_5

important properties, including the fact that all of its Kramers–Moyal coefficients $D^{(n)}(x,t)$ with $n=1$ and $n\geq 3$ vanish. The Wiener process designates a stochastic process whose increments are independent, stationary and normally (Gaussian) distributed.

5.2 The Kramers–Moyal Coefficients of Wiener Process

We first derive the Kramers–Moyal (KM) coefficients $D^{(n)}(w,t)$ of the Wiener process. According to Eq. (5.3), $dW(t)=\Gamma(t)\,dt$ and, therefore,

$$W(t+\Delta t)-W(t)=\int_t^{t+\Delta t}\Gamma(t')\,dt'\ . \tag{5.4}$$

Thus, the first KM coefficient is given by,

$$\begin{aligned} D^{(1)}(w,t) &= \lim_{\Delta t\to 0}\frac{\langle (W(t+\Delta t)-W(t))\rangle\,|_{W(t)=w}}{\Delta t} \\ &= \lim_{\Delta t\to 0}\frac{\left\langle \int_t^{t+\Delta t}\Gamma(t')dt'\right\rangle}{\Delta t} = \lim_{\Delta t\to 0}\frac{\int_t^{t+\Delta t}\left\langle \Gamma(t')\right\rangle dt'}{\Delta t} = 0, \end{aligned} \tag{5.5}$$

in the last step of the derivation we used the the fact that the noise has a zero mean, i.e., $\langle\Gamma(t)\rangle=0$. Similarly to determine the second KM coefficient we need to derive the conditional second moment, given by,

$$\begin{aligned} \left\langle (W(t+\Delta t)-W(t))^2\right\rangle|_{W(t)=w} &= \int_t^{t+\Delta t}dt'\int_t^{t+\Delta t}dt''\left\langle \Gamma(t')\,\Gamma(t'')\right\rangle \\ &= \Delta t \end{aligned} \tag{5.6}$$

and, therefore, the second order KM coefficient is given by,

$$D^{(2)}(w,t)=\frac{1}{2!}\lim_{\Delta t\to 0}\frac{\left\langle (W(t+\Delta t)-W(t))^2\right\rangle|_{W(t)=w}}{\Delta t}=\lim_{\Delta t\to 0}\frac{\Delta t}{2\Delta t}=\frac{1}{2}\ .$$

By a similar procedure one can show that, $D^{(n)}(w,t)=0$ for $n\geq 3$. Therefore, the conditional probability distribution function of the Wiener process satisfies the Fokker–Planck equation and belongs to the class of continuous stochastic processes; see Sect. 4.3.

5.3 Conditional Probability Distribution Function of the Wiener Process

The probability distribution function of Wiener process satisfies the following Fokker–Planck equation with $D^{(1)}(w,t) = 0$ and $D^{(2)}(w,t) = 1/2$,

$$\frac{\partial}{\partial t}\, p(w,t|w_0,t_0) = \frac{1}{2}\frac{\partial^2}{\partial w^2}\, p(w,t|w_0,t_0) \tag{5.7}$$

where the initial condition is $p(w,t_0|w_0,t_0) = \delta(w-w_0)$. Equation (5.7) is a diffusion equation whose solution can be determined using the Laplace transformation, and is given by,

$$p(w,t|w_0,t_0) = (2\pi(t-t_0))^{-1/2}\ \exp\left[-\frac{(w-w_0)^2}{2(t-t_0)}\right] \tag{5.8}$$

for $t > t_0$.

5.4 Statistical Moments of the Wiener Process

Using the conditional PDF of the Wiener process, Eq. (5.8), we show that statistical moments of $W(t)$ are [1]

$$\langle W(t)\rangle = w_0$$

$$\left\langle (W(t)-W_0)^2\right\rangle = t - t_0$$

$$\left\langle (W(t)-W_0)^{2k-1}\right\rangle = 0 \qquad k = 1,2,\ldots$$

$$\left\langle (W(t)-W_0)^{2k}\right\rangle = \int_{-\infty}^{\infty} p(x = w-w_0)\, x^{2k} dx$$

$$= \frac{2}{\sqrt{2\pi(t-t_0)}}\int_0^{\infty} \exp\left(-\frac{x^2}{2(t-t_0)}\right)\, x^{2k}\, dx$$

---

[1] The nth statistical moment of a stochastic variable $x(t)$ with probability $p(x)$ is defined as the expected or mean value of x to the nth power, i.e., $\langle x^n\rangle$. The statistical moments may be determined from the generating function $G(u) = \langle \exp(ux)\rangle = \int \exp(ux)p(x)dx$ using the relation, $\langle x^n\rangle = \frac{\partial^n}{\partial (u)^n} G(u)|_{u=0}$. For the PDF given by Eq. (5.8), the generating function is $G(u) = \exp(\frac{1}{2}(t-t_0)u^2)$ for the variable $x = w - w_0$.

$$= \frac{(2(t-t_0))^k}{\sqrt{\pi}} \int_0^\infty \exp(-z)\; z^{k-\frac{1}{2}}\, dz$$

$$= \frac{\Gamma(k+1/2)}{\Gamma(1/2)}\, 2^k\, (t-t_0)^k \qquad k=1,2,\ldots \tag{5.9}$$

where $\Gamma(x)$ is the Euler gamma function.

5.5 Markov Property of the Wiener Process

The Wiener process is a Markov process. This can be shown by the fact that the conditional probability distribution function Eq. (5.8) satisfies the Chapman–Kolmogorov equation. Moreover, the presence of the first-order time derivative and the absence of memory in the noise $\Gamma(t)$ in Eq. (5.3) confirm the Markov nature of the Wiener process. Let us examine the Chapman–Kolmogorov equation, which means that,

$$\begin{aligned}
p(w_3, t_3|w_1, t_1) &\stackrel{?}{=} \int_{-\infty}^{\infty} p(w_3, t_3|w_2, t_2)\; p(w_2, t_2|w_1, t_1) dw_2 \\
&= \int_{-\infty}^{\infty} \frac{dw_2}{2\pi\sqrt{(t_3-t_2)(t_2-t_1)}} \exp\left(-\frac{(w_3-w_2)^2}{2(t_3-t_2)} - \frac{(w_2-w_1)^2}{2(t_2-t_1)}\right) \\
&= \int_{-\infty}^{\infty} \frac{du}{2\pi\sqrt{(t_3-t_2)(t_2-t_1)}} \\
&\times \exp\left(-\left(\frac{1}{2(t_2-t_1)} + \frac{1}{2(t_3-t_2)}\right)u^2 + \frac{w_3-w_1}{t_3-t_2}\, u - \frac{(w_3-w_1)^2}{2(t_3-t_2)}\right)
\end{aligned}$$

where $u = w_2 - w_1$ and $t_3 > t_2 > t_1$. Completing the square and performing the gaussian integral over u gives,

$$p(w_3, t_3|w_1, t_1) = \frac{1}{2\pi\sqrt{(t_3-t_2)(t_2-t_1)}} \sqrt{\frac{\pi}{a}} \exp\left(\frac{b^2}{4a} - c\right) \tag{5.10}$$

where $a = \frac{(t_3-t_1)}{2(t_3-t_2)(t_2-t_1)}$, $b = \frac{w_3-w_1}{t_3-t_2}$, $c = \frac{(w_3-w_1)^2}{2(t_3-t_2)}$ and $b^2/4a - c = -(w_3-w_2)^2/2(t_3-t_2)$. Finally, we determine the intended result,

$$p(w_3, t_3|w_1, t_1) = \frac{1}{\sqrt{2\pi(t_3-t_1)}} \exp\left(-\frac{(w_3-w_1)^2}{2(t_3-t_1)}\right).$$

5.6 Independence of Increments of the Wiener Process

Using the Markov property of the Wiener process, we write the r-point joint probability distribution (PDF) as,

$$p_r(w_r, t_r; w_{r-1}, t_{r-1}; \ldots; w_1, t_1) = \prod_{i=1}^{r-1} p(w_{i+1}, t_{i+1}|w_i, t_i)\ p(w_1, t_1)$$

and, using Eq. (5.8) for conditional PDF, the r-point joint PDF, $p_r(w_r, t_r; \ldots; w_1, t_1)$ is written as,

$$p_r(w_r, t_r; w_{r-1}, t_{r-1}; \ldots; w_1, t_1) =$$

$$\prod_{i=1}^{r-1} \left[2\pi(t_{i+1} - t_i)\right]^{-1/2} \exp\left\{-\frac{(w_{i+1} - w_i)^2}{2(t_{i+1} - t_i)}\right\} p(w_1, t_1). \tag{5.11}$$

Here, $t_r > t_{r-1} > \cdots > t_1$. If we define the increment $\Delta w_i = w(t_{i+1}) - w(t_i)$ and a time lag, $\Delta t_i = t_{i+1} - t_i$, the r-point joint PDF is factorize as,

$$p_r(w_r, t_r; w_{r-1}, t_{r-1}; \ldots; w_1, t_1) =$$

$$\prod_{i=1}^{r-1} [2\pi\Delta t_i]^{-1/2} \exp\left\{-\frac{(\Delta w_i)^2}{2\Delta t_i}\right\} p(w_1, t_1). \tag{5.12}$$

It is evident from (5.12) that the increments of the Wiener process are independent processes.

5.7 The Correlation Function of the Wiener Process

The correlation function of the Wiener process is given by,

$$\langle W(t)\ W(t')\rangle = min(t, t') \tag{5.13}$$

where $min(t, t')$ is the minimum of (t, t'). The correlation function is the result of the independent increment and the zero mean properties of the Wiener processes. The zero mean property implies that $\langle W(t)\rangle = \langle W(t')\rangle = 0$. We also set $W(0) = 0$.

To prove (5.13), we consider the case $t' < t$, and write $W(t) = W(t') + (W(t) - W(t'))$ as independent increments, then

$$\langle W(t)\ W(t')\rangle = \langle [W^2(t') + W(t')\ (W(t) - W(t'))]\rangle$$

$$= \langle W^2(t')\rangle + \langle [W(t')\,(W(t) - W(t'))]\rangle$$

$$= \langle W^2(t')\rangle + \langle W(t')\rangle\,\langle (W(t) - W(t'))\rangle$$

$$= t' + 0 = t' \tag{5.14}$$

where we have used $\langle W^2(t')\rangle = t'$ [see Eq. (5.9)]. In case $t < t'$ we can show that the correlation function is equal to t to finally arrive at (5.13).

5.8 Wiener Process Is Not Differentiable

The Wiener process has a zero drift and a constant diffusion coefficient, and all of its higher-order Kramers–Moyal coefficients with $n > 2$ vanish. Therefore, its probability distribution satisfies the Fokker–Planck equation and, thus, has continuous trajectory; see Sect. 4.3. It is not, however, differentiable. This means that there is a finite probability

$$Prob\{\lim_{h\to 0^+} \left|\left[W(t+h) - W(t)\right]/h\right| > k\}$$

for any fixed finite k and $h > 0$. This can be shown by interchanging the limit with the probability operations, since time is deterministic, and use the normal distribution of the increment $W(t+h) - W(t)$, so that,

$$Prob\left\{\lim_{h\to 0^+} \left|\left[W(t+h) - W(t)\right]/h\right| > k\right\}$$
$$= \lim_{h\to 0^+} Prob\left\{\left|\left[W(t+h) - W(t)\right]/h\right| > k\right\}$$
$$= \lim_{h\to 0^+} Prob\left\{\left|\left[W(t+h) - W(t)\right]\right| > kh\right\} = \lim_{h\to 0^+}\left[\frac{2}{\sqrt{2\pi h}}\int_{kh}^{\infty} \exp(-\frac{w^2}{2h})dw\right]$$
$$= \lim_{h\to 0^+}\left[\frac{2}{\sqrt{2\pi}}\int_{k\sqrt{h}}^{\infty} \exp(-\frac{v^2}{2})dv\right]$$
$$= \left[\frac{2}{\sqrt{2\pi}}\int_{0}^{\infty} \exp(-\frac{v^2}{2})dv\right] = 1. \tag{5.15}$$

We note that the error is $\frac{2}{\sqrt{2\pi}}\int_0^{k\sqrt{h}} \exp(-\frac{v^2}{2})dv \le \frac{2}{\sqrt{2\pi}}\int_0^{k\sqrt{h}} dv = \frac{2}{\sqrt{2\pi}}k\sqrt{h} << 1.$

Problems

5.1 Wiener process is not differentiable

The correlation function of Wiener process is given by $\langle W(t)\ W(t')\rangle = min(t, t')$. Using the definition in Sect. 4.2.1, show that Wiener process is not mean-square differentiable.

5.2 Higher-order Kramers–Moyal coefficients of Wiener process

Show that Wiener process has vanishing Kramers–Moyal coefficients for $n \geq 3$

$$D^{(n)}(w, t) = 0.$$

5.3 Statistical moments of the Wiener process

Compute

(a) $\langle W(t)\ |W(s) = W_s\rangle$,

(b) $\langle W^2(s)\ W(t)\rangle$
for $t > s$.

5.4 Statistical moments of the integral of the Wiener process

Define $Z(t) = \int_0^t W(s)ds$ and show that

(a) $\langle Z(t)\rangle = 0$,

(b) $var(Z(t)) = \frac{t^3}{3}$.

5.5 Statistical moments generating function

Let W be a Wiener process. Show that,

$$\left\langle e^{uW(t)}\right\rangle = e^{\frac{1}{2}u^2\langle W^2(t)\rangle} = e^{\frac{1}{2}u^2 t}$$

with $W(0) = 0$.

5.6 Conditional probability distribution function of the Wiener process

Fill in the details in the derivation of Eq. (5.8).

5.7 Scaling and increments

Show, for the Wiener process, that,

(a) for $\lambda > 0$, $W(t) \overset{\mathrm{d}}{=} \lambda^{-1/2} W(\lambda t)$

(b) for $0 \leq s \leq t$, $W(t) - W(s) \overset{\mathrm{d}}{=} W(t-s)$.

where $\overset{\mathrm{d}}{=}$ stands for equality of the distributions.

5.8 The Feynman–Kac formula

Given a function $f(x)$, define

$$u(x,t) = \langle f(x + W(t)) \rangle$$

and show that this is solution of the diffusion equation

$$\frac{\partial u}{\partial t} = \frac{1}{2}\frac{\partial^2 u}{\partial x^2}, \qquad u(x,0) = f(x).$$

5.9 The Feynman–Kac formula

Given a function $f(x)$, define

$$u(x,t) = \left\langle \left(f(x + W(t)) \, \exp\left(\int_0^t c(x + W(s)) ds \right) \right) \right\rangle$$

and show that this is solution of following partial differential equation:

$$\frac{\partial u}{\partial t} = \frac{1}{2}\frac{\partial^2 u}{\partial x^2} + c(x)u, \qquad u(x,0) = f(x).$$

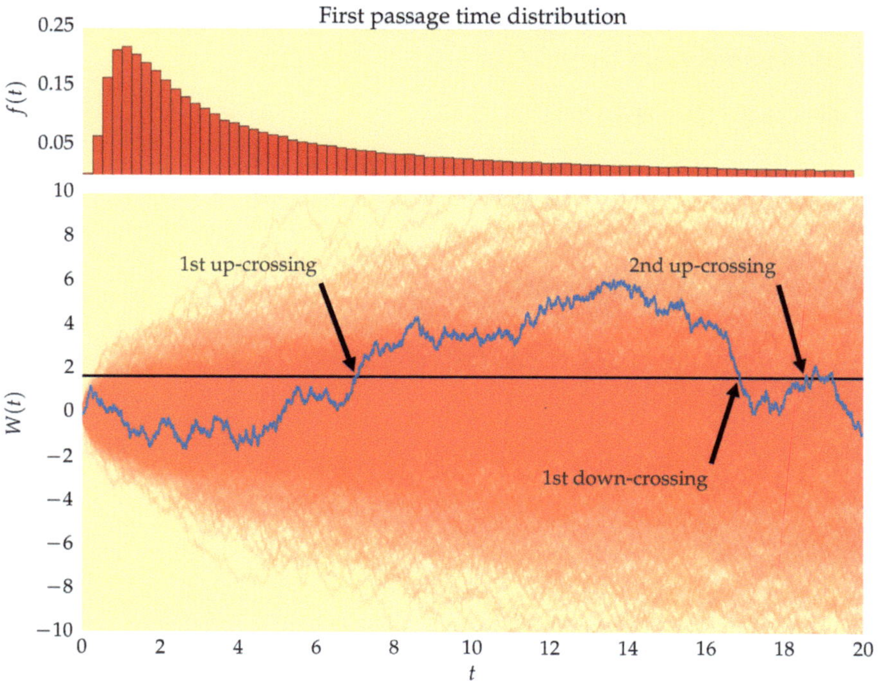

Fig. 5.1 Ensembles of the Wiener processes. The first and second up-crossing points are located in the figures. In the upper panel the first-passage time density $f(t)$ for the Wiener trajectories is plotted

5.10 First-passage time

Let $W(t)$ be a Wiener process. For $a > 0$, define the first passage time

$$\tau_a = \inf\{t : W(t) \geq a\};$$

prove that probability of having $\tau_a \leq t$ may be written as

$$P(\tau_a \leq t) = 2P(W(t) > a).$$

5.11 First-passage time distribution of Wiener process

In Fig. 5.1 an ensemble of the Wiener processes is given. Since the Wiener process is a stochastic process, each of its realization (which starts at $t_0 = 0$ from w_0) reaches a given barrier $x = a$ for the first time at a different time. Therefore, it has a first-passage time probability density $f(t)$.

(a) Show that $f(t)$ is given by:

$$f(t) = \frac{|a - w_0|}{\sqrt{2\pi t^3}} \exp\left\{-\frac{(a - w_0)^2}{2t}\right\}.$$

(b) Show that mean first-passage time $\langle t \rangle$ is infinite.

Chapter 6
Stochastic Integration, Itô and Stratonovich Calculi

In this chapter Itô and Stratonovich calculi are introduced and we prove the Itô lemma and describe Itô calculus for multiplicative noise. Finally Itô-Taylor expansion will be given for white noise-driven Langevin dynamics.

6.1 Stochastic Integration

The Fokker–Planck equation describes the time dependence of marginal probability distribution function $p(x, t)$ and conditional probability distribution function $p(x, t|x', t')$. Now we would like to write down a dynamical equation governing the stochastic process x itself, where its marginal and conditional PDFs satisfy the Fokker–Planck equation. To write down the corresponding stochastic equation i.e. the Langevin equation, we need to know stochastic integration and Itô and Stratonovich calculi [1].

Let us back to the definition of an integral, following the exposition of Gardiner [1], which is given by,

$$\int_0^t f(t')\, dt' = \lim_{n\to\infty} \sum_{i=1}^{n} f(\tau_i)\, (t_i - t_{i-1}) \tag{6.1}$$

where time τ_i is in the interval $[t_{i-1}, t_i]$. For the case that integration measure is differential of a function, e.g. $g(t)$, we define the Riemann–Stieltjes integral as,

$$\int_0^t f(t')\, dg(t') = \lim_{n\to\infty} \sum_{i=1}^{n} f(\tau_i)\, (g(t_i) - g(t_{i-1})) \tag{6.2}$$

for a smooth measure $g(t)$ this limit converges to a unique value independent of the choice of where τ_i taken in the interval $[t_{i-1}, t_i]$ [1].

M. R. Rahimi Tabar, *Analysis and Data-Based Reconstruction of Complex Nonlinear Dynamical Systems*, Understanding Complex Systems,
https://doi.org/10.1007/978-3-030-18472-8_6

Now suppose $g(t) \equiv W(t)$ be a Wiener process so that, its time derivative $\dot{W}(t)$ is delta-correlated as,

$$\langle \dot{W}(t)\ \dot{W}(t') \rangle = \delta(t - t') \tag{6.3}$$

in any interval in real line. The Wiener process is continues, but is not differentiable, see Chap. 5. Therefore we are dealing with the integral that its measure is not smooth. In addition, because of stochastic nature of integral measure for each realization of the trajectory $W(t)$, the integral itself is a stochastic quantity.

Discretization of interval $[t_0, t]$ to n parts and assuming that in each piece τ_i belongs to the interval $t_{i-1} \leq \tau_i \leq t_i$, we define the integral as

$$J = \int_{t_0}^{t} f(t')\, dW(t') = \lim_{n\to\infty} S_n = \lim_{n\to\infty} \sum_{i=1}^{n} f(\tau_i) \left[W(t_i) - W(t_{i-1}) \right]. \tag{6.4}$$

J is a stochastic quantity and one can define an ensemble averaging over different realizations of Wiener process as,

$$\langle J \rangle = \left\langle \int_{t_0}^{t} f(t')\, dW(t') \right\rangle = \lim_{n\to\infty} \sum_{i=1}^{n} \left\langle f(\tau_i)\ \left[W(t_i) - W(t_{i-1}) \right] \right\rangle. \tag{6.5}$$

These type of integrals are very sensitive for choosing of τ_i in the interval $t_{i-1} \leq \tau_i \leq t_i$. To see this, we assume that $f(t) \equiv W(t)$ and Eq. (6.5) can be written as,

$$\begin{aligned}
\langle J \rangle &= \left\langle \int_{t_0}^{t} W(t')\, dW(t') \right\rangle = \lim_{n\to\infty} \sum_{i=1}^{n} \langle W(\tau_i)\ [W(t_i) - W(t_{i-1})] \rangle \\
&= \lim_{n\to\infty} \sum_{i=1}^{n} \{ \langle W(\tau_i)\ W(t_i) \rangle - \langle W(\tau_i)\ W(t_{i-1}) \rangle \} \\
&= \lim_{n\to\infty} \sum_{i=1}^{n} \{ \min(\tau_i, t_i) - \min(\tau_i, t_{i-1}) \} = \lim_{n\to\infty} \sum_{i=1}^{n} \{ \tau_i - t_{i-1} \}
\end{aligned} \tag{6.6}$$

where in the last step we used the Eq. (5.13). Assuming $\tau_i = \alpha t_i + (1 - \alpha) t_{i-1}$ with $0 \leq \alpha \leq 1$, we find,

$$\langle J \rangle = (t - t_0)\alpha \tag{6.7}$$

which means that for given α we have different solutions for the mean integral $\langle J \rangle$, which can vary from 0 to $(t - t_0)$.

There are two standard choice for $f(\tau_i)$ in (6.4), which are suggested by Itô and Stratonovich and are as follows,

$$f(\tau_i) = f(t_{i-1}) \qquad \Rightarrow \mathit{It\hat{o}\,calculus}$$
$$f(\tau_i) = \frac{f(t_i)+f(t_{i-1})}{2} \qquad \Rightarrow \mathit{Stratonovich\,calculus}$$

the descriptions looks arbitrary but we know when to use which calculus, see Sect. 6.2.1.

In general to define the stochastic integral like $\int_0^t f(t')\,dW(t')$, there are two different definitions, Itô and Stratonovich. In Itô description the integral is given by;

$$(I)\int_{t_0}^{t} f(t')\,dW(t') = ms - \lim_{n\to\infty}\left\{\sum_{i=1}^{n} f(t_{i-1})\,[W(t_i) - W(t_{i-1})]\right\}, \tag{6.8}$$

where mean-square limit ($ms - \lim$) is defined as, $ms - \lim_{n\to\infty} X_n$ is equal to x if,

$$\lim_{n\to\infty}\langle (X_n - x)^2\rangle \to 0. \tag{6.9}$$

To see how this prescription works let us consider $f(t) = W(t)$, be the Wiener process. Therefore the integral will be the following sum,

$$I = \int_{t_0}^{t} W(t')\,dW(t') \to S_n = \sum_{i=1}^{n} W(t_{i-1})\,(W(t_i) - W(t_{i-1})) \tag{6.10}$$

and I can be written as,

$$I = \sum_{i=1}^{n} W(t_{i-1})\,\Delta W(t_i) = \frac{1}{2}\sum_{i=1}^{n}\left[(W(t_{i-1}) + \Delta W(t_i))^2 - W(t_{i-1})^2 - (\Delta W(t_i))^2\right] \tag{6.11}$$

the first two terms in the r.h.s of Eq. (6.11) are complete differential and then we find,

$$I = \frac{1}{2}[W(t)^2 - W(t_0)^2] - \frac{1}{2}\sum_{i=1}^{n}(\Delta W(t_i))^2. \tag{6.12}$$

The first two terms in Eq. (6.12) will be fixed by choosing the initial and final times t_0 and t, but the last term will change for different ΔW_is. Therefore we should calculate the $ms - \lim$ of last term, i.e.

$$\left\langle \sum_{i=1}^{n}(\Delta W(t_i))^2 \right\rangle = \sum_{i=1}^{n}\langle (W(t_i) - W(t_{i-1}))^2\rangle$$
$$= \sum_{i=1}^{n}(t_i - t_{i-1}) = t - t_0 \tag{6.13}$$

where in the last step we used the Eq. (5.9). Now we should check that Eq. (6.9) holds or not, then we have to show that the following expression is vanishing in the limit $n \to \infty$,

$$\left\langle \left[\sum_{i=1}^{n} (W(t_i) - W(t_{i-1}))^2 - (t - t_0) \right]^2 \right\rangle = \left\langle \sum_{i=1}^{n} (W(t_i) - W(t_{i-1}))^4 \right\rangle$$
$$+2 \sum_{j=1} \sum_{i>j} \left\langle (W(t_i) - W(t_{i-1}))^2 \, (W(t_j) - W(t_{j-1}))^2 \right\rangle$$
$$-2(t - t_0) \sum_{i=1}^{n} \left\langle (W(t_i) - W(t_{i-1}))^2 \right\rangle + (t - t_0)^2. \tag{6.14}$$

We note that Wiener process $W(t_i) - W(t_{i-1})$ has Gaussian distribution and the increments are independent, therefore,

$$\left\langle (W(t_i) - W(t_{i-1}))^2 \, (W(t_j) - W(t_{j-1}))^2 \right\rangle = (t_i - t_{i-1}) \, (t_j - t_{j-1})$$
$$\left\langle (W(t_i) - W(t_{i-1}))^4 \right\rangle = 3 \left\langle (W(t_i) - W(t_{i-1}))^2 \right\rangle^2 = 3(t_i - t_{i-1})^2. \tag{6.15}$$

Combining the Eqs.(6.14) and (6.15), we find,

$$\left\langle \left[\sum_{i=1}^{n} (W(t_i) - W(t_{i-1}))^2 - (t - t_0) \right]^2 \right\rangle = 2 \sum_{i=1}^{n} (t_i - t_{i-1})^2$$
$$+ \sum_{j=1} \sum_{i>j} \left[(t_i - t_{i-1}) - (t - t_0) \right] \left[(t_j - t_{j-1}) - (t - t_0) \right]$$
$$= 2 \sum_{i=1}^{n} (t_i - t_{i-1})^2. \tag{6.16}$$

In Eq. (6.16) the expression $\sum_{i=1}^{n} (t_i - t_{i-1})^2$ approaches to zero in the limit $n \to \infty$ as $\mathcal{O}(\Delta t)^2$. Therefore we conclude that

$$ms - \lim_{n \to \infty} \sum_{i=1}^{n} (W(t_i) - W(t_{i-1}))^2 = t - t_0. \tag{6.17}$$

Summing up all of the results, we find the stochastic integral in the Itô description as,

$$(I) \int_{t_0}^{t} W(t') \, dW(t') = \frac{1}{2} \left[W^2(t) - W^2(t_0) \right] - \frac{1}{2}(t - t_0). \tag{6.18}$$

The Stratonovich definition of stochastic integral $\int_{t_0}^{t} f(t')dW(t')$ is given by,

$$(S)\int_{t_0}^{t} f(t')\,dW(t') = ms - \lim_{n\to\infty}\left\{\sum_{i=1}^{n}\frac{f(t_i)+f(t_{i-1})}{2}\left[W(t_i)-W(t_{i-1})\right]\right\}. \tag{6.19}$$

Similar calculation for integral (6.10) i.e. with $f(t) \equiv W(t)$, gives us,

$$(S)\int_{t_0}^{t} W(t')\,dW(t') = \frac{1}{2}[W^2(t) - W^2(t_0)]. \tag{6.20}$$

Considering the Eqs. (6.18) and (6.20), shows that normal rules of calculus don't apply to Itô integral, but is applicable to Stratonovich description. For arbitrary function $f(t)$ in the stochastic integrals like $\int_{t_0}^{t} f(t')\,dW(t')$, there is not specific relation between Itô and Stratonovich descriptions.

6.2 Nonanticipating Function and Itô Lemma

The function $f(t)$ in the integral $\int_{t_0}^{t} f(t')\,dW(t')$ is called nonanticipating function if for $s > t$, $f(t)$ and $W(s) - W(t)$ be independent. Now we can present the Itô lemma that states for nonanticipating function $f(t)$,

$$\int_{t_0}^{t} f(t')\,\left[dW(t')\right]^{N+2} = \begin{cases} \int_{t_0}^{t} f(t')\,dt' & N = 0 \\ 0 & N > 0 \end{cases} \tag{6.21}$$

which means that $dW^2 \equiv dt$ and $dW^{2+N} \equiv 0$ for $N > 0$, up to order $\mathcal{O}(\mathrm{d}t)$. To prove the Itô lemma, let us start the case $N = 0$ and define,

$$\begin{aligned} I &= \lim_{n\to\infty}\left\langle\left[\sum_{i=1}^{n} f(t_{i-1})(\Delta W(t_i)^2 - \Delta t_i)\right]^2\right\rangle \\ &= \lim_{n\to\infty}\left\{\left\langle\sum_{i=1}^{n} f(t_{i-1})^2(\Delta W(t_i)^2 - \Delta t_i)^2\right.\right. \\ &\quad \left.\left. + 2\sum_{j=1}\sum_{i<j} f(t_{i-1})f(t_{j-1})(\Delta W(t_i)^2 - \Delta t_i)(\Delta W(t_j)^2 - \Delta t_j)\right\rangle\right\}. \end{aligned} \tag{6.22}$$

For Wiener process we have $\langle \Delta W(t_i)^2\rangle = \Delta t_i$ and $\langle(\Delta W(t_i)^2 - \Delta t_i)^2\rangle = 2\Delta t_i^2$ and the increments are independent. Therefore for nonanticipating function $f(t)$ we have,

$$\left\langle f(t_{i-1})^2(\Delta W(t_i)^2 - \Delta t_i)\right\rangle = \left\langle f(t_{i-1})^2\right\rangle\left\langle(\Delta W(t_i)^2 - \Delta t_i)\right\rangle \tag{6.23}$$

and one finds,

$$I = 2 \lim_{n\to\infty} \left[\sum_{i=1}^{n} \langle (f(t_{i-1}))^2 \rangle \Delta t_i^2 \right] \sim \mathcal{O}(\Delta t)^2 \to 0 \tag{6.24}$$

and we come to conclusion that,

$$ms - \lim_{n\to\infty} \left\{ \sum_{i=1}^{n} f(t_{i-1})\ \Delta W(t_i)^2 - \sum_{i=1}^{n} f(t_{i-1})\ \Delta t_i \right\} = 0 \tag{6.25}$$

which means that,

$$(I) \int_{t_0}^{t} f(t')\ [dW(t')]^2 = \int_{t_0}^{t} f(t')\ dt' \quad . \tag{6.26}$$

For $N > 0$, carrying out similar calculations and using higher moments of Wiener increments $\langle \Delta W^k(t_i) \rangle$ for $k = 2, 3, \ldots$ one can prove the Eq. (6.21). In practice when we are doing the Itô calculus, then will replace dW^2 with dt and dW^{2+N} for $N > 0$ with zero, up to oder $\mathcal{O}(\mathrm{d}t)$. We will use the notations $dW^2 \stackrel{\mathrm{dt}}{=} dt$ and $dW^{2+N} \stackrel{\mathrm{dt}}{=} 0$ for $N > 0$, throughout the book.

6.2.1 *Itô or Stratonovich*

We note that the white noise in (5.1), (5.2) and (6.3) is an idealisation and every real fluctuating forcing has finite amplitude and timescale. If white-noise is approximation to continuously fluctuating noise with finite memory (much shorter than dynamical timescales), appropriate representation is Stratonovich. This is due to the Wong-Zakai theorem [2, 3] that states, for a (deterministic) physical differential equation introducing a realistic *multiplicative* noise, with correlation function $\langle \Gamma(t)\Gamma(t') \rangle \sim \alpha/2 \exp(-\alpha|t - t'|)$, then, in the limit $\alpha \to \infty$, this will converge to a stochastic differential equation with white noise in the Stratonovich interpretation. If white noise approximates set of discrete pulses with finite separation to which system responds, then Itô representation will be appropriate.

6.3 Integration of Polynomial and Examples of Itô Calculus

Here we provide some examples for Itô calculus. In the following examples we will use the relations $dW^2 \stackrel{\mathrm{dt}}{=} dt$ and $dW^{2+N} \stackrel{\mathrm{dt}}{=} 0$ for $N > 0$.

Example 6.1 Let us start with computing $d[W(t)]^n$ which can be written as,

$$d[W(t)]^n = [W(t) + dW(t)]^n - [W(t)]^n = nW^{n-1}dW + \frac{n(n-1)}{2}W^{n-2}dW^2 + \mathcal{O}(dW)^3 + \cdots \tag{6.27}$$

up to orders of $\mathcal{O}(\mathrm{d}t)$ and $\mathcal{O}(dW)$, we find

$$d[W(t)]^n = nW^{n-1}dW + \frac{n(n-1)}{2}W^{n-2}dt. \tag{6.28}$$

Integration over t both sides of Eq. (6.28) for $n = 2$, one gets,

$$\int_{t_0}^{t} d[W(t')]^2 = 2\int_{t_0}^{t} W(t')\,dW(t') + \int_{t_0}^{t} dt. \tag{6.29}$$

This gives us,

$$\int_{t_0}^{t} W(t')\,dW(t') = \frac{1}{2}[W(t)^2 - W(t_0)^2] - \frac{1}{2}(t - t_0) \tag{6.30}$$

which is the Itô integral Eq. (6.18).

Example 6.2 As an another example, in similar way we find the following relation for differential of exponential function,

$$de^W = e^{W(t)+dW(t)} - e^{W(t)} = e^{W(t)}\left[e^{dW(t)} - 1\right] = e^{W(t)}\left[dW(t) + \frac{1}{2}dW(t)^2\right] = e^{W(t)}\left[dW(t) + \frac{1}{2}dt\right]. \tag{6.31}$$

Example 6.3 As an another example for any function $f(W, t)$, its differential is given by,

$$df = \frac{\partial f}{\partial t}dt + \frac{\partial f}{\partial W}dW + \frac{1}{2}\frac{\partial^2 f}{\partial t^2}dt^2 + \frac{1}{2}\frac{\partial^2 f}{\partial W^2}dW^2 + \cdots = \left[\frac{\partial f}{\partial t} + \frac{1}{2}\frac{\partial^2 f}{\partial W^2}\right]dt + \frac{\partial f}{\partial W}dW + \mathcal{O}(dW)^{3/2} + \cdots. \tag{6.32}$$

6.4 Itô Calculus for Multiplicative Noise and Itô-Taylor Expansion

Let us consider the following Langevin equation,

$$dx(t) = a(x,t)\, dt + b(x,t)\, dW(t) \tag{6.33}$$

with arbitrary smooth functions $a(x,t)$ and $b(x,t)$. For non-constant $b(x,t)$ the term $b(x,t)\, dW(t)$ in (6.33) is a multiplicative noise. Here the Wiener process $W(t)$ is given in terms of the white noise $\Gamma(t)$ as $dW(t) = \Gamma(t)\, dt$, where Γ is a zero mean with gaussian distribution function. Differential of arbitrary infinitely differentiable function $f(x)$ will be,

$$\begin{aligned} df(x) &= f(x+dx) - f(x) \\ &= f'(x)\, dx + \frac{1}{2} f''(x)\, (dx)^2 + \mathcal{O}(dx)^3 + \cdots \end{aligned} \tag{6.34}$$

here $f'(x)$ and $f''(x)$ are the first and second derivatives of $f(x)$ with respect to x.

Using the Langevin equation (6.33), up to orders of $\mathcal{O}(dt)$ and $\mathcal{O}(dW)$ reads,

$$\begin{aligned} df(x) &= f'(x) \left[a(x,t)\, dt + b(x,t)\, dW(t) \right] \\ &+ \frac{1}{2} f''(x)\, b^2(x,t)\, dW^2 \\ &= \left[a(x,t)\, f'(x) + \frac{1}{2} f''(x)\, b^2(x,t) \right] dt \\ &+ f'(x)\, b(x,t)\, dW(t) \end{aligned} \tag{6.35}$$

where we replaced dW^2 with dt. In the next chapter we will use Eq. (6.35) to show the equivalence between Langevin approach and Fokker–Planck equation. Defining

$$\begin{aligned} \mathcal{L}^0 &= a(x,t) \frac{\partial}{\partial x} + \frac{1}{2} b^2(x,t) \frac{\partial^2}{\partial x^2} \\ \mathcal{L}^1 &= b(x,t) \frac{\partial}{\partial x} \end{aligned} \tag{6.36}$$

Equation (6.35) becomes

$$df[x(t)] = \mathcal{L}^0 f[x(t)] dt + \mathcal{L}^1 f[x(t)] dW(t) \tag{6.37}$$

that means,

$$f[x(t)] = f[x(t_0)] + \int_{t_0}^{t} \mathcal{L}^0 f[x(s)]ds + \int_{t_0}^{t} \mathcal{L}^1 f[x(s)]dW(s). \tag{6.38}$$

In Eq. (6.38), if we choose the $f(x)$ as x, $a(x,t)$ and $b(x,t)$ then find

$$\begin{aligned} x(t) &= x(t_0) + \int_{t_0}^{t} a[x(s)]ds + \int_{t_0}^{t} b[x(s)]dW(s) \\ a[x(t)] &= a[x(t_0)] + \int_{t_0}^{t} \mathcal{L}^0 a[x(s)]ds + \int_{t_0}^{t} \mathcal{L}^1 a[x(s)]dW(s) \\ b[x(t)] &= b[x(t_0)] + \int_{t_0}^{t} \mathcal{L}^0 b[x(s)]ds + \int_{t_0}^{t} \mathcal{L}^1 b[x(s)]dW(s). \end{aligned} \tag{6.39}$$

Substituting $a[x(t)]$ and $b[x(t)]$ in $x(t)$ leads to

$$\begin{aligned} x(t) = x(t_0) &+ \int_{t_0}^{t} \left\{ a[x(t_0)] + \int_{t_0}^{s_1} \mathcal{L}^0 a[x(s_2)]ds_2 + \int_{t_0}^{s_1} \mathcal{L}^1 a[x(s_2)]dW(s_2) \right\} ds_1 \\ &+ \int_{t_0}^{t} \left\{ b[x(t_0)] + \int_{t_0}^{s_1} \mathcal{L}^0 b[x(s_2)]ds_2 + \int_{t_0}^{s_1} \mathcal{L}^1 b[x(s_2)]dW(s_2) \right\} dW(s_1). \end{aligned} \tag{6.40}$$

We note that $\mathcal{L}^0 a = aa' + \frac{1}{2}b^2 a''$, $\mathcal{L}^0 b = ab' + \frac{1}{2}b^2 b''$, $\mathcal{L}^1 a = ba'$ and $\mathcal{L}^1 b = bb'$. The essence of the expansion is to use the substitution repeatedly $a[x(t)]$ and $b[x(t)]$ in Eq. (6.40) to obtain constant integrands in higher and higher order terms. Eq. (6.40) is the Itô-Taylor expansion for the process (6.33). Once we have the Itô-Taylor expansion, we can construct numerical integration schemes for the process, see Chap. 14.

Problems

6.1 Nonanticipating function

Show the following relations hold for given nonanticipating functions $f(t)$ and $g(t)$,

$$(a) \quad \left\langle \int_{t_0}^{t} f(t')\, dW(t') \right\rangle = 0$$

$$(b) \quad \left\langle \int_{t_0}^{t} f(t')\, dW(t') \int_{t_0}^{t} g(t'')\, dW(t'') \right\rangle = \int_{t_0}^{t} dt' \langle f(t')\, g(t') \rangle .$$

6.2 Itô and Stratonovich interpretations

For Langevin equation

$$\frac{dx}{dt} = a(x,t) + b(x,t)\, \Gamma(t)$$

where $\Gamma(t)$ is Gaussian white noise, show that

(a) with Itô interpretation

$$\langle b(x,t)\, \Gamma(t) \rangle = 0.$$

(b) with Stratonovich interpretation

$$\langle b(x,t)\Gamma(t) \rangle = \frac{1}{2} \langle b'(x,t) b(x,t) \rangle$$

where $b'(x,t) = \frac{\partial}{\partial x} b(x,t)$.

6.3 Itô and Stratonovich interpretations

For the following Langevin equation,

$$\frac{dx(t)}{dt} = -x(t)\, \Gamma(t), \qquad x(0) = x_0$$

derive $\langle x(t) \rangle$ in Itô and Stratonovich senses.

6.4 Itô's lemma

Let $W(t)$ be a Wiener process. Compute $(dx)^2$ using Itô's lemma for:

$$dx(t) = dt + x(t)\, dW(t).$$

6.5 Stochastic differential equation

Let $W(t)$ be a Wiener process. Compute stochastic differential equation, i.e. $dz(t)$, followed by:

(a) $z(t) = (x(t))^2$, where $dx(t) = \mu x(t)dt + \sigma x(t)dW(t)$,

(b) $z(t) = 3 + t + e^{W(t)}$,

(c) $z(t) = e^{x(t)}$, where $dx(t) = \mu dt + \sigma dW(t)$,

with constant μ and σ.

6.6 Itô's lemma

Using the Itô's lemma, calculate the following double integral and show that

$$\int_{t_0}^{t} \int_{t_0}^{s_1} dW(s_2)dW(s_1) = \frac{1}{2}[W(t) - W(t_0)]^2 - \frac{1}{2}(t - t_0)$$

6.7 Itô's integral

Calculate the following multiple Itô integral and show that

$$\int_0^t dW(s_1) \int_0^{s_1} dW(s_2) \cdots \int_0^{s_{n-1}} W(s_n)dW(s_n) = \frac{1}{(n+1)!} H_{n+1}(W(t))$$

where $H_n(x)$ is the nth order Hermit polynomial with Rodrigues' formula

$$H_n(x) = e^{\frac{1}{2}x^2} \left(\frac{d}{dx}\right)^n e^{-\frac{1}{2}x^2} \quad .$$

6.8 Itô-Taylor expansion

Show that solution of the Langevin equation

$$dx(t) = a(x,t)\, dt + b(x,t)\, dW(t)$$

can be written as

$$x(t) = x(t_0) + a[x(t_0)] \int_{t_0}^{t} ds_1 + b[x(t_0)] \int_{t_0}^{t} dW(s_1)$$
$$+ b[x(t_0)]b'[x(t_0)] \int_{t_0}^{t} \int_{t_0}^{s_1} dW(s_2)dW(s_1) + R$$

where R is remainder of order $\mathcal{O}(dt^{3/2})$ and higher if $dt = t - t_0$ is small. Keeping the first three and four terms in above expansion will give the Euler-Maruyama and Milstein schemes, respectively.

References

1. C.W. Gardiner, *Handbook of Stochastic Methods* (Springer, Berlin, 1983)
2. E. Wong, M. Zakai, Ann. Math. Statist. **36**, 1560 (1965)
3. E. Wong, M. Zakai, Internat. J. Engrg. Sci. **3**, 213 (1965)

Chapter 7
Equivalence of Langevin and Fokker–Planck Equations

In this chapter we show the equivalence between the Langevin approach and the Fokker–Planck equation, and derive the equation for the statistical moments of the process whose dynamics is described by the Langevin equation.

7.1 Probability Distribution Functions of Langevin Dynamics

We aim to show that the probability distribution function of stochastic processes whose dynamics is given by the Langevin equation

$$dx(t) = a(x,t)dt + b(x,t)\, dW(t) \tag{7.1}$$

satisfies the Fokker–Planck equation, Eq. (3.15). As shown by Eq. (6.35), any arbitrary infinitely differentiable function f of the process x has the differential,

$$\begin{aligned} df(x) &= f(x+dx) - f(x) \\ &= f'(x)\, dx + \frac{1}{2} f''(x)\, (dx)^2 + \mathcal{O}(dx)^3 + \cdots . \end{aligned}$$

Suppose that the dynamics of $x(t)$ is given by the Langevin equation, Eq. (7.1). Therefore, we find that

$$\begin{aligned} df(x) &= f'(x)\ \left(a(x,t)dt + b(x,t)\, dW(t)\right) \\ &\quad + \frac{1}{2} f''(x)\, (a(x,t)dt + b(x,t)\, dW)^2 + \mathcal{O}(dx)^3 \end{aligned} \tag{7.2}$$

M. R. Rahimi Tabar, *Analysis and Data-Based Reconstruction of Complex Nonlinear Dynamical Systems*, Understanding Complex Systems,
https://doi.org/10.1007/978-3-030-18472-8_7

using the relation, $dW^2 \stackrel{\text{dt}}{=} dt$, up to first orders of dt and dW, we find,

$$df(x) = \left\{ f'(x)\, a(x,t) + \frac{1}{2} f''(x)\, b^2(x,t) \right\} dt \\ + f'(x)\, b(x,t)\, dW(t). \tag{7.3}$$

The time derivative of f is given by

$$\frac{df(x)}{dt} = f'(x)\, a(x,t) + \frac{1}{2} f''(x)\, b^2(x,t) \\ + f'(x)\, b(x,t)\, \Gamma(t) \tag{7.4}$$

where $\Gamma(t)$ is zero mean Gaussian white noise. Ensemble averaging both side of Eq. (7.4) over the distribution of Γ and assuming that b is a nonanticipating function, give us,

$$\left\langle \frac{df(x)}{dt} \right\rangle = \left\langle f'(x)\, a(x,t) + \frac{1}{2} f''(x)\, b^2(x,t) \right\rangle + 0 \tag{7.5}$$

where we used the zero-mean property of the noise $\Gamma(t)$. The left side of Eq. (7.5) is written as,

$$\left\langle \frac{df(x)}{dt} \right\rangle = \frac{d}{dt} \langle f(x) \rangle = \int dx\; f(x)\; \partial_t p(x,t). \tag{7.6}$$

where $p(x,t)$ is the portability distribution function of x, therefore

$$\int dx\; f(x)\; \partial_t p(x,t) = \int dx\; \left\{ f'(x) a(x,t) + \frac{1}{2} f''(x) b^2(x,t) \right\}\; p(x,t). \tag{7.7}$$

Simple partial integration, once in the first term and twice in the second term, gives

$$\int dx\; f(x)\; \partial_t p(x,t) \\ = \int dx f(x) \left\{ -\partial_x \left(a(x,t)\; p(x,t) \right) + \frac{1}{2} \partial^2_{xx} \left(b^2(x,t)\; p(x,t) \right) \right\} \tag{7.8}$$

where we used the fact that $p(x,t)$ and its derivative with respect to x approach zero in the limit, $x \to \pm\infty$. Now, for an arbitrary functions f, we find the Fokker–Planck equation

$$\partial_t\; p(x,t) = -\partial_x\; \left(a(x,t)\; p(x,t) \right) + \frac{1}{2} \partial^2_{xx}\; \left(b^2(x,t)\; p(x,t) \right). \tag{7.9}$$

To derive Eq. (7.9) we have used the Itô interpretation, which leads the PDF of white noise-driven Langevin dynamics that satisfies the Fokker–Planck equation with the drift coefficient (compare with 3.15)

$$D^{(1)}(x,t) = a(x,t)$$

and diffusion coefficient

$$D^{(2)}(x,t) = \frac{1}{2} b^2(x,t).$$

It can also be shown that one can derive the *same* Fokker–Planck equation in the Stratonovich interpretation, if one rewrites the Langevin equation as,

$$dx(t) = \left\{ a(x,t) - \frac{1}{2} b(x,t)\, b'(x,t) \right\} dt + b(x,t)\, dW(t). \tag{7.10}$$

where $b'(x,t) = \partial b(x,t)/\partial x$. This means that we can go from the Itô to the Stratonovich interpretation by appropriately modifying the drift term. The correction term, $-\frac{1}{2} b(x,t) b'(x,t)$, is often called *noise-induced* drift. We note that for constant multiplication function $b(x,t) = b_0$, there is no difference between the two interpretation. In what follows we will use the Itô interpretation—thanks to the Itô lemma that provides us with a powerful tool—to deal with stochastic processes analytically.

7.2 Equation for Statistical Moments Based on the Langevin Equation

Suppose that the dynamics of a stochastic variable x is given by a Langevin equation:

$$dx(t) = a(x,t)\, dt + b(x,t)\, dW(t).$$

We are interested in deriving a differential equation for the statistical moments, $\langle x^n \rangle$, based on the Langevin equation. From Eq. (7.5) we know that the expectation of an arbitrary differentiable function f satisfies

$$\left\langle \frac{df(x)}{dt} \right\rangle = \left\langle f'(x)\, a(x,t) + \frac{1}{2} f''(x)\, b^2(x,t) \right\rangle,$$

if we apply this to $f = x^n$ with $n \geq 1$, we find

$$\frac{d}{dt} \langle x^n \rangle = n \left\langle x^{n-1}\, a(x,t) \right\rangle + \frac{1}{2} n(n-1) \left\langle x^{n-2}\, b^2(x,t) \right\rangle \tag{7.11}$$

which gives the equations for the statistical moments with $n \geq 1$. One can also derive (7.11) directly from the Fokker–Planck equation, i.e., Eq. (7.9), by multiplying its both sides by x^n and integrating over the variable x.

We note that for the general nonlinear functions a and b, the equation for the nth-order moment depends on the moments of higher order $m > n$. Thus, in order to actually compute the required moments, we would need to integrate an infinite number of moment equations. This problem is known as the *closure problem*. Statistical moments may be approximated by using the moment closure methods that which typically are based on replacing the higher-order statistical moments with suitable approximations or truncation of the equations after a specific term. For example, one can set the statistical moments above a certain order to zero, or approximate them with their steady-state values [1, 2].

7.3 Existence of Solutions to Langevin Equation

One important issue in differential equations is the question of whether their solution exists and, if it does, whether it is unique. The conditions that are required for the existence and uniqueness of the solution for the Langevin equation, Eq. (7.1), in the time interval $[t_0, T]$ are [1, 3],

(i) Lipschitz condition: it means that a K exists such that

$$|a(x,t) - a(y,t)| + |b(x,t) - b(y,t)| \leq K\, |x - y|$$

for all x, y and t in the interval $[t_0, T]$,

(ii) Growth condition: a K exists such that for all t in the interval $[t_0, T]$

$$|a(x,t)|^2 + |b(x,t)|^2 \leq K^2 \left(1 + |x|^2\right).$$

Under such conditions there will be a unique nonanticipating solution $x(t)$ in the range $[t_0, T]$. The Lipschitz condition is indeed a smoothness condition and, in practice, every Langevin equation satisfies this condition.

Problems

7.1 Stratonovich interpretation

Starting from the Langevin equation, Eq. (7.10), prove that probability density function $p(x,t)$ satisfies the Fokker–Planck equation, Eq. (7.9), with the Stratonovich interpretation.

7.2 Kramers–Moyal coefficients

For the Langevin equation

$$dx(t) = a(x,t)\,dt + b(x,t)\,dW(t)$$

where $W(t)$ is a Wiener process, use the Itô lemma to prove that

(a) $D^{(1)}(x,t) = \lim_{dt\to 0} \frac{1}{dt}\langle (x(t+dt) - x(t))|x(t) = x\rangle = a(x,t)$

(b) $D^{(2)}(x,t) = \lim_{dt\to 0} \frac{1}{2dt}\langle (x(t+dt) - x(t))^2|x(t) = x\rangle = \frac{1}{2}b^2(x,t)$

and for some $\delta > 0$

(c) $\lim_{dt\to 0} \frac{1}{dt}\langle (x(t+dt) - x(t))^{2+\delta}|x(t) = x\rangle = 0.$

which means that higher-order Kramers–Moyal coefficients are $D^{(n)}(x,t) = 0$, for $n \geq 3$.

7.3 Transition probability distribution

For the Langevin equation

$$dx(t) = a(t)\,dt + b(t)\,dW(t)$$

use the corresponding Fokker–Planck equation to prove that the transition (conditional) probability distribution with initial condition $p(y,s|x,s) = \delta(x-y)$ is given by:

$$p(y,t|x,s) = \frac{1}{\sqrt{2\pi \int_s^t b^2(u)du}} \exp\left[-\frac{\left(y - x - \int_s^t a(u)du\right)^2}{2\int_s^t b^2(u)du} \right].$$

7.4 Statistical moments of the first-passage times

Suppose that the process x is the solution of the white noise-driven Langevin equation

$$\frac{dx}{dt} = a(x,t) + b(x,t)\Gamma(t).$$

Since the process x is stochastic, each realization (which starts at $t_0 = 0$ from $x_0 \in [A, B]$) of it reaches the barrier $x_A = A$ or $x_B = B$ for the *first time* at a different passage (exit) time T. Therefore, it has a first passage time probability density $f(T|x_0)$ and its statistical moments, assuming that they exist, are defined as $\tau_n(x_0) \equiv \langle T^n \rangle$. The density $f(T|x_0)$ depends on the starting position x_0 and on the chosen boundary conditions. For example, we can choose (i) reflecting at A and absorbing boundary conditions at B, or (ii) both boundaries to be absorbing. In case (i) we are interested in to study the statistics of the first-passage times to reach the boundary at B, while in the second case one is interested in looking at the first-passage time for reaching either boundary.

Define the survival probability $G(x_0, t)$ of a trajectory $x(t)$ to be within the interval $[A, B]$ at time t, if it started at $x_0 \in [A, B]$ at time $t = 0$.

(a) Show that $G(x_0, t)$ is given in term of the conditional probability distribution function as

$$G(x_0, t) = \int_A^B p(x', t|x_0, 0)dx'.$$

For the systems with time translation invariant, i.e., $a(x, t) = a(x)$ and $b(x, t) = b(x)$, we write

$$p(x', t|x_0, 0) = p(x', 0|x_0, -t).$$

(b) By noting that $p(x', 0|x_0, t)$ satisfies the backward Fokker–Planck equation (see problem 3.3)

$$\frac{\partial}{\partial t}p(x', 0|x_0, t) = -a(x_0)\frac{\partial}{\partial x_0}p(x', 0|x_0, t) - \frac{1}{2}b^2(x_0)\frac{\partial^2}{\partial x_0^2}p(x', 0|x_0, t)$$

show that the survival probability $G(x_0, t)$ for case (i) satisfies the partial differential equation

$$\frac{\partial}{\partial t}G(x_0, t) = a(x_0)\frac{\partial}{\partial x_0}G(x_0, t) + \frac{1}{2}b^2(x_0)\frac{\partial^2}{\partial x_0^2}G(x_0, t)$$

with initial and boundary conditions, $G(x_0, 0) = \begin{cases} 1 & x_0 \in [A, B] \\ 0 & x_0 \notin [A, B] \end{cases}$, absorbing boundary at B, $G(x_0 = B, t) = 0$ and reflecting boundary at A, $\partial_{x_0}G(x_0 = A, t) = 0$. Also $G(x_0, t \to \infty) = 0$.

(c) Argue that

$$f(T|x_0) = -\partial_T G(x_0, T).$$

(d) Show that the statistical moments of the first-passage times, T, i.e. $\tau_n(x_0) = \int_0^\infty T^n f(T|x_0)dt$, satisfies the following Darling and Siegert's recursive relation

$$-n\tau_{n-1}(x) = a(x)\frac{\partial}{\partial x}\tau_n(x) + \frac{1}{2}b^2(x)\frac{\partial^2}{\partial x^2}\tau_n(x)$$

where $x \equiv x_0$ with $\tau_0(x) = 1$ and $\tau_n(A) = \tau_n(B) = 0$, for $n = 1, 2, \ldots$. This yields a hierarchy of equations that can be solved recursively: once we have solved for the nth moment, we can go to moment $n + 1$ to solve it.

7.5 Mean first-passage time

Consider a stochastic process x that evolves within the interval $[A, B]$ according to the white noise-driven Langevin equation

$$\frac{dx(t)}{dt} = -\frac{dU(x)}{dx} + \Gamma(t)$$

where $-\frac{dU(x)}{dx}$ is a force due to the potential $U(x)$. Use the Darling and Siegert's recursive relation to compute the mean first-passage time $\tau_1(x_0)$ for $U(x) = -x$ and $U(x) = \frac{1}{2}x^2$ with the boundary conditions

(a) A reflecting and B absorbing,

(b) both A and B absorbing.

7.6 Kramers' escape time

Consider the stochastic process x that evolves according to the white noise-driven Langevin equation

$$\frac{dx(t)}{dt} = -\frac{dU(x)}{dx} + \epsilon\Gamma(t), \quad x(0) = x_0$$

where $-\frac{dU(x)}{dx}$ is a force due to the potential $U(x)$. The limit $\epsilon \to 0$ is known as the small-noise limit. Assume that $U(x)$ has a single non-degenerate minimum at $x = a$ and maximum at $x = b$, and that $U(x) \to \infty$ as $x \to -\infty$.

(a) Show that mean exit time satisfies the following equation with boundary conditions,

$$\left[\frac{1}{2}\epsilon^2\frac{d^2}{dx^2} - U'(x)\frac{d}{dx}\right]\tau(x) = -1, \quad \tau(b) = 0, \quad \tau_x(-\infty) = 0.$$

(b) Using the boundary conditions show that

$$\tau(x) = -\frac{2}{\epsilon^2}\int_b^x e^{2U(z)/\epsilon^2}\int_{-\infty}^z e^{-2U(y)/\epsilon^2}dydz.$$

(c) Suppose that $x_0 \approx a$ and show, in the small-noise limit, that the mean exit time from the interval $(-\infty, b]$ is given by

$$\tau(x_0) \approx \pi\exp\left\{\frac{2(U(b) - U(a))}{\epsilon^2}\right\}\sqrt{\frac{1}{U''(a)|U''(b)|}}\ .$$

Hint: There is no possibility of exit from $x \to -\infty$. Thus, we impose reflecting boundary condition at $x = -\infty$. Therefore, the boundary conditions in the Darling and Siegert's relation (with $n = 1$) will be $\tau(b) = 0$ and $\tau_x(-\infty) = 0$.

7.7 Novikov–Furutsu–Donsker relation

For a Gaussian-distributed noise $\eta(t)$ with the covariance

$$C(t, t') = \langle\eta(t)\eta(t')\rangle - \langle\eta(t)\rangle\langle\eta(t')\rangle$$

show that correlation between a functional $g[\eta]$ and noise $\eta(t)$ can be written in terms of $C(t, t')$ as

$$\langle \eta(t) g[\eta] \rangle = \langle \eta(t) \rangle \langle g[\eta] \rangle + \int_0^t C(t, t') \left\langle \frac{\delta g[\eta])}{\delta \eta(t')} \right\rangle dt'$$

where $\frac{\delta g[\eta]}{\delta \eta(t')}$ is the functional derivative of $g[\eta]$ with respect to the Gaussian random function $\eta(t')$.

References

1. C.W. Gardiner, *Handbook of Stochastic Methods* (Springer, Berlin, 1983)
2. U. Frisch, *Turbulence: The Legacy of AN Kolmogorov* (Cambridge University Press, 1995)
3. A. Pazy, *Semigroups of Linear Operators and Applications to Partial Differential Equations, Applied Mathematical Sciences 44* (Springer, New York, 1983)

Chapter 8
Example of Stochastic Calculus

In this chapter we present three examples of the Langevin dynamics, namely, the Anderson–Kubo, Ornstein–Uhlenbeck, and Black–Scholes processes. We will use the Itô interpretation that provides us with a powerful tool for dealing with such linear and nonlinear stochastic processes analytically.

8.1 Anderson–Kubo Process

The Anderson–Kubo process is defined as a process with $a(x, t) = 0$ and $b(x, t) = cx$ in the Langevin equation, Eq. (7.1), with constant c [1–3]. Indeed, it is a multiplicative process whose *stochastic solution* is derived here using the Itô calculus.

Making the change of the variable $y = \ln x$, we find the following expansion (up to orders of $\mathcal{O}(\mathrm{d}t)$ and $\mathcal{O}(dW)$),

$$\begin{aligned} dy &= \frac{1}{x}dx - \frac{1}{2x^2}(dx)^2 + \cdots \\ &= c\, dW(t) - \frac{1}{2}c^2 dt \end{aligned} \tag{8.1}$$

where we used the Langevin equation, $dx = cx dW(t)$. The solution of Eq. (8.1) is,

$$y(t) = y(t_0) + c\,(W(t) - W(t_0)) - \frac{1}{2}c^2\,(t - t_0) \tag{8.2}$$

Then, one finds the stochastic solution of the Anderson–Kubo process as follows,

$$x(t) = x(t_0)\, \exp\left\{c(W(t) - W(t_0)) - \frac{1}{2}c^2(t - t_0)\right\}. \tag{8.3}$$

The original version of this chapter was revised: Belated correction has been incorporated. The correction to this chapter is available at https://doi.org/10.1007/978-3-030-18472-8_24

M. R. Rahimi Tabar, *Analysis and Data-Based Reconstruction of Complex Nonlinear Dynamical Systems*, Understanding Complex Systems,
https://doi.org/10.1007/978-3-030-18472-8_8

Using the solution of Anderson–Kubo process, Eq. (8.3), we determine the mean value, as well as correlation function of $x(t)$. We note that the Wiener process $W(t)$ has Gaussian probability distribution function, and that for any zero mean Gaussian process z, one finds[1]

$$\langle \exp(z) \rangle = \exp\left(\frac{\langle z^2 \rangle}{2}\right). \tag{8.4}$$

Using (8.4) we calculate the mean and correlation function (also second statistical moment) of the Anderson–Kubo process as,

$$\begin{aligned} \langle x(t) \rangle &= \langle x(t_0) \rangle \; \exp\left(\frac{1}{2}c^2(t-t_0) - \frac{1}{2}c^2(t-t_0)\right) \\ &= \langle x(t_0) \rangle \end{aligned} \tag{8.5}$$

where $\langle \cdots \rangle$ denotes an average over the Wiener process and the initial conditions. The correlation function will then be,

$$\begin{aligned} &\langle x(t)x(s) \rangle \\ &= \langle x^2(t_0) \rangle \left\langle \exp\left\{ c\,[W(t) + W(s) - 2W(t_0)] - \frac{1}{2}c^2(t+s-2t_0) \right\} \right\rangle . \end{aligned}$$

Using (8.4), gives us

$$\begin{aligned} &\langle x(t)x(s) \rangle \\ &= \langle x^2(t_0) \rangle \; \exp\left\{ \frac{1}{2}c^2 \left[\langle (W(t) + W(s) - 2W(t_0))^2 \rangle - (t+s-2t_0) \right] \right\} \\ &= \langle x^2(t_0) \rangle \; \exp\left\{ \frac{1}{2}c^2 \left[t+s-2t_0 + 2 \; \min(t,s) - (t+s-2t_0) \right] \right\} \\ &= \langle x^2(t_0) \rangle \; \exp\left\{ c^2 \; \min(t-t_0, s-t_0) \right\} \end{aligned} \tag{8.6}$$

where we used Eqs. (5.9) and (5.13). By setting $t = s$ we determine the second statistical moment of the Anderson–Kubo process as,

$$\langle x^2(t) \rangle = \langle x^2(t_0) \rangle \exp\left\{ c^2(t-t_0) \right\}.$$

See Sect. 8.3 for all the statistical moments of the Anderson–Kubo process [3].

---

[1] The average of exponential function is $\langle \exp(z) \rangle = \int dz \frac{1}{\sqrt{2\pi \langle z^2 \rangle}} \exp(-\frac{z^2}{2\langle z^2 \rangle}) \exp(z)$, with the Gaussian integral giving the relation (8.4).

8.2 Ornstein–Uhlenbeck Process

The Ornstein–Uhlenbeck (OU) process is the only process that is simultaneously Gaussian, Markov and stationary. It is given by a Langevin equation with $a(x, t) = -kx$ and $b(x, t) = \sqrt{2D}$, with a real k and positive constant D. Defining $y = x\exp(kt)$, we have

$$dy = dx\ \exp(kt) + x\,d(\exp(kt)) + \mathcal{O}(dxdt) + \mathcal{O}(dx)^2 + \mathcal{O}(dt)^2$$

where the coefficient of the term $\mathcal{O}(dx)^2$ (which is second order derivative of x with respect to x) is zero. Using the Langevin equation for dx up to the first-order in both dt and dW, we find,

$$dy = \sqrt{2D}\ \exp(kt)\ dW(t)$$

which gives us the stochastic solution of Ornstein–Uhlenbeck process:

$$x(t) = x(0)\ \exp(-kt) + \sqrt{2D}\int_0^t \exp\left(-k(t-t')\right)\ dW(t'). \tag{8.7}$$

The mean and correlation function of the OU-process are given by,[2]

$$\langle x(t)\rangle = \langle x(0)\rangle\ \exp(-kt) \tag{8.8}$$

and

$$\begin{aligned}\langle x(t)\,x(s)\rangle = &\left[\mathrm{var}\{x(0)\} - \frac{D}{k}\right]\ \exp\left(-k(t+s)\right)\\ &+\frac{D}{k}\exp\left(-k|t-s|\right)\end{aligned} \tag{8.9}$$

where $\mathrm{var}\{x(0)\}$ is the variance of the initial values of x. For finite and positive k, in the long-time limit only the last term in Eq. (8.9) survives, and in the limit, $t, s \to \infty$ we find, $\mathrm{var\{x(t)\} = D/k}$.

In general, the *correlation time scale* of a process whose correlation function decays as $C(t-s) = \langle x(t)\,x(s)\rangle \sim \exp(-k|t-s|)$, is defined as,

$$\tau = \frac{1}{\mathrm{var}(x)}\int_0^\infty dt'\ C(t'), \tag{8.10}$$

[2]The correlation function is $\langle x(t)x(s)\rangle = \mathrm{var}\{x(0)\}\exp(-k(t+s)) + 2D\langle\int_0^t \exp(-k(t-t'))\,dW(t')\int_0^s \exp(-k(s-s'))dW(s')\rangle = \mathrm{var}\{x(0)\}\exp(-k(t+s)) + 2D\int_0^{\min(t,s)}\exp(-k(t+s-2s'))dt' = \left[\mathrm{var}\{x(0)\} - \frac{D}{k}\right]\exp(-k(t+s)) + \frac{D}{k}\exp(-k|t-s|)$.

from which we find $\tau = 1/k$ for the OU-process. Using the drift and diffusion coefficients of the OU process, i.e., $D^{(1)}(x) = a(x) = -kx$ and $D^{(2)}(x) = b^2(x)/2 = D$, we write down the corresponding Fokker–Planck equation,

$$\frac{\partial p(x, t|x', t')}{\partial t} = k\frac{\partial}{\partial x}\left(x\, p(x, t|x', t')\right) + D\frac{\partial^2}{\partial x^2}\, p(x, t|x', t') \tag{8.11}$$

with the initial condition $p(x, t'|x', t') = \delta(x - x')$. Equation (8.11) may be solved by taking the Fourier transform with respect to x,

$$p(x, t|x', t') = \frac{1}{2\pi}\int e^{ilx} p(l, t|x', t') dl$$

which results in the following equation,

$$\frac{\partial p(l, t|x', t')}{\partial t} = -kl\frac{\partial p(l, t|x', t')}{\partial l} - Dl^2 p(l, t|x', t'). \tag{8.12}$$

The solution of Eq. (8.12) with the initial condition $p(l, t|x', t') = \exp(ilx')$ is,

$$p(l, t|x', t') = \exp\left[-ilx'e^{-k(t-t')} - \frac{Dl^2}{2k}\left(1 - e^{-2k(t-t')}\right)\right]. \tag{8.13}$$

By applying the inverse Fourier transform, one finally obtains the solution of the Fokker–Planck equation describing the Ornstein–Uhlenbeck process as,

$$p(x, t|x', t') = \sqrt{\frac{k}{2\pi D\left(1 - e^{-2k(t-t')}\right)}}\exp\left\{-\frac{k\left(x - x'e^{-k(t-t')}\right)^2}{2D\left(1 - e^{-2k(t-t')}\right)}\right\}. \tag{8.14}$$

In the limit $k \to 0$ we obtain the Gaussian distribution for a Wiener process (expand exponentials for small k and set $D = 1/2$). For positive k and large time difference $k(t - t') >> 1$, the OU process has a stationary distribution given by,

$$p_{st}(x) = \sqrt{\frac{k}{2\pi D}}\exp\left\{-\frac{kx^2}{2D}\right\}. \tag{8.15}$$

We note that for $k \leq 0$ no stationary solution exists.

Next, we determine the nth statistical moment of the OU process, $M_n = \int x^n p(x, t)dx$. Let $n \geq 2$ and write down the Fokker–Planck equation (8.11) for $p(x, t)$ (by multiplying Eq. (8.11) by $p(x', t')$ and integrating over x'). Multiplying the corresponding Fokker–Planck equation by x^n and integrating by parts (once on the first term on the right side and twice on the second), we obtain:

$$\frac{d}{dt}\int x^n p(x, t)dx = -kn\int x^n p(x, t)dx + Dn(n - 1)\int x^{n-2} p(x, t)dx$$

or, equivalently

$$\frac{d}{dt}M_n = -knM_n + Dn(n-1)M_{n-2}. \tag{8.16}$$

This is a first-order linear inhomogeneous differential equation whose solution (time-dependent statistical moments) is:

$$M_n(t) = M_n(0)e^{-knt} + Dn(n-1)\int_0^t M_{n-2}(s)\, e^{-kn(t-s)} ds. \tag{8.17}$$

The stationary statistical moments of the OU process are determined directly from the stationary PDF, Eq. (8.15), as follows;

$$\begin{aligned}\lim_{t\to\infty} M_n(t) = \langle x^n \rangle &= \sqrt{\frac{k}{2\pi D}} \int_{-\infty}^{\infty} x^n \exp\left\{-\frac{kx^2}{2D}\right\} dx \\ &= \begin{cases} 1 \times 3 \cdots (n-1) \left(\frac{D}{k}\right)^{n/2} & n\ even \\ 0 & n\ odd \end{cases} .\end{aligned}$$

8.3 Black–Scholes Process, or Geometric Brownian Motion

The Black–Scholes process, also called a geometric Brownian motion, is given by the following Langevin equation with $a(x,t) = \lambda x$ and $b(x,t) = \mu x$, where λ and μ are constant,

$$dx(t) = \lambda\, x(t)\, dt + \mu\, x(t)\, dW(t). \tag{8.18}$$

In the Itô description the drift and diffusion coefficients are $D^{(1)}(x) = a(x,t) = \lambda x$ and $D^{(2)}(x) = 1/2\, b^2(x,t) = 1/2\, \mu^2 x^2$, respectively. By defining $y = \ln(x)$ one finds,

$$\begin{aligned} dy &= \frac{1}{x}\, dx - \frac{1}{2x^2}\, d^2x + \cdots \\ &= \lambda\, dt + \mu\, dW(t) - \frac{1}{2x^2}\, (\lambda\, x\, dt + \mu\, x\, dW(t))^2 + \cdots . \end{aligned} \tag{8.19}$$

Up to the first orders in dt and $dW(t)$, we obtain,

$$dy = \left(\lambda - \frac{\mu^2}{2}\right) dt + \mu\, dW. \tag{8.20}$$

Integration of (8.20) and retaining $y = \ln(x)$ yields the stochastic solution of Black–Scholes process:

$$x(t) = x(0)\ \exp\left\{\left(\lambda - \frac{\mu^2}{2}\right)\ (t - t_0) + \mu\ (W(t) - W(t_0))\right\}. \tag{8.21}$$

The mean value and the correlation function of the Black–Scholes process are determined using the Anderson–Kubo process by a simple transformation,

$$c \to \mu$$
$$-\frac{1}{2}c^2 \to \left(\lambda - \frac{\mu^2}{2}\right).$$

In finance λ and μ are known as *interest rate* and *volatility*, respectively. From Eq. (7.11) we obtain an equation for the nth moment of the Black–Scholes process, given by

$$\frac{d}{dt}M_n(t) = \left(\lambda n + \frac{1}{2}\mu^2 n(n-1)\right) M_n(t) \tag{8.22}$$

for $n \geq 1$. The solution of this equation is:

$$M_n(t) = M_n(0)\exp\left\{\left(\lambda + \frac{1}{2}\mu^2(n-1)\right) nt\right\}. \tag{8.23}$$

For $n = 1$, we determine $M_1(t)$ directly from (8.21):

$$\begin{aligned} M_1(t) &= M_1(0)\exp\left\{\left(\lambda - \frac{1}{2}\mu^2\right)t\right\}\langle \exp\{\mu W(t)\}\rangle \\ &= M_1(0)\exp\left\{\left(\lambda - \frac{1}{2}\mu^2\right)t\right\}\exp\left\{\frac{1}{2}\mu^2 t\right\} \\ &= M_1(0)\exp\{\lambda t\} \end{aligned} \tag{8.24}$$

here, for simplicity, we set $W(0) = 0$ and used the relation, $\langle \exp\{\mu W(t)\}\rangle = \exp\{\frac{1}{2}\mu^2\langle W^2(t)\rangle\} = \exp\{\frac{1}{2}\mu^2 t\}$ for the Wiener process. We note that the nth moment might diverge as $t \to \infty$, depending on the values of λ and μ:

$$\lim_{t\to\infty} M_n(t) = \begin{cases} 0 & \frac{\mu^2}{2} < -\frac{\lambda}{n-1} \\ M_n(0) & \frac{\mu^2}{2} = -\frac{\lambda}{n-1} \\ +\infty & \frac{\mu^2}{2} > -\frac{\lambda}{n-1} \end{cases}.$$

Problems

8.1 Stationary solution

Fill in the details in the derivation of Eq. 8.15.

8.2 Black–Scholes process

For the Langevin equation

$$dx(t) = \lambda x(t)dt + \mu x(t)dW(t), \qquad x(0) = x_0.$$

where λ and μ are real constants. Show that the solution cannot change sign.

8.3 Lamperti transform

The Lamperti transform is a change of variables that allows one-dimensional Langevin equation with multiplicative noise

$$dx = a(x,t)dt + b(x,t)dW(t)$$

to be rewritten as another Langevin equation with additive noise. Let $y = h(x)$ for some function h to be determined.

(a) Using the Eq. (6.35), show that

$$dy = \mathcal{L}h(x)dt + h'(x)b(x)dW(t)$$

where $\mathcal{L} = a(x,t)\frac{\partial}{\partial x} + \frac{1}{2}b^2(x,t)\frac{\partial^2}{\partial x^2}$. To achieve constant diffusion, we require $h'(x)b(x) = b_0$ for some constant b_0.

(b) Show that

$$\mathcal{L}h(x) = b_0\left[\frac{a(x,t)}{b(x,t)} - \frac{b'(x,t)}{2}\right]$$

and y satisfies the following Langevin equation with additive noise,

$$dy = A(y)dt + b_0 dW(t),$$

where $A(y)$ is

$$A(y) = \frac{b_0\, a(h^{-1}(y),t)}{b(h^{-1}(y),t)} - \frac{1}{2}b_0\, b'(h^{-1}(y),t).$$

(c) For the following Langevin equation (Cox–Ingersoll–Ross)

$$dx = (a_0 + a_1 x)dt + b_0\sqrt{x}dW(t)$$

with constants a_0, a_1 and b_0, eliminate $\sqrt{x}$ with $y = h(x) = 2\sqrt{x}$ and show that y satisfies the Langevin equation,

$$dy(t) = \left[\frac{2a_0}{y} + \frac{a_1 y}{2} - \frac{b_0^2}{2y}\right] dt + b_0 \, dW(t).$$

8.4 Lamperti transform

Suppose that $x(t)$ satisfies the Itô stochastic dynamics

$$dx = f(x)dt + \sqrt{a_0 + a_1 x + a_2 x^2} \, dW(t)$$

for some $a_0, a_1, a_2 > 0$. Apply the Lamperti transform $y = h(x)$ and show that,

$$dy = \frac{f(x)}{\sqrt{a_0 + a_1 x + a_2 x^2}} dt - \frac{1}{4} \frac{a_1 + 2a_2 x}{a_0 + a_1 x + a_2 x^2} dt + dW(t)$$

where $x = h^{-1}(y)$ and

$$y = h(x) = \frac{1}{\sqrt{a_2}} \ln \left(\frac{a_1}{2\sqrt{a_2}} + \sqrt{a_2} x + \sqrt{a_0 + a_1 x + a_2 x^2} \right)$$

where

$$x = h^{-1}(y) = \frac{\left(e^{y\sqrt{a_2}} - \frac{a_1}{2\sqrt{a_2}} \right)^2 - a_0}{a_1 + 2\sqrt{a_2} \left(e^{y\sqrt{a_2}} - \frac{a_1}{2\sqrt{a_2}} \right)} .$$

To prove that x and y are one-to-one functions, show $\frac{dy}{dx} > 0$ (for the values of x that argument of logarithm in y be positive).

8.5 Stochastic differential equation

Show that the following nonlinear (Itô) stochastic differential (Langevin) equation,

$$dx(t) = dt + 2\sqrt{x(t)} \, dW(t), \qquad x(0) = x_0$$

has the stochastic solution

$$x(t) = \left(W(t) + \sqrt{x_0} \right)^2 .$$

8.6 Stochastic differential equation

For class of the following (Itô) stochastic differential equations, there exists an analytical formula for their solution:

$$dx(t) = f(t, x(t))dt + c(t)\, dW(t), \qquad x(0) = x_0.$$

where $f(t, x(t)) \in \mathbb{R}$ and $c(t) \in \mathbb{R}$. Show that the solution can be derived as follows:

$$x(t) = y(t) + F(t)$$

where

$$dy(t) = f(t, y(t) + F(t))dt, \qquad y(0) = x_0$$

$$dF(t) = c(t)dW(t),$$

with $F(0) = 0$. The stochastic differential equation for $F(t)$ can be integrated, i.e., $F(t) = \int_0^t c(s)dW(s)$.

8.7 Stochastic differential equation

For the class of the following nonlinear (Itô) stochastic differential equations, there exists an analytical formula as their solution:

$$dx(t) = f(t, x(t))dt + c(t)x(t)\, dW(t), \qquad x(0) = x_0.$$

where $f(t, x(t)) \in \mathbb{R}$ and $c(t) \in \mathbb{R}$. Show that the solution can be derived as follows:

$$x(t) = F^{-1}(t)y(t)$$

where

$$dF(t) = F(t)c^2(t)dt - F(t)c(t)\, dW(t) \quad \text{with } F(0) = 1 \text{ and}$$

$$dy(t) = F(t)f(t, F^{-1}(y(t)))dt.$$

8.8 Stochastic differential equation

Calculate the analytical (Itô) stochastic solution for,

$$dx(t) = \frac{dt}{x(t)} + \alpha x(t)\, dW(t), \qquad x(0) = x_0.$$

with $\alpha \in \mathbb{R}$. Use the steps in problem (8.7) to check the following relations:

$$F(t) = e^{\frac{1}{2}\alpha^2 t - \alpha W(t)}$$

$$dy(t) = \frac{F(t)}{F^{-1}(t)y}dt = \frac{F^2(t)}{y}dt$$

$$dy(t)y(t) = F^2(t)dt, \qquad \frac{1}{2}y^2(t) = \int_0^t F^2(s)ds + c_0$$

$$y(t) = \left(x_0^2 + 2\int_0^t e^{\alpha^2 s - 2\alpha W(s)}ds\right)^{\frac{1}{2}}$$

$$x(t) = e^{-\frac{1}{2}\alpha^2 t + \alpha W(t)}\left(x_0^2 + 2\int_0^t e^{\alpha^2 s - 2\alpha W(s)}ds\right)^{\frac{1}{2}}$$

where c_0 is a real constant.

8.9 Stochastic differential equation

Solve the following (Itô) stochastic differential equations for $x(t)$ and determine $\langle x(t)\rangle$ and $\mathrm{var}(x(t))$:

$$(a) \qquad dx(t) = \left(a\sqrt{x(t)} + b^2/4\right)dt + b\sqrt{x(t)}\,dW(t)$$

$$(b) \qquad dx(t) = \left(ax^2(t) + b^2x^3(t)\right)dt + bx^2(t)\,dW(t)$$

and calculate $\langle x(t)\rangle$ from

$$(c) \qquad dx(t) = \left(ax^{3/4}(t) + \frac{3}{8}b^2x^{1/2}(t)\right)dt + bx^{3/4}(t)\,dW(t)$$

where a and b are real constants, and $x(0) = x_0 > 0$. (Hint: Using Lamperti transform, seek a transformation $y = f(x)$ for some f such that $y(t)$ satisfies a constant diffusion coefficient.)

References

1. R. Kubo, J. Phys. Soc. Jpn. **9**, 935 (1954)
2. P.W. Anderson, J. Phys. Soc. Jpn. **9**, 316 (1954)
3. A. Brissaud, U. Frisch, J. Math. Phys. **15**, 524 (1974)

Chapter 9
Langevin Dynamics in Higher Dimensions

This chapter describes and discusses Langevin dynamics and the Fokker–Planck equation in higher dimension, and discrete-time evolution and discrete-time approximation of stochastic evolution equations. We close the chapter with calculations of short-time propagators of d-dimensional Fokker–Planck equation.

9.1 d-Dimensional Langevin Dynamics

In this section we generalize the Langevin dynamics to higher dimensions. We consider a class of systems that are described by a multivariate state vector $\mathbf{x}(t)$, contained in a d-dimensional state space $\{\mathbf{x}\}$. The evolution of the state vector $\mathbf{x}(t)$ is assumed to be governed by a deterministic part to which a stochastic term is added:

$$\frac{d}{dt}\mathbf{x}(t) = \mathbf{N}\big(\mathbf{x}, t\big) + \mathbf{F}\big(\mathbf{x}, t\big). \tag{9.1}$$

Here, $\mathbf{N}$ denotes quite generally a nonlinear function that depends on the stochastic variable (vector) $\mathbf{x}(t)$ and, additionally, may explicitly depend on time t. In the following we shall always adopt the Itô interpretation, implying that the drift vector $\mathbf{D}^{(1)}(\mathbf{x}, t)$ coincides with the nonlinear vector field $\mathbf{N}(\mathbf{x}, t)$. Note, however, that the time t may also be considered as a general variable and replaced by, for example, a scale variable r, as described by, for example in [1]. Because the function $\mathbf{N}$ can be nonlinear, those systems that exhibit chaotic time evolution in the deterministic case are also included in the class of stochastic processes (9.1) by formally taking, $\mathbf{F} = 0$.

The term $\mathbf{F}\big(\mathbf{x}, t\big)$ fluctuates on a fast time scale. We assume that the d components F_i can be represented in the form

$$F_i\left(\mathbf{x}, t\right) = \sum_{j=1}^{d} g_{ij}\left(\mathbf{x}, t\right)\ \Gamma_j(t) \tag{9.2}$$

M. R. Rahimi Tabar, *Analysis and Data-Based Reconstruction of Complex Nonlinear Dynamical Systems*, Understanding Complex Systems,
https://doi.org/10.1007/978-3-030-18472-8_9

where the noise $\Gamma_j(t)$ terms are considered to be random functions whose statistical characteristics are well-defined. It is evident that these properties determine significantly the dynamical behavior of the state vector $\mathbf{x}(t)$. Therefore, using Eqs. (9.1) and (9.2) with the Itô interpretation, the d-dimensional Langevin dynamics is given by

$$\frac{d}{dt}\mathbf{x}(t) = \mathbf{N}\big(\mathbf{x}, t\big) + \mathbf{g}\big(\mathbf{x}, t\big)\, \boldsymbol{\Gamma}(t). \tag{9.3}$$

9.2 The Fokker–Planck Equation in Higher Dimensions

The conditional probability distribution $p(\mathbf{x}, t|\mathbf{x}', t')$, is obtained from the solution of the Fokker–Planck equation (also known as the second Kolmogorov equation, [2]):

$$\begin{aligned}\frac{\partial}{\partial t} p(\mathbf{x}, t\,|\,\mathbf{x}', t') = &-\sum_{i=1}^{d} \frac{\partial}{\partial x_i}\Big[D_i^{(1)}(\mathbf{x}, t)\, p(\mathbf{x}, t\,|\,\mathbf{x}', t')\Big] \\ &+ \frac{1}{2}\sum_{i,j=1}^{d} \frac{\partial^2}{\partial x_i \partial x_j}\Big[D_{ij}^{(2)}(\mathbf{x}, t)\, p(\mathbf{x}, t\,|\,\mathbf{x}', t')\Big].\end{aligned} \tag{9.4}$$

Here, $\mathbf{D}^{(1)}$ and $\mathbf{D}^{(2)}$ are the drift vector and the diffusion matrix, respectively. By considering the Itô definitions of the stochastic integrals, the coefficients $\mathbf{D}^{(1)}$ and $\mathbf{D}^{(2)}$ of the Fokker–Planck equation, Eq. (9.4), and the functionals $\mathbf{N}$ and $\mathbf{g}$ of the Langevin equation, Eq. (9.3), are related by

$$D_i^{(1)}(\mathbf{x}, t) = N_i(\mathbf{x}, t), \tag{9.5}$$

$$D_{ij}^{(2)}(\mathbf{x}, t) = \sum_{l=1}^{d} g_{il}(\mathbf{x}, t)\, g_{jl}(\mathbf{x}, t). \tag{9.6}$$

They are defined according to

$$\begin{aligned} D_i^{(1)}(\mathbf{x}, t) &= \lim_{\tau\to 0} \frac{1}{\tau}\big\langle x_i(t+\tau) - x_i\big\rangle\Big|_{\mathbf{x}(t)=\mathbf{x}} \\ &= \lim_{\tau\to 0} \frac{1}{\tau}\int d\mathbf{x}'\, p(\mathbf{x}', t+\tau|\mathbf{x}, t)\, (x_i' - x_i) \end{aligned} \tag{9.7}$$

$$\begin{aligned} D_{ij}^{(2)}(\mathbf{x}, t) &= \lim_{\tau\to 0} \frac{1}{\tau}\big\langle (x_i(t+\tau) - x_i)\, (x_j(t+\tau) - x_j)\big\rangle\Big|_{\mathbf{x}(t)=\mathbf{x}} \\ &= \lim_{\tau\to 0} \frac{1}{\tau}\int d\mathbf{x}'\, p(\mathbf{x}', t+\tau|\mathbf{x}, t)\, (x_i' - x_i)\, (x_j' - x_j). \end{aligned} \tag{9.8}$$

These expressions demonstrate that the drift vector and the diffusion matrix are determined as the first and second moments of the conditional probability distributions $p(\mathbf{x}', t+\tau|\mathbf{x}, t)$ in the small-time limit, nonparametrically, i.e., directly from the time series, see Chaps. 16–21.

9.3 The Kramers–Moyal Expansion in Higher Dimensions

For a d-component Markov process $x_j(t)$, $j = 1, 2, \ldots, d$, the Kramers–Moyal (KM) expansion is given by the following multivariate expansion,

$$\frac{\partial}{\partial t} p(\mathbf{x}, t \mid \mathbf{x}', t') = \sum_{m=1}^{\infty} \frac{(-1)^m}{m!} \sum_{j_1, j_2, \ldots}^{d} \frac{\partial^m}{\partial x_{j_1} \cdots \partial x_{j_m}} \left[M^{(m)}_{j_1,\ldots,j_m}(\mathbf{x}, t)\, p(\mathbf{x}, t \mid \mathbf{x}', t') \right]$$

where the KM-coefficients $M^{(m)}_{j_1,\ldots,j_m}$ are,

$$M^{(m)}_{j_1,\ldots,j_m}(\mathbf{x}, t) = \lim_{\tau\to 0} \frac{1}{\tau} \left\langle \prod_{\nu=1}^{m} [x_{j_\nu}(t+\tau) - x_{j_\nu}(t)] \right\rangle \bigg|_{\mathbf{x}(t)=\mathbf{x}}.$$

9.4 Discrete Time Evolution

It is convenient to consider the temporal evolution (9.3) of the state vector $\mathbf{x}(t)$ on a time scale that is large compared to the time scale of the fluctuations $\boldsymbol{\Gamma}(t)$. As we shall briefly indicate below, a stochastic process related to the evolution (9.3) is modeled by stochastic evolution laws that relate the state vectors $\mathbf{x}(t)$ at times $t_i,\ t_{i+1} = t_i + \tau,\ t_{i+2} = t_i + 2\tau, \ldots$, for small but finite values of τ. Here, we deal with the class of proper Langevin processes that are defined by the following discrete time evolutions.

9.4.1 Proper Langevin Equations: White Noise

The discrete time evolution of a proper Langevin process—the discrete version of the temporal evolution (9.3)—is given by[1]

---

[1] To understand the coefficient $\sqrt{\tau}$ in the last term of Eq. (9.9), in the integration of Langevin equation, we consider one-dimensional Langevin equation as,

$$\frac{dx}{dt} = a(x) + b(x)\Gamma(t)$$

$$\mathbf{x}(t_{i+1}) = \mathbf{x}(t_i) + \mathbf{D}^{(1)}[\mathbf{x}(t_i), t_i]\tau + \mathbf{g}(\mathbf{x}(t_i), t_i)\sqrt{\tau}\boldsymbol{\eta}(t_i) \tag{9.9}$$

where $\mathbf{D}^{(1)}[\mathbf{x}(t_i), t_i]$ and $\boldsymbol{\eta}(t_i)$ are the drift vector and a fluctuating vectorial quantity characterized by a Gaussian distribution with zero mean, $\langle \boldsymbol{\eta}(t_i) \rangle = 0$,

$$p(\boldsymbol{\eta}) = p(\eta_1, \eta_2, \cdots, \eta_d) = \frac{1}{(\sqrt{2\pi})^d} \exp\left(-\frac{\boldsymbol{\eta}^2}{2}\right) = \frac{1}{(\sqrt{2\pi})^d} \exp\left(-\frac{1}{2}\sum_{\alpha=1}^{d} \eta_\alpha^2\right). \tag{9.10}$$

Furthermore, their components are statistically independent for different times

$$\langle \eta_\alpha(t_i)\eta_\beta(t_j) \rangle = \delta_{\alpha\beta}\delta_{ij}. \tag{9.11}$$

The discrete Langevin, Eq. (9.9), must be considered in the limit $\tau \to 0$.

9.5 Discrete-Time Approximation of Stochastic Evolution Equations

In order to motivate the discrete-time approximations, Eq. (9.9), we integrate the white noise-driven Langevin equation over a finite but small time increment τ:

$$\begin{aligned} \mathbf{x}(t+\tau) &= \mathbf{x}(t) + \int_t^{t+\tau} dt' \mathbf{D}^{(1)}(\mathbf{x}, t') + \int_t^{t+\tau} dt' \mathbf{g}(\mathbf{x}, t')\boldsymbol{\Gamma}(t') \\ &\approx \mathbf{x}(t) + \tau \mathbf{D}^{(1)}(\mathbf{x}, t) + \int_t^{t+\tau} dt' \mathbf{g}(\mathbf{x}, t')\boldsymbol{\Gamma}(t'). \end{aligned} \tag{9.12}$$

The time interval τ is chosen to be larger than the time scale of the fluctuations $\Gamma_j(t)$. It involves the rapidly fluctuating quantities $\Gamma_j(t)$, and is represented by a stochastic integral, as discussed in Chap. 6. If we assume the matrix $\mathbf{g}$ to be independent of the time t and the state vector $\mathbf{x}(t)$, we arrive at the integrals

$$d\mathbf{W}(t, \tau) = \int_t^{t+\tau} dt' \boldsymbol{\Gamma}(t'). \tag{9.13}$$

---

where Γ is a zero-mean Gaussian white noise with intensity 1. We know from Wiener process that, $dW = \Gamma(t)\, dt$ and its first and second order incremental moments are, $\langle (W(t+\tau) - W(t)) \rangle = 0$, and $\langle (W(t+\tau) - W(t))^2 \rangle = \tau$. Therefore, dW and Γ have dimensions $\sqrt{\tau}$ and $1/\sqrt{\tau}$, respectively. This means that we can replace Γ with $\eta/\sqrt{\tau}$, where η is an uncorrelated Gaussian noise with unit variance to obtain the following discretized equation with time step τ of the Langevin equation

$$x(t+\tau) = x(t) + a(x)\tau + b(x)\sqrt{\tau}\eta.$$

This is equivalent replacing $dW = W(t+\tau) - W(t)$ with $\sqrt{\tau}\eta$. In the next chapter we show that a similar argument holds for the Lévy-driven Langevin equation where, using the relation $\langle |dL| \rangle = \tau^{1/\alpha}$, we will replace the Lévy noise with $\eta\tau^{1/\alpha}$, with η being a Lévy uncorrelated noise.

These are the quantities for which a statistical characterization can be derived. By looking at Eq. (9.12), we encounter the difficulty that the integrals over the noise forces may involve functions of the state vector within the time interval $(t, t+\tau)$. The interpretation of such integrals for wildly fluctuating stochastic quantities $\mathbf{\Gamma}(t)$ is difficult. As we have shown, the simplest way is to formulate an interpretation of such terms that lead to different interpretations of the stochastic evolution equation, Eq. (9.3). Once again, similar to one dimension, we formulate the widely used definitions due to Itô and Stratonovich, which are the higher-dimensional generalization of stochastic integrals.

In the Itô sense, the integral is interpreted as

$$\int_t^{t+\tau} dt' \mathbf{g}(\mathbf{x}, t') \mathbf{\Gamma}(t') = \mathbf{g}(\mathbf{x}, t) \int_t^{t+\tau} dt' \mathbf{\Gamma}(t') = \mathbf{g}(\mathbf{x}, t) d\mathbf{W}(t, \tau) \quad . \tag{9.14}$$

The *Stratonovich* definition is

$$\begin{aligned}\int_t^{t+\tau} dt' \mathbf{g}(\mathbf{x}, t') \mathbf{\Gamma}(t') &= \mathbf{g}\left(\frac{\mathbf{x}(t+\tau)+\mathbf{x}(t)}{2}, t+\frac{\tau}{2}\right) \int_t^{t+\tau} dt' \boldsymbol{\Gamma}(t') \\ &= \mathbf{g}\left(\frac{\mathbf{x}(t+\tau)+\mathbf{x}(t)}{2}, t+\frac{\tau}{2}\right) d\mathbf{W}(t, \tau).\end{aligned} \tag{9.15}$$

Since from experiments one obtains the probability distributions of the stochastic processes that are related to the stochastic Langevin equations, we are free to choose a certain interpretation of the process.

9.6 Short-Time Propagators of d-Dimensional Fokker–Planck Equation

It is straightforward to determine the short-time propagator from the finite-time approximation of the Langevin equation (9.3), Eq. (9.9). We shall denote the propagators by $p(\mathbf{x}_{j+1}, t+\tau|\mathbf{x}_j, t)$ and consider the case of Gaussian noise. From finite-time interpretation of the Langevin equation (9.9) we find,

$$\boldsymbol{\eta}(t_i) = \frac{1}{\tau^{1/2}} \left[\mathbf{g}(\mathbf{x}_i, t_i)\right]^{-1} \left[\mathbf{x}_{i+1} - \mathbf{x}_i - \tau \mathbf{D}^{(1)}(\mathbf{x}_i, t_i)\right]. \tag{9.16}$$

The variables $\boldsymbol{\eta}(t_i)$ are Gaussian random vectors with probability distribution as given by (9.10) and are related to the noise $\mathbf{\Gamma}$ via the relation, $\mathbf{\Gamma}(t_i) = \sqrt{\tau}\boldsymbol{\eta}(t_i)$. In turn, relation (9.16) defines the transition probability distribution

$$p(\mathbf{x}_{i+1}, t_{i+1}|\mathbf{x}_i, t_i)\, d\mathbf{x}_{i+1} = p[\boldsymbol{\Gamma}(t_i) = \boldsymbol{\Gamma}]\, J(\mathbf{x}_i, t_i)\, d\mathbf{x}_{i+1} \quad , \tag{9.17}$$

where J is the determinant of the Jacobian

$$J_{\alpha\beta} = \frac{\partial \Gamma_\alpha(t_i)}{\partial [\mathbf{x}_{i+1}]_\beta} \quad , \tag{9.18}$$

and $\left[\mathbf{g}\right]^{-1}$ denotes the inverse of the matrix $\mathbf{g}$ (which is assumed to exist). In what follows it is convenient to use the so-called diffusion matrix (9.6), i.e., $\mathbf{D}^{(2)}(\mathbf{x}_i, t_i)$

$$\mathbf{D}^{(2)}(\mathbf{x}_i, t_i) = \mathbf{g}(\mathbf{x}_i, t_i)\, \mathbf{g}^T(\mathbf{x}_i, t_i). \tag{9.19}$$

We are now able to explicitly state the short-time propagator of the Langevin process:

$$p(\mathbf{x}_{i+1}, t_{i+1}|\mathbf{x}_i, t_i) = \frac{1}{\sqrt{(2\pi\tau)^d \,\mathrm{Det}\left[\mathbf{D}^{(2)}(\mathbf{x}_i, t_i)\right]}} \exp\left[-\frac{S(\mathbf{x}_{i+1}, \mathbf{x}_i, t_i, \tau)}{2}\right] \tag{9.20}$$

with $\tau = t_{i+1} - t_i$. Here, $\mathrm{Det}[\mathbf{D}^{(2)}]$ is the determinant of $d \times d$ matrix $\mathbf{D}^{(2)}(\mathbf{x}_i, t_i)$, and we have defined the quantity $S(\mathbf{x}_{i+1}, \mathbf{x}_i, t_i, \tau)$ according to

$$S(\mathbf{x}_{i+1}, \mathbf{x}_i, t_i, \tau) =$$

$$\tau \left[\frac{\mathbf{x}_{i+1} - \mathbf{x}_i}{\tau} - \mathbf{D}^{(1)}(\mathbf{x}_i, t_i)\right] \left[\mathbf{D}^{(2)}(\mathbf{x}_i, t_i)\right]^{-1} \left[\frac{\mathbf{x}_{i+1} - \mathbf{x}_i}{\tau} - \mathbf{D}^{(1)}(\mathbf{x}_i, t_i)\right]. \tag{9.21}$$

As we see, the short-time propagator (9.20), which yields the transition probability density from state $\mathbf{x}_i$ to state $\mathbf{x}_{i+1}$ in the finite but small-time interval τ is a normal distribution. Equation (9.20) is the d-dimensional generalisation of (3.18) for short-term propagator of the Langevin dynamics in one dimension [3].

Problems

9.1 Short-term propagator of d-dimensional Langevin dynamics

Fill in the details in the derivation of Eq. (9.20).

9.2 The Stratonovich interpretation

Prove that with Stratonovich interpretation the drift vector and diffusion matrix are related to $\mathbf{N}$ and $g_{ij}(\mathbf{x}(t), t)$ by

$$D_i^{(1)}(\mathbf{x}, t) = N_i(\mathbf{x}, t) + \sum_{l=1}^{d} g_{il}(\mathbf{x}, t) \frac{\partial}{\partial x_l} g_{il}(\mathbf{x}, t),$$

$$D_{ij}^{(2)}(\mathbf{x}, t) = \sum_{l=1}^{d} g_{il}(\mathbf{x}, t)\, g_{jl}(\mathbf{x}, t).$$

9.3 The diffusion matrix

By construction, diffusion matrix given by Eq. (9.8) is a real and symmetric. Show that it is also positive semidefinite, which means that for an arbitrary column vector a

$$\sum_{i=1}^{d}\sum_{j=1}^{d} D_{ij}^{(2)} a_i a_j = a^{\mathrm{T}} D^{(2)} a \geq 0, \quad \forall\, a_1, \ldots, a_d \in \mathbb{R}$$

where a^{T} denotes the transpose of a.

9.4 The correlation functions

Consider a particular case for the drift vector and diffusion matrix [4], given by,

$$D_i^{(1)}(\mathbf{x}, t) = -A_{ij}(t)\, x_j$$

$$D_{ij}^{(2)}(\mathbf{x}, t) = D_{ij}^{(2)}(t) \quad .$$

Show that in the Itô sense the first and second statistical moments satisfy the following coupled first order differential equations,

$$\frac{d}{dt}\langle x_i\rangle = -A_{ij}(t)\,\langle x_j\rangle$$

$$\frac{d}{dt}\langle x_i x_j\rangle = -A_{il}(t)\,\langle x_l x_j\rangle - \langle x_i x_l\rangle\,\left(\mathbf{A}^{\mathrm{T}}\right)_{lj} + D_{ij}^{(2)}(t)$$

where $\mathbf{A}^{\mathrm{T}}$ is the transpose of $\mathbf{A}$. Repeated indices are implicitly summed over.

9.5 Two-dimensional stochastic differential equation and nonlinear equations

Consider the coupled nonlinear (Itô) stochastic differential equations,

$$\frac{dx_1}{dt} = a(1 - x_1^2 - x_2^2)x_1 - bx_2 + \sqrt{2k}\Gamma_1(t) = f_1(x_1, x_2) + \sqrt{2k}\Gamma_1(t),$$

$$\frac{dx_2}{dt} = a(1 - x_1^2 - x_2^2)x_2 + bx_1 + \sqrt{2k}\Gamma_2(t) = f_2(x_1, x_2) + \sqrt{2k}\Gamma_2(t)$$

with zero mean Gaussian noises $\Gamma_i(t)$ and

$$\langle \Gamma_i(t)\Gamma_j(t')\rangle = \delta_{ij}\delta(t - t')$$

where a, b and k are positive constants.

Show that the Fokker–Planck equation for the joint probability distribution function $p(x_1, x_2; t)$ is:

$$\frac{\partial p}{\partial t} + \frac{\partial}{\partial x_1}[f_1\, p] + \frac{\partial}{\partial x_2}[f_2\, p] - k\left(\frac{\partial^2 p}{\partial x_1^2} + \frac{\partial^2 p}{\partial x_2^2}\right) = 0.$$

Write down the Fokker–Planck equation in the polar coordinates (r, θ), and show that the radial-dependent $(\partial p/\partial\theta = 0)$ stationary distribution will be,

$$p(r, t \to \infty) = \frac{\sqrt{a/k}}{\pi^{3/2}} \frac{e^{-a/4k}}{1 + erf(\sqrt{a/4k})} \exp\left[\frac{a}{k}\left(\frac{r^2}{2} - \frac{r^4}{4}\right)\right],$$

where $r^2 = x_1^2 + x_2^2$ and $erf(z) = \frac{2}{\sqrt{\pi}} \int_0^z e^{-s^2} ds$.

References

1. R. Friedrich, J. Peinke, M. Sahimi, M.R. Rahimi Tabar, Phys. Rep. **506**, 87 (2011)
2. A.N. Kolmogorov, Math. Ann. **104**, 415 (1931)
3. H. Risken, *The Fokker-Planck Equation* (Springer, Berlin, 1989)
4. J. Honerkamp, *Stochastic Dynamical Systems : Concepts, Numerical Methods, Data Analysis* (Wiley-VCH, Weinheim, 1993)

Chapter 10
Lévy Noise-Driven Langevin Equation and Its Time Series–Based Reconstruction

In this chapter, we introduce Lévy noise-driven Langevin and fractional Fokker–Planck equations, and derive short-time propagator of the Lévy-driven processes. Then, we provide the details of the limit theorems for the Wiener and Lévy processes. Finally, non-parametric reconstruction of the Lévy-driven Langevin dynamics from time series is described.

10.1 Langevin Equation with Lévy Noise

As we saw in Chap. 9, *normal* diffusion processes under the influence of an external force field are often described by a Langevin equation (with the Itô interpretation) for the d-dimensional state variable $\mathbf{x}(t)$

$$d\mathbf{x}(t) = \mathbf{D}^{(1)}\big(\mathbf{x}, t\big)dt + \mathbf{g}\big(\mathbf{x}, t\big)d\mathbf{W}(t). \tag{10.1}$$

where the trajectory of the stochastic process $\mathbf{x}(t)$ is determined by a deterministic drift $\mathbf{D}^{(1)}$ and a stochastic part $\mathbf{g}$. Here $d\mathbf{W}$ stands for an infinitesimal d-dimensional Brownian motion, i.e., an infinitesimal Wiener process. The d-dimensional driving noise source $\mathbf{\Gamma}(t)$ of the Wiener process $d\mathbf{W} = \mathbf{\Gamma}(t)dt$ are zero mean δ-function correlated Gaussian white noises with

$$\langle \Gamma_i(t) \rangle = 0$$

$$\langle \Gamma_i(t)\Gamma_j(t') \rangle = \delta_{ij}\delta(t - t'). \tag{10.2}$$

Many physical and biological systems are described by models like Eq. (10.1); see Chap. 22. As we will see in Chaps. 16–21, the unknown deterministic drift $\mathbf{N}$ and the stochastic part $\mathbf{g}$ can be determined directly from measured time series.

M. R. Rahimi Tabar, *Analysis and Data-Based Reconstruction of Complex Nonlinear Dynamical Systems*, Understanding Complex Systems,
https://doi.org/10.1007/978-3-030-18472-8_10

The assumption of Gaussianity of the statistics of the noise $\mathbf{\Gamma}(t)$ is not necessary. One can consider fluctuating forces with non-Gaussian probability distributions. In this case the probability distributions must be characterized by the higher-order moments or, more explicitly, by their cumulants (or the tails of the probability distribution function). Although there are many heavy-tailed alternatives to the Gaussian law, such as the Student's t, hyperbolic, normal inverse Gaussian, or truncated stable distributions, there is one good reason for modeling noise $\mathbf{\Gamma}(t)$ with stable distributions. Indeed, they are supported by the generalized Central Limit Theorem that states that stable laws are the only possible limit distributions for properly normalized and centered sums of independent, identically distributed random variables. Stable distributions accommodate the heavy tails and asymmetry, and they often provide a very good fit to empirical data. In particular, they are valuable models for data sets covering extreme events, such as market crashes, natural catastrophes, etc.

In this section, the class of Langevin systems will be extended to the larger class of Langevin-like systems where the Gaussian white noise is replaced by the more general Lévy white noise [1, 2]. Stochastic Lévy processes constitute an interesting generalisation of normal diffusion processes. The typical length of Lévy flights grows according to,

$$L \sim t^{1/\alpha} \qquad \alpha \in (0, 2], \tag{10.3}$$

where $\alpha = 2$ represents the normal diffusion. This property has made Lévy flights natural candidates for the description of enhanced or suppressed diffusion.

The distribution of Lévy noise has a long-range algebraic tails corresponding to large but infrequent steps, the so-called *rare events*. Such events become more important and more likely in Lévy flights than in Brownian motion. The superdiffusive characteristics of Lévy flights have recently been used to model a broad variety of physical processes. Some examples are the description of anomalous transport in one dimension with absorbing boundary [3], modeling of anomalous diffusion at liquid surfaces (the so-called bulk-mediated surface diffusion) where the molecules execute Lévy walks on the surface [4], in the porous media [5–9] and eye lenses [10]. Enhanced diffusion has been reported in systems of polymer-like breakable micelles [11]. Turbulence has been investigated under the point of view of Lévy flights [12]. Even wandering of albatrosses has been modeled by Lévy flights [13]; see [14] for a recent review of the subject.

In the following the theoretical model of a generalization of the Langevin equation for normal diffusion processes, Eq. (10.1), will be presented. For dynamic systems that are describable by such an evolution equation, a method of analysis will be presented that provides us with a direct data-driven tool to formulate model equations for the dynamics of the system.

Instead of Eq. (10.1), the following nonlinear Langevin-like differential equation for a stochastic vector $\mathbf{x}(t)$ is assumed to describe the dynamics of the systems of interest,

$$d\mathbf{x}(t) = \mathbf{D}^{(1)}\big(\mathbf{x}, t\big)dt + \mathbf{g}\big(\mathbf{x}, t\big)\, d\mathbf{L}_{\alpha}^{\gamma,\beta,\mu}(t). \tag{10.4}$$

Here, $d\mathbf{L}_\alpha^{\gamma,\beta,\mu}(t)$ denotes an infinitesimal d-dimensional α-stable Lévy motion with,[1]

$$d\mathbf{L}_\alpha^{\gamma,\beta,\mu}(t) = \boldsymbol{\eta}_\alpha^{\gamma,\beta,\mu}(t)\, dt^{1/\alpha} \tag{10.5}$$

See Sect. 10.5 for the fractional power of dt in Eq. (10.5). Here, the d stochastically independently chosen components $(\boldsymbol{\eta}_\alpha^{\gamma,\beta,\mu})_i(t)$ of the Lévy noise creating such a motion are characterized by the four parameters: Lévy stability, or index of stability $\alpha \in (0, 2]$, also called the tail index, tail exponent or characteristic exponent, and scale parameter $\gamma \geq 0$, skewness parameter $\beta \in [-1, 1]$ and center or location parameter $\mu \in \mathbb{R}$. For $\alpha = 2$, $\gamma = 0.5$, $\beta = 0$ and $\mu = 0$, Eq. (10.4) reduces to the original Langevin equation, Eq. (10.1). For $\alpha < 2$, the variance is infinite and the tails are asymptotically equivalent to a Pareto law, i.e. they exhibit a power-law behavior. For decreasing α the larger deviations of the realization become larger and more frequent.

For simplicity we can assume that the matrix $\mathbf{g}$ is diagonal, i.e. $\mathbf{g}_{ij}\big(\mathbf{x}, t\big) = \mathbf{g}_{ii}\big(\mathbf{x}, t\big)\delta_{ij}$.

10.1.1 Lévy Noise

In general, the Lévy noise $y \equiv \eta_\alpha^{\gamma,\beta,\mu}(t)$ is not defined by its probability density, but rather by its statistical moment generating function. The probability density distribution exists and is continuous but, with some exceptions, it is not known in closed form. The characteristic or generating function of each Lévy noise component, i.e., the Fourier transform of the probability density distribution $p_\alpha^{\gamma,\beta,\mu}(y)$ of each component, has the following form [15]:

$$\begin{aligned}\langle e^{iqy}\rangle &= \int_{-\infty}^{\infty} p_\alpha^{\gamma,\beta,\mu}(y)\, e^{iqy} dx \\ &= \begin{cases} \exp\left\{-\gamma|q|^\alpha\left(1 - i\beta\, \mathrm{sgn}(q)\, \tan\left(\frac{\pi\alpha}{2}\right)\right) + i\mu q\right\} & \alpha \neq 1 \\ \exp\left\{-\gamma|q|\left(1 + i\frac{2}{\pi}\beta\, \mathrm{sgn}(q)\, \ln|q|\right) + i\mu q\right\} & \alpha = 1. \end{cases}\end{aligned} \tag{10.6}$$

Here $\mathrm{sgn}(q)$ is the sign function of q. Equation (10.6) is the most general form of a characteristic function of a stable process. In the limits $\gamma = 1$ and $\beta = \mu = 0$ the

[1] A probability distribution $f(x)$ is called stable (such as, for example, Lévy α-stable distribution) if a linear combination of two identically distributed independent random variables x_1 and x_2 has the same distribution, up to the location (mean) and scale parameters. This means that for positive constants a_1 and a_2, the constants $a > 0$ and $b \in \mathbb{R}$ exist so that

$$a_1x_1 + a_2x_2 \stackrel{\mathrm{d}}{=} ax + b$$

where $\stackrel{\mathrm{d}}{=}$ stands for equality of distributions. A stochastic process $\{x(t), t \in T\}$ is stable if all its finite-dimensional distributions are stable.

probability distribution function is given by,

$$p_\alpha^{\gamma=1,\beta=0,\mu=0}(y) = \frac{1}{\pi}\int_0^\infty \exp(-|q|^\alpha)\cos(qy)dq \tag{10.7}$$

where a series expansion of Eq. (10.7), valid for large arguments ($|y| \gg 1$), is given by,

$$p_\alpha^{\gamma=1,\beta=0,\mu=0}(|y|) = -\frac{1}{\pi}\sum_{k=1}^{n}\frac{(-1)^k}{k!}\frac{\Gamma(\alpha k+1)}{|y|^{\alpha k+1}}\sin\left(\frac{k\pi\alpha}{2}\right) + \mathcal{O}\left(|y|^{-\alpha(n+1)-1}\right).$$

with $\Gamma(\cdots)$ being the Euler Γ function. The leading term for large values of $|y|$ is given by

$$p_\alpha^{\gamma=1,\beta=0,\mu=0}(|y|) \sim \frac{\Gamma(\alpha+1)\sin(\frac{\pi\alpha}{2})}{\pi|y|^{(1+\alpha)}}, \qquad |y| \to \infty \quad . \tag{10.8}$$

The asymptotic behavior for large values of $|y|$ is a power law, which results in the divergence of all the statistical moments $\langle |y|^n \rangle$ with $n \geq \alpha$. In particular, all Lévy stable processes with $\alpha < 2$ have infinite variance. For small y one finds

$$p_\alpha^{\gamma=1,\beta=0,\mu=0}(|y|) \sim \frac{\Gamma(\frac{1}{\alpha})}{\pi\alpha}, \qquad |x| \to 0$$

which implies that $p_\alpha^{\gamma=1,\beta=0,\mu=0}(|y| \to 0) \to \infty$ as $\alpha \to 0$ and, therefore, the center of the probability distribution density becomes sharper and, according to Eq. (10.8), its tails become fatter.

Explicit densities with $\mu = 0$ are known for some cases, Gaussian distribution, $\alpha = 2$, $\beta = 0$, Cauchy distribution, $p(y) = 1/\pi(1+y^2)$ for $\alpha = 1$, $\beta = 0$ and the Lévy-Smirnov distribution with $\alpha = 1/2$, $\beta = 1$ whose explicit form is,

$$p(y) = \begin{cases} \frac{1}{\sqrt{2\pi}} y^{-3/2}\exp(-\frac{1}{2y}) & y \geq 0 \\ 0 & y < 0. \end{cases} \tag{10.9}$$

For $\alpha = 1/3$, $\beta = 1$ the distribution is $1/((2\pi/3^{7/6})\ y)^{3/2} K_{1/3}[2^{5/2}/(3^{9/4}\sqrt{y})]$, where $K_\nu(u)$ is the modified Bessel function of the second kind. In addition, for $\alpha = 2/3$, $\beta = 0$ the distribution is

$$p(y) = \frac{1}{6}\sqrt{\frac{3}{\pi y^2}}\exp\left(\frac{2}{27y^2}\right) W_{-1/2,1/6}\left(\frac{4}{27y^2}\right)$$

where $W(u)$ is the Whittaker function. Finally, the Holtsmark distribution represents the limit $\alpha = 3/2$ and $\beta = 0$.

10.1.2 Fractional Fokker–Planck Equations

The finite-time propagators or conditional probability distributions of stochastic processes containing the Lévy noise lead to the fractional diffusion equations. For a discussion of this topic we refer the reader to [16–18]; see also Sect. 10.3. For example, consider one-dimensional Lévy-noise driven Langevin equation as,

$$dx(t) = D^{(1)}(x)dt + dL_{\alpha}^{\gamma,\beta,\mu},$$

where the noise is understood in the sense of the Itô interpretation. The noise $L_{\alpha}^{\gamma,\beta,\mu}(t)$ is the uncorrelated Lévy process with the stability index $(0 < \alpha \leq 2)$. The cumulant expansion of the characteristic function, truncated at the order α, with scale parameter γ produces the following fractional Fokker–Planck equation [19],

$$\frac{\partial}{\partial t} p(x,t|x',t') = -\frac{\partial}{\partial x}\left(D^{(1)}(x)\, p(x,t|x',t')\right) + \gamma \frac{\partial^{\alpha}}{\partial |x|^{\alpha}} p(x,t|x',t')$$
$$+ \gamma\beta \tan\left(\frac{\pi\alpha}{2}\right) \frac{\partial}{\partial x} \frac{\partial^{\alpha-1}}{\partial |x|^{\alpha-1}} p(x,t|x',t')$$

where we need to deal with partial differentiations with fractional order α and $\alpha - 1$. In Sect. 10.3 we provide the short-time propagator of the Langevin dynamics driven by the Lévy processes, if we do not have the Fokker–Planck equation itself.

10.2 Discrete Time Approximation of Langevin Equations: Lévy Noise-Driven

A more general class is formed by the discrete-time evolution laws [1],

$$\mathbf{x}(t_{i+1}) = \mathbf{x}(t_i) + \mathbf{D}^{(1)}(\mathbf{x}(t_i), t_i)\tau + \mathbf{g}(\mathbf{x}(t_i), t_i)\tau^{1/\alpha}\boldsymbol{\eta}_{\alpha}^{\gamma,\beta,\mu}(t_i) \tag{10.10}$$

where $\mathbf{D}^{(1)}(\mathbf{x}(t_i), t_i)$ and $\boldsymbol{\eta}_{\alpha}^{\gamma,\beta,\mu}(t_i)$ are the drift vector and a random quantity, distributed according to the Lévy stable law, characterized by the Lévy parameters α, γ, β and μ. We set the mean $\mu = 0$ and the scale parameter $\gamma = 1$. See Sects. 9.4.1 and 10.5, for the origin of the factor $\tau^{1/\alpha}$ in Eq. (10.10).

As we pointed out earlier, the Gaussian distribution represents a special limit of the class of the Lévy stable distributions with $\alpha = 2$ and $\beta = 0$. Formally, one has, $0 < \alpha \leq 2$. For almost all the applications, however, it seems reasonable to choose, $1 < \alpha \leq 2$, in order for the first moment of the noise source to be finite. The consideration of this type of statistics for the noise variables η is based on the central-limit theorem, as discussed in Sect. 10.5.

The discrete Langevin equation, Eq. (9.9), and the generalized Langevin equation, Eq. (10.10), must be considered in the limit $\tau \to 0$. They are the basis of all the further treatments. A central point is that, if one assumes the noise sources to be independent of the variable $\mathbf{x}(t_i)$, the discrete-time evolution equations define a Markov process, the generator of which, i.e., its conditional probability distribution or short time propagator, is established on the basis of Eqs. (9.9) and (10.10).

10.3 Short-Time Propagator of the Lévy Noise-Driven Langevin Processes

It is now straightforward to determine the short-time propagator of the Lévy noise-driven Langevin processes. We must replace the Gaussian distribution by the (multivariate) Lévy distribution $p_\alpha^{\gamma,\beta,\mu}(\eta)$. As a consequence, we obtain the conditional probability, i.e., the short-time propagator, of the Lévy processes in d-dimensions:

$$p(\mathbf{x}_{i+1}, t_i + \tau|\mathbf{x}_i, t_i) = \frac{1}{\tau^{d/\alpha}\mathrm{Det}\left[\mathbf{g}(\mathbf{x}_i, t_i)\right]} p_\alpha^{\gamma,\beta,\mu} \left\{\frac{1}{\tau^{1/\alpha}} \left[\mathbf{g}(\mathbf{x}_i, t_i)\right]^{-1} \left[\mathbf{x}_{i+1} - \mathbf{x}_i - \tau \mathbf{D}^{(1)}(\mathbf{x}_i, t_i)\right]\right\} \tag{10.11}$$

where from Eq. (10.10) we used

$$\eta_\alpha^{\gamma,\beta,\mu}(t_i) = \left\{\frac{1}{\tau^{1/\alpha}} \left[\mathbf{g}(\mathbf{x}_i, t_i)\right]^{-1} \left[\mathbf{x}_{i+1} - \mathbf{x}_i - \tau \mathbf{D}^{(1)}(\mathbf{x}_i, t_i)\right]\right\}.$$

The steps of proof of Eq. (10.11) are similar to what we did in Sect. 9.6. The importance of this approach is that it allows us to have an approximation of $p(\mathbf{x}_{i+1}, t_i + \tau|\mathbf{x}_i, t_i)$, even when we do not have the Fokker–Planck equation itself.

10.4 Joint Probability Distribution and Markovian Properties

Due to the statistical independence of the random vectors components $\eta(t_i)$ and $\eta(t_j)$ for $i \neq j$, we obtain the n-point joint probability distribution as a product of the distributions $p(\eta)$:

$$p(\eta_n, \ldots, \eta_1) = p(\eta_n)p(\eta_{n-1}) \cdots p(\eta_1). \tag{10.12}$$

Here, we omitted α, γ, and β, as well as the μ-dependence of the noise η to enhance readability. Furthermore, we observe that under the assumption that the random vector $\eta(t_i)$ is independent of the variables $\mathbf{x}(t_j)$ for all $j \leq i$ and using Eq. (10.11), we construct the n-point joint probability distribution

$$p(\mathbf{x}_n, t_n; \cdots; \mathbf{x}_1, t_1) = p(\mathbf{x}_n, t_n | \mathbf{x}_{n-1}, t_{n-1}) \cdots p(\mathbf{x}_2, t_2 | \mathbf{x}_1, t_1) p(\mathbf{x}_1, t_1).$$

This is the definition of a Markov process, however. We note that, the transition probabilities are the short-time propagators, i.e., the representation (10.11) is valid in the short-time limit, $\tau = t_n - t_{n-1} \to 0$. This stems from the fact that Eq. (10.10) is valid for very small but finite τ. The probability distribution (10.11) is the discrete approximation of the *path integral representation* of the stochastic process under consideration [20]; see problem (3.2).

The statistical description of the Langevin equation based on the n-points joint probability distribution leads to its representation in terms of the conditional probability distributions. This representation is the definition of a Markov process. Due to the assumptions on the statistics of the fluctuating forces, various processes arise. If the fluctuating forces are assumed to be Gaussian, the short-time propagator is Gaussian and, as a consequence, is solely defined by the drift vector and the diffusion matrix. If the fluctuating forces are assumed to be Lévy-distributed, more complex short-time propagators arise, i.e., (10.11).

We note that the Markovian property, i.e., the fact that the propagator $p(\mathbf{x}_i, t_i |\mathbf{x}_{i-1} t_{i-1})$ does not depend on the states $\mathbf{x}_k$ at times $t_k < t_{i-1}$, is usually violated by many physical systems because the noise sources become correlated for small time differences t_M, a point which has already been emphasized in the famous work of Albert Einstein on the Brownian motion, which is one of the first works on stochastic processes [21]. This time scale is called the Markov-Einstein time scale, and appears to be a highly interesting quantity in analysing complex time series [22–24]. In Chap. 16 we will provide some methods to estimate the Markov-Einstein time scale from measured time series.

10.5 Limit Theorems, and Wiener and Lévy Processes

In the following we shall discuss the possibilities of characterizing the stochastic integrals

$$L(t+\tau) - L(t) = \int_t^{t+\tau} dt' \Gamma(t') \tag{10.13}$$

where $\Gamma(t)$ is a zero mean rapidly fluctuating quantity. In order to characterize the properties of this noise, one can resort to the limit theorems of statistical mechanics [25]. The central-limit theorem states that if the quantities Γ_j, $j = 1, \ldots, n$ are statistically independent variables with zero mean and variance σ^2, then the sum

$$\frac{1}{\sqrt{n}}\sum_{j=1}^{n}\Gamma_j=\Gamma \tag{10.14}$$

tends to a Gaussian random variable with variance σ^2 for large values of n. The limiting probability distribution $p(\Gamma)$ is then a Gaussian distribution with variance σ^2:

$$p(\Gamma)=\frac{1}{\sqrt{2\pi\sigma^2}}\exp(-\Gamma^2/2\sigma^2). \tag{10.15}$$

There exists a generalization of the central-limit theorem that is applicable to random variables whose second statistical moments do not exist. It states that the distribution of the sum over identically distributed random variables with zero mean, Γ_j [26, 27]

$$\frac{1}{n^{1/\alpha}}\sum_{j=1}^{n}\Gamma_j\rightarrow\eta_\alpha \tag{10.16}$$

for large values of n tends to a random variable η_α that is distributed according to the Lévy-stable distribution $p_\alpha(\eta)$. Here, we ignored superscripts γ, β and μ in order to enhance the readability.

In order to evaluate the integral (10.13) using the limit theorems, it is convenient to represent the stochastic force $\Gamma(t)$ as a sum over N_τ δ-kicks occurring at discrete times t_j

$$\Gamma(t)=\sum_j\Gamma_j\,(\Delta t)^{1/\alpha}\delta(t-t_j)\quad , \tag{10.17}$$

where Δt is the time difference between the occurrence of the two kicks. Then, we obtain

$$\begin{aligned} dL(t,\tau) &= \sum_{j,t_j\in\tau}\Gamma_j\,(\Delta t)^{1/\alpha} \\ &= (N_\tau\Delta t)^{1/\alpha}\frac{1}{N_\tau^{1/\alpha}}\sum_{j,t_j\in\tau}\Gamma_j=\tau^{1/\alpha}\eta_\alpha(t) \end{aligned} \tag{10.18}$$

where N_τ is the number of kicks in the interval τ and in the limit $N_\tau\rightarrow\infty$, $\frac{1}{N_\tau^{1/\alpha}}\sum_{j,t_j\in\tau}\Gamma_j$ tends to a stable random variable. Thus, for the case of $\alpha=2$, the stochastic variable $dL(t,\tau)$ will be the increment of the Wiener processes and may be represented by

$$dW(t,\tau)=\sqrt{\tau}\eta(t) \tag{10.19}$$

where $\eta(t)$ is a random variable, distributed according to a Gaussian distribution. For the more general case, $dL(t,\tau)$ is a stochastic variable

$$dL(t,\tau) \equiv d\mathbf{L}_{\alpha}^{\gamma,\beta,\mu}(t) = \tau^{1/\alpha}\boldsymbol{\eta}_{\alpha}^{\gamma,\beta,\mu}(t) \tag{10.20}$$

where the distribution of $\eta_{\alpha}^{\gamma,\beta,\mu}$ is the Lévy distribution.

10.6 Reconstruction of Lévy Noise-Driven Langevin Dynamics by Time Series

In this section we present a time series–based, i.e., data-driven non-parametric method for Lévy noise-driven Langevin dynamics, and show that the Lévy stability index α, as well as the nonlinear deterministic and stochastic parts of the dynamics, together with their uncertainties, are estimated from a time series [1].

Let us assume that the Langevin equation with the Lévy noise are given by equation,

$$\mathbf{x}(t_{i+1}) = \mathbf{x}(t_i) + \mathbf{D}^{(1)}[\mathbf{x}(t_i), t_i]\,\tau + \mathbf{g}(\mathbf{x}(t_i), t_i)\,\tau^{1/\alpha}\boldsymbol{\eta}_{\alpha}^{\gamma=1,\beta=0,\mu=0}(t_i) \tag{10.21}$$

for small τ, where we fixed all the parameters of the Lévy noise except α. Setting the skewness parameter $\beta = 0$ ensures that the Lévy distribution is symmetric, while $\mu = 0$ implies that the Lévy noise has zero mean. We also consider the Lévy noise with scale parameter $\gamma = 1$ and remove any scaling dependence in the function $\mathbf{g}(\mathbf{x})$. For the extracting the deterministic and stochastic parts from fluctuating data, first, it is assumed that the underlying dynamics is described by an iteration like Eq. (10.21). At the end, after all analysis results are obtained, the assumption can be verified by some self-consistency checks.

Because we took the case $\beta = \mu = 0$, the probability density of $\boldsymbol{\eta}_{\alpha}^{\gamma=1,\beta=0,\mu=0}$ is a zero-mean symmetric function. Therefore, conditional averaging the Eq. (10.21) over various realization of the Lévy noise, and taking into account the non-anticipating property of $\mathbf{g}(\mathbf{x})$, yields

$$\mathbf{D}^{(1)}(\mathbf{x}) = \lim_{\tau\to 0}\frac{1}{\tau}\langle(\mathbf{x}(t+\tau) - \mathbf{x}(t))\rangle|_{\mathbf{x}(t)=\mathbf{x}} \quad . \tag{10.22}$$

If stationarity is given, then the deterministic and stochastic functions $\mathbf{D}^{(1)}(\mathbf{x})$ and $\mathbf{g}(\mathbf{x})$ are not explicitly time dependent. Therefore, the ensemble average in Eq. (10.22) is estimated as the conditional temporal average over the entire time series. For a non-stationary time series, see Sect. 17.1 for estimating the time-dependence of $\mathbf{D}^{(1)}(\mathbf{x}, t)$.

For stationary time series, the state space of the process is discretized, and the conditional average must be calculated for every $\mathbf{x}$ (with some binning) separately. To estimate the stochastic part $g(\mathbf{x})$, the Lévy index α must be determined, first. Therefore, the following conditional average is considered and approximated for small τ by

$$T_i^{(2)}(\mathbf{x}, \tau) = \langle |\mathbf{x}_i(t+\tau) - \mathbf{x}_i(t) - \mathbf{D}_i^{(1)}(\mathbf{x})\tau|\rangle|_{\mathbf{x}(t)=\mathbf{x}}$$

$$\simeq g_{ii}(\mathbf{x})\tau^{1/\alpha} \int_{-\infty}^{\infty} |\eta|\, p_\alpha^{(\gamma=1,\beta=0,\mu=0)}(\eta)\, d\eta$$

$$= g_{ii}(\mathbf{x})\tau^{1/\alpha} F(\alpha). \tag{10.23}$$

The logarithm of Eq. (10.23) is given by

$$\ln\left[T_i^{(2)}(\mathbf{x}, \tau)\right] = \ln\left[g_{ii}(\mathbf{x})F(\alpha)\right] + \frac{1}{\alpha}\ln(\tau). \tag{10.24}$$

Using Eq. (10.24) we estimate both α and the function $g_{ii}(\mathbf{x})$. To estimate α, we calculate numerically the function $T_i^{(2)}(\mathbf{x}, \tau)$ for several small values of τ, and then plot it for fixed $\mathbf{x}$ in a log − log manner over τ, after which α is determined as the inverse slope of a fitted straight line. In theory, for each $\mathbf{x}$ and i the value of α should be received by this method and the average value α is used as an estimate for the Lévy index. The standard deviation of distribution for α may be taken as uncertainty of index.

Knowing the parameter α allows us to estimate the expression $F(\alpha)$ via simulating the Lévy noise with the estimated α with $(\alpha, \gamma = 1, \beta = 0, \mu = 0)$, and taking the average of the absolute value of the realization. The realization $\eta_\alpha^{\gamma=1,\beta=0,\mu=0}$ of Lévy noise is then constructed as follows [28].[2]

A realization r of a uniformly distributed random variable in the interval $[-\pi/2, \pi/2]$ is taken and, independently, a realization v of an exponential random variable with mean 1 is also generated. The noise $\eta_\alpha^{\gamma=1,\beta=0,\mu=0}$ will be [2],

$$\eta_\alpha^{\gamma=1,\beta=0,\mu=0} = \begin{cases} \dfrac{\sin(\alpha r)}{\cos^{1/\alpha}(r)} \left\{\dfrac{\cos((1-\alpha)r)}{v}\right\}^{(1-\alpha)/\alpha} & \alpha \neq 1 \\ \tan(r) & \alpha = 1. \end{cases} \tag{10.25}$$

Having α and $F(\alpha)$ enable us to determine the stochastic part $g_{ii}(\mathbf{x})$

$$g_{ii}(\mathbf{x}) = \lim_{\tau\to 0} \frac{[T_i^{(2)}(\mathbf{x}, \tau)]}{\tau^{1/\alpha}F(\alpha)}. \tag{10.26}$$

The procedure concerning the calculation of the limit $\tau \to 0$ in each bin is analogous to the determination of the deterministic part [1]. See Sect. 16.7 for possible self-consistency check of the constructed Lévy noise-driven Langevin dynamics, Eq. (10.21), from time series.

---

[2]In addition, one can use fast Fourier transform that must be applied to the characteristic function, Eq. (10.6), or by utilizing a direct numerical integration.

Problems

10.1 Tails of the Lévy distribution function

Fill in the details in the derivation of Eq. (10.8) from Eq. (10.7).

10.2 Short-time propagator

Fill in the details in the derivation of Eq. (10.11).

10.3 Lévy noise

Use Eq. (10.25) to generate 10^6 Lévy noise with $\alpha = 0.5, 1, 1.5$ and check that the tails of their probability distribution functions are consistent with Eq. (10.8).

10.4 Lévy process

Use Eq. (10.25) to generate $n = 10^3$ Lévy noise with $\alpha = 0.5, 1, 1.9$ and compute

$$\Delta L_k = L(t_{k+1}) - L(t_k) = (t_{k+1} - t_k)^{1/\alpha} \eta_\alpha^{\gamma=1,\beta=0,\mu=0}$$

for $k = 1, 2, \ldots, n$ and with $L(t_1) = 0$ and $t_{k+1} - t_k = 0.001$. Plot the discretized trajectories of Lévy processes, which are given by $L(t_i) = \sum_{k=1}^{i} \Delta L_k$.

References

1. S. Siegert, R. Friedrich, Phys. Rev. E **64**, 041107 (2001)
2. S. Borak, W. Hardle, R. Weron, Stable distributions, SFB 649 Discussion Paper **008**, 1–28 (2005)
3. G. Zumofen, J. Klafter, Phys. Rev. E **51**, 2805 (1995)
4. O.V. Bychuk, B. OŚhaughnessy, Phys. Rev. Lett. **74**, 1795 (1995)
5. M. Sahimi, Hydrodynamic dispersion near the percolation threshold: scaling and probability densities. J. Phys. A **20**, L1293–L1298 (1987)
6. M. Sahimi, A.O. Imdakm, The effect of morphological disorder on hydrodynamic dispersion in flow through porous media. J. Phys. A **21**, 3833–3870 (1988)
7. M. Sahimi, Dispersion in porous media, continuous-time random walks, and percolation. Phys. Rev. E **85**, 016316 (2012)
8. M. Sahimi, H. Rassamdana, A. Mehrabi, *Fractals in Porous Media: From Pore to Field Scale*, ed. by F. Family, P. Meakin, B. Sapoval, R. Wood. Fractal Aspects of Materials, MRS Proceedings, vol. 367 (1995), pp. 203–214
9. S. Stapf, R. Kimmich, R.-O. Seitter, Phys. Rev. Lett. **75**, 2855 (1995)
10. J. Bodurka, R.-O. Seitter, R. Kimmich, A. Gutsze, J. Chem. Phys. **107**, 5621 (1997)
11. A. Ott, J.P. Bouchaud, D. Langevin, W. Urbach, Phys. Rev. Lett. **65**, 2201 (1990)
12. G. Zumofen, A. Blumen, J. Klafter, M.F. Shlesinger, J. Stat. Phys. **54**, 1519 (1989)
13. G.M. Viswanathan, V. Afanasyev, S.V. Buldyrev, E. Murphy, P.A. Prince, H.E. Stanley, Nature **381**, 413 (1996)
14. V. Zaburdaev, S. Denisov, J. Klafter, Rev. Mod. Phys. **87**, 483 (2015)

15. A. Kyprianou, *Introductory Lectures on Fluctuations of Lévy Processes with Applications* (Springer, Berlin, 2006)
16. D. Schertzer, M. Larchevéque, J. Duan, V.V. Yanovsky, S. Lovejoy, J. Math. Phys. **42**, 200 (2001)
17. V.V. Yanovsky, A.V. Chechkin, D. Schertzer, A.V. Tur, Phys. A **282**, 13 (2000)
18. R. Metzler, J. Klafte, Phys. Rep. **339**, 1 (2000)
19. S.I. Denisov, P. Hänggi, H. Kantz, Europhys. Lett. **85**, 40007 (2009)
20. H. Risken, *The Fokker-Planck Equation* (Springer, Berlin, 1989)
21. A. Einstein, Über die von der molekularkinetischen Theorie der Wärme geforderte Bewegung von in ruhenden Flüssigkeiten suspendierten Teilchen. Ann. Phys. **17**, 549 (1905)
22. R. Zeller, J. Friedrich, J. Peinke, Europhys. Lett. **41**, 153 (1998)
23. C. Renner, J. Peinke, R. Friedrich, J. Fluid Mech. **433**, 383 (2001)
24. M.R. Rahimi Tabar, M. Sahimi, F. Ghasemi, K. Kaviani, M. Allamehzadeh, J. Peinke, M. Mokhtari, M. Vesaghi, M.D. Niry, A. Bahraminasab, S. Tabatabai, F. Fayazbakhsh, M. Akbari, Modelling *Critical and Catastrophic Phenomena in Geoscience: A Statistical Physics Approach.* Lecture Notes in Physics, vol. 705, p. 281, (Springer, Berlin, 2007)
25. B.V. Gnedenko, A.N. Kolmogorov, *Limit Distributions of Sums of Independent Random Variables* (Addison-Wesley, Boston, 1954)
26. R. Friedrich, J. Peinke, M. Sahimi, M.R. Rahimi Tabar, Phys. Rep. **506**, 87 (2011)
27. J.P. Nolan, *Stable Distributions: Models for Heavy-Tailed Data* (Springer, New York, 2016)
28. G. Samorodnitzky, M. Taqqu, *Stable Non-Gaussian Random Processes* (Chapman & Hall, New York, 1994)

Chapter 11
Stochastic Processes with Jumps and Non-vanishing Higher-Order Kramers–Moyal Coefficients

In this chapter we study stochastic processes in the presence of jump discontinuity, and discuss the meaning of non-vanishing higher-order Kramers–Moyal coefficients. We describe in details the stochastic properties of Poisson jump processes. We derive the statistical moments of the Poisson process and the Kramers–Moyal coefficients for pure Poisson jump events. Growing evidence shows that continuous stochastic modeling (white noise-driven Langevin equation) of time series of complex systems should account for the presence of discontinuous jump components [1–6]. Such time series have some distinct important characteristics, such as heavy tails and occasionally sudden large jumps. Nonparametric (data-based) modeling of time series with jumps provides an attractive way of conducting research and gaining intuition of such processes.

Many theoretical models have been developed in the context of mathematical finance to describe discontinuous components [7]. Among of them, jump-diffusion modeling provides a theoretical tool and has received great attention in the literature due to the existence of nonparametric estimation procedure, which we will discuss in details in Chaps. 20 and 21. It allows us study the potential nonlinearities in the drift, in the diffusive volatility, and in the intensity of the discontinuous jump component [1, 6].

As we will see in Chaps. 12 and 13, from a theoretical point of view, the second Kramers–Moyal conditional moment of the process can be expressed as the sum of the infinitesimal conditional volatility of the diffusion component and the conditional second moment of the jump part. In addition jumps play a significant role when all the KM conditional moments of order higher than two are not vanishing. In particular higher KM conditional moments would carry information about the probability of arrival (jump rate) and the features of the distribution of the jump size.

The focus in this chapter is introducing stochastic tools for investigation of time series with discontinuous jump components. We will start with the meaning of non-vanishing higher order KM coefficients and its impact on the continuity condition. Similarly to the role of the Wiener process in the Langevin modeling, the Pois-

M. R. Rahimi Tabar, *Analysis and Data-Based Reconstruction of Complex Nonlinear Dynamical Systems*, Understanding Complex Systems,
https://doi.org/10.1007/978-3-030-18472-8_11

son jump process plays an essential role in jump-diffusion modeling. Therefore we present stochastic properties of the Poisson jump process, such as its statistical moments, waiting time distribution, etc.

11.1 Continuous Stochastic Processes

As we have mentioned already in the Introduction, systems under the influence of random forcing or in the presence of non-linear interactions with other systems can behave in a very complex stochastic manner [8]. The governing process can be either stochastic continuous or discontinuous processes. For stochastic continuous Markov processes we have seen in Chap. 7, that the dynamics of process x driven by white noise can be written as (using Itô interpretation),

$$dx(t) = D^{(1)}(x,t)dt + \sqrt{2D^{(2)}(x,t)}dW(t), \tag{11.1}$$

where $\{W(t), t \geq 0\}$ is a scalar Wiener (Brownian) motion, $D^{(1)}(x,t)$ and $D^{(2)}(x,t)$ are the deterministic drift and the diffusion functions (the first and second Kramers–Moyal coefficients).

As we have shown in Chap. 4, in Eq. (11.1) $x(t)$ are *continuous* diffusion processes. The first and second order Kramers–Moyal (KM) coefficients can be determined non-parametrically, i.e., directly form measured time series. For the continuous diffusion process Eq. (11.1) (for infinitesimal dt) we have:

$$\begin{aligned} \langle (x(t+dt) - x(t))^1|_{x(t)=x}\rangle &= D^{(1)}(x,t)dt \\ \langle (x(t+dt) - x(t))^2|_{x(t)=x}\rangle &= 2D^{(2)}(x,t)dt \\ \langle (x(t+dt) - x(t))^{2+s}|_{x(t)=x}\rangle &= 0 \end{aligned} \tag{11.2}$$

with $s > 0$. Here KM coefficients are given by $D^{(m)}(x) = \lim_{dt\to 0} \frac{1}{m!dt} K^{(m)}(x)$, where $K^{(m)}(x) = \langle (x(t+dt) - x(t))^m|_{x(t)=x}\rangle$. Remind that $D^{(m)}(x)$ is related to $M^{(m)}(x)$ as $D^{(m)}(x) = M^{(m)}(x)/m!$

As we have seen in Chap. 4, using the conditional probability distribution, a continuous process $x(t)$ satisfies the following continuity condition, given some $\delta > 0$,

$$C(t) = \lim_{dt\to 0} \frac{Prob[|\Delta x(t)| > \delta\ |_{x(t)=x}]}{dt} = 0 \tag{11.3}$$

where $\Delta x(t) = x(t+dt) - x(t)$, uniformly for all x and t.

11.2 Non Vanishing Higher Order Kramers–Moyal Coefficients and the Continuity Condition

In this section, we show that the non-vanishing higher order KM coefficients $n > 2$ will be indicator of the noncontinuity of underlying stochastic processes. Let us prove the following relation for the continuity condition in terms of Kramers–Moyal coefficients,

$$\lim_{dt\to 0} \frac{Prob[|\Delta x| > \delta \,|_{x(t)=x}]}{dt} \le \frac{M^{(m)}(x)}{\delta^m} \tag{11.4}$$

where $M^{(m)}(x) = \lim_{dt\to 0} \frac{1}{dt}\langle |x(t+dt) - x(t)|^m |_{x(t)=x}\rangle$ (note the absolute value in the conditional moments). This means that KM coefficients provide an upper limit for continuity condition.

Proof

To prove (11.4), let us consider the order m conditional moment as,

$$\langle |x(t+dt) - x(t)|^m\rangle|_{x(t)=x} = \langle |u - x|^m\rangle|_{x(t)=x} \tag{11.5}$$

where $x(t + dt) = u$. We can write the conditional moments in terms of short-time propagator as:

$$\begin{aligned}\langle |u - x|^m\rangle|_{x(t)=x} &= \int_{-\infty}^{\infty} du\ |u - x|^m p(u, t + dt|x, t)\\ &= \int_{-\infty}^{x-\delta} du\ |u - x|^m p(u, t + dt|x, t) + \int_{x-\delta}^{x+\delta} du\ |u - x|^m p(u, t + dt|x, t)\\ &+ \int_{x+\delta}^{\infty} du\ |u - x|^m p(u, t + dt|x, t).\end{aligned} \tag{11.6}$$

Taking into account the positivity of $p(u, t + dt|x, t)$ we can write the following inequality,

$$\langle |u - x|^m\rangle|_{x(t)=x} \ge \int_{|u-x|>\delta} du\ |u - x|^m p(u, t + dt|x, t) \tag{11.7}$$

where we have ignored the second term in r.h.s of Eq. (11.6). Also using $|u - x|^m > \delta^m$, we have:

$$\langle |u - x|^m\rangle|_{x(t)=x} \ge \delta^m \int_{|u-x|>\delta} du\ p(u, t + dt|x, t) \tag{11.8}$$

finally dividing Eq. (11.8) by dt and in the limit of $dt \to 0$ we find,

$$\lim_{dt \to 0} \frac{1}{dt} \langle |u - x|^m \rangle|_{x(t)=x} \geq \delta^m \lim_{dt \to 0} \frac{1}{dt} \int_{|u-x|>\delta} du\ p(u, t + dt|x, t) \quad (11.9)$$

which means that

$$\lim_{dt \to 0} \frac{Prob[|\Delta x| > \delta\ |_{x(t)=x}]}{dt} \leq \frac{M^{(m)}(x)}{\delta^m}. \quad (11.10)$$

We note that the moment condition (11.10) with non-vanishing coefficients for $m = 1$ and $m = 2$, will be for any $m \geq 3$. Therefore any vanishing KM coefficients, specially fourth-order $M^{(4)}(x)$ guarantee that the process is statistically continues (noting to the Pawula theorem), otherwise the non-vanishing KM coefficients provide upper limit for continuity condition. To check the continuity condition for given time series, depending on the problem formulation, the higher KM moment condition (11.10) may be easier to demonstrate than estimating the tail of the probability distribution. In the next sections we show that in the presence of jump events in the process, inequality in Eq. (11.10) changes to equality.

11.3 Stochastic Properties of Poisson Process

Similar to the role of Wiener process for continuous processes, Poisson process plays an important role in modeling of time series with jumps. We will assume that jump events are rare and can be modeled via a Poisson process. Therefore is this section we aim to derive the statistical properties of Poisson process that will be useful in determining the functions and parameters in the jump-diffusion modeling in Chap. 12.

The Poisson process $J(t)$ generate positive jumps of integer magnitude and we will see that it is a discontinuous process (with definition in Eq. 11.3). For this reason, the Poisson process is also called a counting process or point process. The analytical calculations for the Poisson process are more severe than for the Wiener process. This is due to the fact that the singularities of the Poisson process arise at the zeroth order with the value of $J(t)$ jumping, while for Wiener process $W(t)$ arise at the first order derivative. However, the jumps of the Poisson process have a modeling benefit over the Wiener process which is useful for applications with extreme jump events.

In summary, Poisson process $J(t)$ is a discontinuous process and satisfies the following properties, (following the exposition of Hanson [9]):

- $J(t)$ has unit jumps, this means that if the value of $J(t)$ jumps at time $T_k > 0$, then

$$J(T_k^+) = J(T_k^-) + 1$$

where $J(T_k^+)$ and $J(T_k^-)$ denote the limit from the right and the left, respectively. Therefore $J(t)$ is discontinuous, increasing and has instantaneous jumps.

- $J(t)$ is right-continuous, since

$$J(t^+) = J(t) \geq J(t^-), \qquad t > 0 \quad .$$

- $J(t)$ has independent increments, since the Poisson increments

$$\Delta J(t_i) = J(t_i + \Delta t_i) - J(t_i)$$

are mutually independent for all t_i on non-overlapping time intervals defined such that $t_i \geq 0$, $t_{i+1} = t_i + \Delta t_i$ and any $\Delta t_i > 0$ for $i = 0, \ldots, n$ so that $t_i < t_{i+1}$ for $i = 0, \ldots, n$. Noting that $J(t_i) = J(0) + \sum_{j=0}^{i-1} \Delta J(t_j)$, depending on all preceding increments, recalling that $J(0) = 0$ at $t_0 = 0$, i.e.,

$$Prob\left[\Delta J(t_i) \leq j_i; \Delta J(t_j) \leq j_j\right] = Prob[\Delta J(t_i) \leq j_i]\; Prob[\Delta J(t_j) \leq j_j]$$

if $i \neq j$, such that there is no overlap in the time intervals $(t_i, t_{i+1}]$ and $(t_j, t_{j+1}]$. Note that $\Delta J(t_i)$, as a forward increment is independent of $J(t_i)$ and recall that $\Delta J(t_i) = J(t_i + \Delta t_i) - J(t_i)$ is associated with the time interval $[t_j, t_j + \Delta t_j)$, open on the right since the process $J(t_i)$ is right continuous.

- $J(t)$ is a stationary process, since the distribution of the increment $\Delta J(t) = J(t + \Delta t) - J(t)$ is independent of t.

In terms of the counting process this means that for fixed $s > 0$, $J(s + t) - J(s)$ (the number of arrivals during the first t time units after time s) has the same distribution as $J(t)$ (the number of arrivals during the first t time units), and is independent of $\{J(u) : 0 \leq u \leq s\}$ (the counting process up to time s). This illustrates the stationary and independent increments properties. In particular, $J(s)$ is independent of $J(s + t) - J(s)$ for all $s, t \geq 0$. It is helpful to note that $J(t)$ is synonymous with the increment $J(t) - J(0)$.

- $J(t)$ is a Markov process, since

$$Prob[J(t + \Delta t) = k|J(s), s \leq t] = Prob[J(t + \Delta t) = k\;|J(t)]$$

for any $t \geq 0$, any $\Delta t > 0$.

- $J(t)$ is Poisson distributed with mean $\mu = \lambda t$ and variance $\sigma^2 = \lambda t$, with $t > 0$, i.e.,

$$Prob[J(t) = k] = p_k(\lambda t) = \frac{(\lambda t)^k}{k!} \exp(-\lambda t) \tag{11.11}$$

for integer values $k = 0, 1, 2, \ldots$, with constant (jump rate) $\lambda > 0$ and for $t \geq 0$.

- By the stationarity property of the Poisson process increment $\Delta J(t) = J(t + \Delta t) - J(t)$ has the same discrete distribution as $J(t)$ in (11.11), so the parameter λt will change to $\lambda \Delta t$, i.e

$$Prob[\Delta J(t) = k] = p_k(\lambda \Delta t) = \frac{(\lambda \Delta t)^k}{k!} \exp(-\lambda \Delta t)$$

for $k = 0, 1, \cdots$ and for $\Delta t \geq 0$.
The simulation of the simple Poisson process $J(t)$ is usually based upon simulating the time between jumps, the inter-arrival time $T_{k+1} - T_k$, since the inter-arrival time can be shown to be exponentially distributed.

- Exponential distribution of time between jumps.
Let $J(t)$ be a simple Poisson process with fixed jump rate $\lambda > 0$ and let T_j denote the jth jump time, then the distribution of the inter-jump time $\Delta T_j = T_{j+1} - T_j$ for $j = 0, 1, \ldots$ conditioned on T_j (with defining T_0 =0), is

$$Prob[\Delta T_j \leq \Delta t \,|T_j\,] = 1 - \exp(-\lambda \Delta t) \simeq \lambda \Delta t.$$

provided $\lambda \Delta t \ll 1$.
To prove this, we use the basic idea that the probability of the time between jumps $\Delta T_j = T_{j+1} - T_j$ less than Δt, conditioned on the prior jump time T_j, will be the same as the probability that there be at least one jump in the time interval. In turn this is the same as one minus the probability that there are no jumps in the time interval, i.e.,

$$\begin{aligned} Prob[\Delta T_j \leq \Delta t \,|T_j\,] &= 1 - Prob[\Delta T_j > \Delta t \,|T_j\,] \\ &= 1 - Prob[\Delta J(T_j) = 0 \,|T_j\,]. \end{aligned} \tag{11.12}$$

By the stationary property of the simple Poisson process $J(t)$ the probability of the difference does not depend on the common time T_j, but on the difference in time ΔT_j,

$$\begin{aligned} Prob[\Delta T_j \leq \Delta t \,|T_j\,] &= 1 - Prob[J(\Delta t) - J(0) = 0\,] \\ &= 1 - Prob[J(\Delta t) = 0] = 1 - p_0(\lambda \Delta t) \\ &= 1 - \exp(-\lambda \Delta t) \simeq \lambda \Delta t \end{aligned} \tag{11.13}$$

provided $\lambda \Delta t \ll 1$.

- If $J(t)$ is a Poisson process, then,

$$Cov[J(t), J(s)] = \langle J(t)J(s)\rangle - \langle J(t)\rangle\langle J(s)\rangle = \lambda \, min(t, s). \quad (11.14)$$

The proof of this relation is similar to what we have done for Wiener process. This is an application of the independent increment property of Poisson or Markov jump processes, also demonstrating how application of independent increments rely on the zero mean properties. For the Poisson process the zero mean property comes from using the Poisson deviation or centered Poisson processes $\hat{J}(t) = J(t) - \lambda t$ where, $\langle J(t)\rangle = \lambda t$, so that $\langle \hat{J}(t)\rangle = 0$. Consider $s < t$ and write,

$$\begin{aligned} Cov[J(t), J(s)] &= \langle \hat{J}(t)\hat{J}(s)\rangle = \langle \hat{J}(s)^2 + \hat{J}(s)(\hat{J}(t) - \hat{J}(s))\rangle \\ &= \langle \hat{J}(s)^2\rangle + \langle \hat{J}(s)(\hat{J}(t) - \hat{J}(s))\rangle \\ &= var[\hat{J}(s)] + \langle \hat{J}(s)\rangle\langle(\hat{J}(t) - \hat{J}(s))\rangle \\ &= \lambda s + 0 = \lambda s. \end{aligned} \quad (11.15)$$

where we used $var[\hat{J}(s)] = \lambda s$. In the case of $t < s$ the result will be λt in similar way, therefore we find $Cov[J(t), J(s)] = \lambda \, min(t, s)$.

11.4 Statistical Moments of Poisson Process

The averages for the integer powers of the Poisson increment follow from the mean over the Poisson distribution (11.11) and summed by differentiation of the exponential series. The statistical moment of increments can be written as,

$$\begin{aligned} \langle(\Delta J)^k\rangle &= \exp(-\lambda \Delta t) \sum_{l=0}^{\infty} \frac{(\lambda \Delta t)^l l^k}{l!} \\ &= \left[e^{-z} \left(z \frac{d}{dz} \right)^k e^z \right]_{z=\lambda \Delta t} \end{aligned} \quad (11.16)$$

Table 11.1 Statistical moments of Poisson increments

k	$\langle(\Delta J)^k\rangle$	$\langle(\Delta J - \lambda\Delta t)^k\rangle$
0	1	–
1	$\lambda\Delta t$	0
2	$\lambda\Delta t(1+\lambda\Delta t)$	$\lambda\Delta t$
3	$\lambda\Delta t(1+3\lambda\Delta t+(\lambda\Delta t)^2)$	$\lambda\Delta t$
4	$\lambda\Delta t(1+7\lambda\Delta t+6(\lambda\Delta t)^2+(\lambda\Delta t)^3)$	$\lambda\Delta t(1+3\lambda\Delta t)$

for $k = 0, 1, \ldots$, where $z = \lambda\Delta t$. This result can be shown by induction from the definition of statistical moments.[1] Some statistical moments of Poisson increments can be found from Eq. (11.16) and summarised in Table 11.1.

The results presented in table I, can be applied to other expected statistical moments, for example,

$$var[\Delta J] = \langle(\Delta J - \lambda\Delta t)^2\rangle$$
$$= \langle(\Delta J)^2\rangle - 2\lambda\Delta t\langle\Delta J\rangle + (\lambda\Delta t)^2\langle 1\rangle = \lambda\Delta t. \qquad (11.17)$$

The presented results can also be applied to statistical moments of Poisson differential process as an increment process, $dJ(t) = J(t+dt) - J(t)$, by replacing Δt by dt and neglecting terms of $\mathcal{O}(dt)^2$. Statistical moments of $J(t)$ also follow by replacing Δt by t except the higher powers of t would not be negligible compared to the first power, unless t is small.

11.5 The Process in Presence of Jumps, Pure Poisson Process as an Example, Kramers–Moyal Coefficients

We can imagine that adding jump processes to diffusion process invalidates the continuity condition (11.3). Indeed the inequality in Eq. (11.10) changes to equality for the processes that have jumps. As an simple example consider the homogeneous Poisson process which counts events that occur at a constant jump rate $\lambda > 0$, which is the average number of "events" or "arrivals" that occur per unit time. The number

[1] The statistical moments of Poisson process can be also found using its the generating or characteristic function, where can be written as

$$\langle e^{iuJ_t}\rangle = \sum_{n=0}^{\infty} e^{iun} p(J_t = n) = \sum_{n=0}^{\infty} e^{iun}\frac{(\lambda t)^n}{n!}e^{-\lambda t} = e^{-\lambda t}\sum_{n=0}^{\infty}\frac{(\lambda t e^{iu})^n}{n!}$$

$$= e^{-\lambda t}\left(1 + \frac{(\lambda t e^{iu})}{1!} + \frac{(\lambda t e^{iu})^2}{2!} + \cdots\right) = e^{-\lambda t}e^{(\lambda t e^{iu})} = e^{\{\lambda t(e^{iu}-1)\}}.$$

of events in time interval $(t, t + dt]$ follows a Poisson distribution with associated parameter λdt. The probability to have k jumps in time scale dt is given by,

$$Prob[\Delta x(t) = k] = p_k(\lambda dt) = \frac{e^{-\lambda dt}(\lambda dt)^k}{k!} \qquad k = 0, 1, 2, \ldots. \tag{11.18}$$

For this process the continuity condition reads,

$$\begin{aligned} C(t) &= \lim_{dt \to 0} \frac{Prob[|\Delta x(t)| > \delta \ \ |_{x(t)=x}]}{dt} \\ &= \lim_{dt \to 0} \frac{1}{dt} \sum_{k=1}^{\infty} Prob[\Delta x(t) = k] = \lim_{dt \to 0} \frac{1}{dt} \sum_{k=1}^{\infty} e^{\lambda dt} \frac{(\lambda dt)^k}{k!} \\ &= \lim_{dt \to 0} \frac{1 - e^{-\lambda dt}}{dt} = \lambda. \end{aligned} \tag{11.19}$$

We can show also that all of the conditional moments are equal to jump rate λ. For instance consider the third and fourth order KM coefficients as,

$$\begin{aligned} M^{(3)}(x) &= \lim_{dt \to 0} \frac{1}{dt} \langle |\Delta x(t)|^3 |_{x(t)=x} \rangle \\ &= \lim_{dt \to 0} \frac{1}{dt} \sum_{k=1}^{\infty} Prob[\Delta x(t) = k]\, k^3 = \lim_{dt \to 0} \frac{1}{dt} \sum_{k=1}^{\infty} e^{\lambda dt} \frac{(\lambda dt)^k}{k!} k^3 \\ &= \lim_{dt \to 0} (\lambda^3 dt^2 + 3\lambda^2 dt + \lambda) = \lambda \end{aligned} \tag{11.20}$$

and in a similar way,

$$\begin{aligned} M^{(4)}(x) &= \lim_{dt \to 0} \frac{1}{dt} \langle |\Delta x(t)|^4 |_{x(t)=x} \rangle \\ &= \lim_{dt \to 0} \frac{1}{dt} \sum_{k=1}^{\infty} Prob[\Delta x(t) = k]\, k^4 = \lim_{dt \to 0} \frac{1}{dt} \sum_{k=1}^{\infty} e^{\lambda dt} \frac{(\lambda dt)^k}{k!} k^4 \\ &= \lim_{dt \to 0} (\lambda^4 dt^3 + 6\lambda^3 dt^2 + 7\lambda^2 dt + \lambda) = \lambda. \end{aligned} \tag{11.21}$$

A non-vanishing fourth-order KM coefficient $M^{(4)}(x)$ indicates that (according to the Pawula theorem [10]) for the process with jump events all of the KM coefficients are needed to describe the dynamics of the probability distribution function [11, 12], see also Chap. 19.

Problems

11.1 Exponential distribution of time between jumps for Poisson process

Fill in the details in the derivation of Eq. (11.13).

11.2 Mean inter-jump times

Show that the mean inter-jump times $\tau_k = T_{k+1} - T_k$ for Poisson jump process with constant jump rate λ is given by,

$$\langle \tau_k \rangle = \frac{1}{\lambda} \quad .$$

11.3 Correlation coefficient of Poisson process

Suppose $\{J(t), t \geq 0\}$ is a Poisson process with constant jump rate λ, then show that the correlation coefficient of $J(t)$ and $J(t+s)$ $(s \geq 0)$ is given by,

$$\rho(t,s) = \frac{Cov[J(t), J(t+s)]}{\sqrt{var[J(t)] \times var[J(t+s)]}} = \left\{ \frac{t}{t+s} \right\}^{1/2} .$$

11.4 Statistical moments of Poisson jump process

Show that

$$\langle (\Delta J)^m \rangle = \lambda \Delta t \, (1 + \mathcal{O}(\lambda \Delta t))$$

for $\lambda \Delta t \ll 1$, by induction for $m \geq 1$.

11.5 Poisson zero-one jump law

For Poisson jump process with jump rate λ, show that

$$Prob[\Delta J(t) = 0] = 1 - \lambda \Delta t + \mathcal{O}(\lambda \Delta t)^2$$

$$Prob[\Delta J(t) = 1] = \lambda \Delta t + \mathcal{O}(\lambda \Delta t)^2$$

$$Prob[\Delta J(t) > 1] = \mathcal{O}(\lambda \Delta t)^2$$

$$Prob[(\Delta J(t))^m = \Delta J(t)] = 1 - \frac{1}{2}(\lambda \Delta t)^2 + \mathcal{O}(\lambda \Delta t)^3$$

for $m \geq 2$. For $\lambda \Delta t \ll 1$ we find $Prob[\Delta J(t) > 1] \ll Prob[\Delta J(t) = 1]$. This asymptotic relationship is specific characteristic of Poisson jump process.

11.6 Integral including Poisson jump process

Show that

$$I[J](t) = \int_0^t J(s)dJ(s) \stackrel{\text{ms}}{=} I_{ms}[J](t) \equiv \frac{1}{2}J(t)(J(t)-1)$$

where the mean square limit integral is,

$$I_{ms}[J](t) \stackrel{\text{ms}}{=} ms - lim_{n\to\infty} \sum_{i=0}^{n} J(t_i)\Delta J(t_i).$$

11.7 Wiener and Poisson jump processes

Show that

$$dJ\, dt \stackrel{\text{dt}}{=} 0,$$

$$dJ\, dW \stackrel{\text{dt}}{=} 0,$$

$$(dJ)^m \stackrel{\text{dt}}{=} dJ \stackrel{\text{dt}}{=} \lambda dt$$

for $m \geq 1$, where $W(t)$ and $J(t)$ are independent Wiener and Poisson jump processes, respectively. The symbol $\stackrel{\text{dt}}{=}$ means that the equality with an accuracy $\mathcal{O}(dt)$.

11.8 Poisson jump process with time dependent jump rate $\lambda(t)$

Consider a Poisson jump process with time dependent jump rate $\lambda(t)$ and suppose that the first jump occurs at time $t_0 = 0$.

(a) Show that the waiting time distribution $f(t|t_0 = 0)$ to have an another jump at time t, is given by,

$$f(t|t_0 = 0) = \lambda(t) \exp\left\{-\int_0^t \lambda(t')dt'\right\}.$$

(b) Show that for known $f(t|t_0 = 0)$, the time dependent jump rate $\lambda(t)$ can be found as

$$\lambda(t) = \frac{f(t|t_0 = 0)}{1 - \int_0^t f(t'|t_0 = 0)dt'} \quad .$$

References

1. M. Anvari, K. Lehnertz, M.R. Rahimi Tabar, J. Peinke, Sci. Rep. **6**, 35435 (2016)
2. T. Andersen, L. Benzoni, J. Lund, An empirical investigation of continuous-time equity return models. J. Financ. **57**, 1239 (2002)
3. G. Bakshi, C. Cao, Z. Chen, Empirical performance of alternative option pricing models. J. Financ. **52**, 2003 (1997)
4. D. Duffe, J. Pan, K. Singleton, Transform analysis and asset pricing for affine jump-diffusions. Econometrica **68**, 1343 (2000)
5. S. Das, The surprise element: jumps in interest rates. J. Econom. **106**, 27 (2002)
6. M. Johannes, The statistical and economic role of jumps in continuous-time interest rate models. J. Financ. **59**, 227 (2004)
7. Z. Cai, Y. Hong, Some Recent Developments, in *Nonparametric Finance*, ed. by Q. Li, J.S. Racine (ed.) Nonparametric Econometric Methods (Advances in Econometrics, vol. 25), (Emerald Group Publishing Limited, Bingley, 2009), pp. 379–432
8. R. Friedrich, J. Peinke, M. Sahimi, M.R. Rahimi Tabar, Phys. Rep. **506**, 87 (2011)
9. F.B. Hanson, *Applied Stochastic Processes and Control for Jump-Diffusions: Modeling, Analysis, and Computation* (SIAM, Philadelphia, 2007)
10. R.F. Pawula, Phys. Rev. **162**, 186 (1967)
11. H. Risken, *The Fokker-Planck Equation* (Springer, Berlin, 1989)
12. C.W. Gardiner, *Handbook of Stochastic Methods* (Springer, Berlin, 1983)

Chapter 12
Jump-Diffusion Processes

In this chapter we introduce jump-diffusion processes and provide a theoretical framework that justifies the nonparametric (data-based) extraction of the parameters and functions controlling the arrival of a jump and the distribution of the jump size from the estimated conditional Kramers–Moyal moments. The method and the results are applicable to both *stationary* and *nonstationary* time series in the presence of discontinuous jump components; see Chap. 17.

12.1 Kramers–Moyal Coefficients of Jump-Diffusion Processes

In general, non-vanishing higher order Kramers–Moyal coefficients $m > 2$, indicates that the corresponding time series does not belong to the class of diffusion process. Accordingly, an improvement of the Langevin-type modeling (Eq. (11.1)) is needed. An important generalization of Langevin-type modeling is to include jump processes, with properties that can also be determined from measured time series.

The jump-diffusion process is given by a (Itô) dynamical stochastic equation:

$$dx(t) = N(x,t)dt + \sqrt{D(x,t)}dW(t) + \xi\, dJ(t), \tag{12.1}$$

where $\{W(t), t \geq 0\}$ is a scalar Wiener process, $N(x,t)$ the drift function, $D(x,t)$ the diffusion function, and $J(t)$ is a Poisson jump process. For simplicity, we assumed that jump events are rare and can be modeled via a Poisson process. The jump has rate $\lambda(x)$, can be state dependent and size ξ which we assume it has some symmetric distribution with finite even order statistical moments $\langle \xi^{2j} \rangle$ for $j \geq 1$.

All of the unknown functions and coefficients in Eq. (12.1) can be derived directly from measured time series, as will be shown here. Using Eq. (12.1) one can disentangle the non-linear contributions in a given time series. Indeed, there are three sources for a non-linear dynamics in Eq. (12.1): non-linear dependence on x of the drift

M. R. Rahimi Tabar, *Analysis and Data-Based Reconstruction of Complex Nonlinear Dynamical Systems*, Understanding Complex Systems,
https://doi.org/10.1007/978-3-030-18472-8_12

term $N(x)$, of the multiplicative volatility term $D(x)$, or of the jump rate $\lambda(x)$ with amplitude ξ. In this respect a *linear theory* is defined with a linear drift, a constant diffusion coefficient and a vanishing jump amplitude.

Processes with jumps have been widely used to describe the random evolution of, e.g., neuron dynamics [1–3], of soil moisture dynamics [4], or of financial figures such as stock prices, market indices, and interest rates [5]. A non-parametric procedure for estimating continuous-time diffusion processes was introduced in [6], in a continuous path setting and then extended to mixed jump-diffusion processes in [7, 8], that provide a rigorous treatment of the underlying statistical theory.

Theorem *Kramers–Moyal Coefficients of Jump-Diffusion Processes*

A general jump-diffusion process is given by the dynamical stochastic equation (12.1), where all of the functions and parameters in this modeling can be found non-parametrically from measured time series by estimating the Kramers–Moyal conditional moments as [1]:

$$
\begin{aligned}
K^{(1)}(x,t) &= N(x,t)\,\mathrm{d}t \\
K^{(2)}(x,t) &= [D(x,t) + \lambda(x,t)\langle\xi^2\rangle]\,\mathrm{d}t \\
K^{(j)}(x,t) &= \lambda(x,t)\langle\xi^j\rangle\,\mathrm{d}t, \qquad \textit{for} \quad j > 2
\end{aligned}
\tag{12.2}
$$

where Kramers–Moyal conditional moment are $K^{(j)}(x,t) = \langle (x(t+dt) - x(t))^j|_{x(t)=x}\rangle$. For non-stationary processes all of the functions and parameters are also time dependent, see Chap. 17. Here we omit the t-dependences in Eq. (12.2) to enhance readability.

Proof To prove the relations in Eq. (12.2), we can use the different statistical moments of the Wiener process $\{W(t), t \geq 0\}$ and the jump process $J(t)$. Using the definition of a Wiener process as the integration of a white-noise signal, we know that: $\langle dW(t)\rangle = 0$, $\langle dW(t)^2\rangle = dt$ and $\langle (dW(t))^{m=2k}\rangle = (2dt)^k\Gamma(k+1/2)/\Gamma(1/2)$, see Eq. (5.9).

Also for Poisson-distributed jumps with rate λ, we have found in Chap. 11, the statistical moments of dJ as, for instance for $m = 1, 2, 3$,

$$
\begin{aligned}
\langle (dJ)\rangle &= \lambda dt \\
\langle (dJ)^2\rangle &= \lambda dt(1 + \lambda dt) \\
\langle (dJ)^3\rangle &= \lambda dt(1 + 3\lambda dt + (\lambda dt)^2).
\end{aligned}
$$

In the limit $dt \to 0$ all of the statistical moments of jumps are proportional to λdt. Therefore we can use the following relations in the limit of vanishing dt, $dW^2 \overset{\mathrm{dt}}{=} dt$

(with $\langle dW \rangle = 0$) and $(dW)^m \overset{\text{dt}}{=} 0$ for $m > 2$. In a similar way, in this limit we have $(dJ)^m \overset{\text{dt}}{=} dJ = \lambda dt$ for $m > 1$. The symbol $\overset{\text{dt}}{=}$ means that the equality with an accuracy $\mathcal{O}(dt)$.

Conditional averaging of Eq. (12.1) over three *independent* processes, Wiener, Poisson-distributed jumps and jump amplitude ξ gives:

$$\begin{aligned}
&\langle dx(t)|_{x(t)=x}\rangle \\
&= N(x)dt + \sqrt{D(x)}\langle dW\rangle + \langle \xi dJ\rangle \\
&= N(x)dt + \sqrt{D(x)}\langle dW\rangle + \langle \xi\rangle\langle dJ\rangle \\
&= N(x)dt + 0 + \langle \xi\rangle \lambda(x)dt \\
&= \Big[N(x) + \langle \xi\rangle\lambda(x)\Big]dt,
\end{aligned} \tag{12.3}$$

with assuming $\langle \xi \rangle = 0$, this proves the first relation in Eq. (12.2). Here we used the non-anticipating property of $D(x)$. We also have used the independence of the amplitude of the jumps and of the Poisson process dJ. Similarly, the second conditioned moment of dx leads to:

$$\begin{aligned}
&\langle dx(t)^2|_{x(t)=x}\rangle \\
&= \langle\Big[N(x)dt + \sqrt{D(x)}dW + \xi dJ\Big]^2\rangle \\
&= \langle\Big[N(x)^2dt^2 + D(x)dW^2 + \xi^2 dJ^2 \\
&+ 2N(x)\sqrt{D(x)}dtdW \\
&+ 2N(x)\xi dtdJ + 2\sqrt{D(x)}dW\xi dJ\Big]\rangle
\end{aligned} \tag{12.4}$$

up to order of $\mathcal{O}(dt)$ will be:

$$\begin{aligned}
&\langle dx(t)^2|_{x(t)=x}\rangle \\
&= \langle D(x) + \xi^2\lambda(x)\rangle dt \\
&= \Big[D(x) + \langle \xi^2\rangle\lambda(x)\Big]dt
\end{aligned} \tag{12.5}$$

which proves the second relation in Eq. (12.2). Finally for $k \geq 3$, we find:

$$\langle dx(t)^k|_{x(t)=x}\rangle =$$

$$\sum_{l,m,n}\left\langle\left[A_{l,m,n}(N(x)dt)^l(D(x)dW)^m(\xi dJ)^n\right]\right\rangle \tag{12.6}$$

where $A_{l,m,n} = k!/l!m!n!$, so that $l + m + n = k$. Up to order of $\mathcal{O}(dt)$ we find:

$$\langle dx(t)^k|_{x(t)=x}\rangle = \langle \xi^k\rangle \lambda(x)dt \tag{12.7}$$

which proves the third relation in Eq. (12.2).

Identification of the drift, diffusive, intensity of the jumps and parameters of the distribution of the jump component simply requires choice of an appropriate parametric family for the probability measure of the jump size as well as use of the estimated conditioned moments. Here we discuss two choices for the distribution of the jump component that accommodate different jump behaviours, namely Gaussian and mixed Gaussian jump sizes. Extensions to alternative specifications are straightforward based on our subsequent discussion.

12.2 Gaussian Distributed Jump Amplitude

We can assume that jump amplitude has zero mean Gaussian distribution with variance σ_ξ^2 which can symmetrically jumps up and down, i.e. $\xi \sim \mathcal{N}(0, \sigma_\xi^2)$. We call σ_ξ^2 the jump amplitude that may depends on state variable x and time t. For a Gaussian random variable ξ, we can use the relations

$$\begin{aligned}\langle \xi^{2n}\rangle &= \frac{(2n)!}{2^n n!}\langle \xi^2\rangle^n \\ &= \langle \xi^2\rangle^n \prod_{l=1}^{n}(2l-1) \\ \langle \xi^{2n-1}\rangle &= 0\end{aligned}$$

for $n = 1, 2, \ldots$. Now using the last relation in Eq. (12.2), with $j = 4$ and $j = 6$, we first estimate the jump amplitude $\sigma_\xi^2(x)$ and the jump rate $\lambda(x)$ as:

$$\sigma_\xi^2(x) = \frac{M^{(6)}(x)}{5M^{(4)}(x)}, \qquad \lambda(x) = \frac{M^{(4)}(x)}{3\sigma_\xi^4(x)} \tag{12.8}$$

where $M^{(j)}(x) = \lim_{dt \to 0} \frac{1}{dt} K^{(j)}(x)$. Once the jump components $\sigma_\xi^2(x)$ and $\lambda(x)$ are identified, the second moment $M^{(2)}(x)$ identifies the diffusion function $D(x)$ and the first moment gives us the estimate for the drift function $N(x)$. We can also define the averaged jump amplitude as $\bar{\sigma_\xi}^2 = \frac{1}{n} \sum_{n=1}^{N} \sigma_\xi^2(x_n)$, where have assumed that the state space of the process is discretized and the conditional average is calculated for every x_n (binning with N-bins) separately.

12.3 Mixed Gaussian Jumps—The Variance Gamma Model

The variance gamma model for jump amplitude has been introduced in [9] as a way to capture long tailedness in daily stock returns [10, 11].

Let us assume that $\xi \sim \mathcal{N}(0, \sigma_\xi^2\, U)$, where U has a gamma distribution [8],

$$p(U) = \frac{1}{b\, \Gamma(1/b)} \left(\frac{U}{b}\right)^{1/b\, -1} e^{(-U/b)} \tag{12.9}$$

for $U \geq 0$ and $\Gamma(\cdot)$ is the gamma function and b is a positive parameter. U has mean value equal to 1 and variance equal to b, hence σ_ξ^2 serves as a scale parameter.

What we need in Eq. (12.2) is the statistical moments of $\langle \xi^k \rangle$. Using the gamma distribution i.e. Eq. (12.9), we can write

$$\begin{aligned}
\langle \xi^{2n} \rangle &= \langle\, \langle \xi^{2n}[U] \rangle_\xi \,\rangle_U \\
&= \left\langle \sigma_\xi^{2n} \left(\prod_{l=1}^{n} (2l-1) \right) U^n \right\rangle_U \\
&= \sigma_\xi^{2n} \left(\prod_{l=1}^{n} (2l-1) \right) \langle U^n \rangle_U \\
&= \sigma_\xi^{2n} \left(\prod_{l=1}^{n} (2l-1) \right) \frac{\Gamma(1/b+n)}{\Gamma(1/b)} b^n
\end{aligned} \tag{12.10}$$

and

$$\langle \xi^{2n-1} \rangle = 0 \tag{12.11}$$

for $n = 1, 2, \ldots$ and we have used the statistical moments of gamma function $\langle U^n \rangle = \frac{\Gamma(1/b\, +n)}{\Gamma(1/b)} b^n$.[1] Thus, a possible way to extract the functions and parameters

[1] The generating function of gamma distribution is $\langle \exp(itU) \rangle = (\frac{1}{1-ibt})^{1/b}$. For gamma-distribution

of interest from the estimated Kramers–Moyal conditioned moments is as follows. To estimate the $\sigma_\xi^2(x)$, λ and parameter b, we need at least three equations. Setting $n = 2$ and $n = 3$ in Eqs. (12.10) and (12.2) we find,

$$\sigma_\xi^2(x;\, b) = \frac{1}{5b}\left\{\frac{\Gamma(1/b+2)}{\Gamma(1/b+3)}\right\}\frac{M^{(6)}(x)}{M^{(4)}(x)} = \frac{1}{5}\left(\frac{1}{1+2b}\right)\frac{M^{(6)}(x)}{M^{(4)}(x)} \tag{12.12}$$

implying that $\sigma_\xi^2(x;\, b)$ is expressed as a function of the unknown parameter b and we used the relation $\Gamma(x+1) = x\Gamma(x)$ for gamma function. Another relation comes from the KM coefficients of order 8 and 6, (for $n = 4$ and $n = 3$) i.e.,

$$\sigma_\xi^2(x;\, b) = \frac{1}{7b}\left\{\frac{\Gamma(1/b+3)}{\Gamma(1/b+4)}\right\}\frac{M^{(8)}(x)}{M^{(6)}(x)} \quad . \tag{12.13}$$

Equality of two equations for $\sigma_\xi^2(x;\, b)$ gives the following equation for parameter b,

$$\frac{M^{(8)}(x)}{M^{(6)}(x)} - \frac{7}{5}\frac{M^{(6)}(x)}{M^{(4)}(x)}\left\{\frac{\Gamma(1/b+2)\Gamma(1/b+4)}{\Gamma^2(1/b+3)}\right\} = 0 \tag{12.14}$$

which gives,

$$b(x) = \left\{\frac{M^{(8)}(x)}{M^{(6)}(x)} - \frac{7}{5}\frac{M^{(6)}(x)}{M^{(4)}(x)}\right\} \Big/ \left\{\frac{21M^{(6)}(x)}{5M^{(4)}(x)} - 2\frac{M^{(8)}(x)}{M^{(6)}(x)}\right\}. \tag{12.15}$$

$$p(U, \alpha = 1/b, \beta = b) = \frac{1}{\beta^\alpha \Gamma(\alpha)} U^{\alpha-1}\exp(-U/\beta)$$

for $U > 0$ and $\alpha, \beta > 0$ we have,

$$\langle \exp(itU)\rangle = \int_0^\infty e^{-zU} p(U, \alpha, \beta) dU = \frac{1}{\beta^\alpha \Gamma(\alpha)} \int_0^\infty U^{\alpha-1} e^{-(z+1/\beta)U} dU$$

where $z = -iU$. Defining $y = (z + 1/\beta)U$ we find,

$$\langle \exp(itU)\rangle = \frac{(z+1/\beta)^{-\alpha}}{\beta^\alpha \Gamma(\alpha)} \int_0^\infty e^{-y} y^{\alpha-1} dy = (1+z\beta)^{-\alpha}$$

where we have used the $\Gamma(\alpha) = \int_0^\infty e^{-y} y^{\alpha-1} dy$.

In general b can be a x dependent parameter and to have an averaged $\hat{b}$ we can define $\hat{b} = \frac{1}{n}\sum_{n=1}^{N} b(x_n)$, where again we have assumed that the state space of the process is discretized and the conditional average is calculated for every x_n (in N-bins) separately. It is now straightforward to compute

$$\sigma_\xi^2(x;\hat{b}) = \frac{1}{5}\left(\frac{1}{1+2\hat{b}}\right)\frac{M^{(6)}(x)}{M^{(4)}(x)} \tag{12.16}$$

and obtain the jump rate from Eq. (12.2) as,

$$\lambda(x) = \frac{M^{(4)}(x)}{3\,\sigma_\xi^4(x;\hat{b})\,(1+\hat{b})} \tag{12.17}$$

and finally using the Eq. (12.2) the diffusion function and drift coefficient can be obtained.

12.4 Jump-Drift Process, Stochastic Solution, Example

Consider the multiplicative process as

$$dx = \alpha x\,dt + \beta x\,dJ \tag{12.18}$$

with α and β are real constants and dJ is Poisson process with jump-rate λ. Since dJ takes only the values 0 and 1 then $(dJ)^m \overset{\text{dt}}{=} dJ$ for $m > 1$. Also $dJdt \overset{\text{dt}}{=} 0$, where

$$\langle dtdJ\rangle = dt\langle dJ\rangle = \lambda(dt)^2 \overset{\text{dt}}{=} 0.$$

Using $(dJ)^m \overset{\text{dt}}{=} dJ$ we have

$$\begin{aligned} e^{kdJ} &= \sum_{n=0}^{\infty}\frac{(kdJ)^n}{n!} \overset{\text{dt}}{=} 1 + dJ\sum_{n=1}^{\infty}\frac{k^n}{n!} \\ &= 1 + dJ[e^k - 1] \end{aligned} \tag{12.19}$$

or, alternatively (with $\beta = [e^k - 1]$),

$$1 + \beta dJ \overset{\text{dt}}{=} e^{\ln[1+\beta]dJ}. \tag{12.20}$$

Therefore we can write,

$$
\begin{aligned}
x(t+dt) = x(t) + dx &\overset{\text{dt}}{=} (1 + \alpha dt + \beta dJ)x(t) \\
&\overset{\text{dt}}{=} (1 + \alpha dt)(1 + \beta dJ)x(t) \\
&\overset{\text{dt}}{=} e^{\alpha dt + \ln[1+\beta]dJ} x(t). \qquad (12.21)
\end{aligned}
$$

The stochastic solution of $x(t)$, iterating Eq. (12.21), is therefore

$$
\begin{aligned}
x(t) &= x(0) \, \exp\left[\alpha \int_0^t dt + \ln(\beta + 1) \int_0^t dJ(s)\right] \\
&= x(0) e^{\alpha t + \ln(1+\beta) J(t)} \\
&= x(0) e^{\alpha t} (1 + \beta)^{J(t)} \qquad (12.22)
\end{aligned}
$$

with $J(0) = 0$.

Problems

12.1 Jump-drift process

Consider the following jump-drift process

$$dx(t) = x(t) \, [a(t)dt + b(t)dJ]$$

with initial condition $x(0) = x_0 > 0$. Here $a(t)$ and $b(t)$ are drift and jump-amplitude coefficients. Also $J(t)$ is Poisson jump with jump rate $\lambda(t)$.

Transform $x(t)$ to $y(t)$ via $y(t) = \ln(x(t))$ and show that,

(a) $\langle dy(t) \rangle = [a(t) + \lambda(t) \ln(1 + b(t))]dt$

and infinitesimal variance is

(b) $var[dy(t)] = \lambda(t) \ln^2(1 + b(t))dt$.

For constant coefficients $a(t) = a_0$, $b(t) = b_0$ and $\lambda(t) = \lambda_0$, show that the stochastic solution is,

(c) $x(t) = x(t_0) \exp[a_0(t - t_0)](1 + b_0)^{(J(t) - J(t_0))}$

and its mean and variance are given by,

(d) $\langle x(t) \rangle = x_0 \exp[(a_0 + \lambda_0 b_0)(t - t_0)]$

(e) $var[x(t)] = \langle x(t) \rangle^2 \left(\exp[\lambda_0 b_0^2 (t - t_0)] - 1\right)$.

12.2 Jump-drift process

Solve the following jump-drift equation for $x(t)$ and mean:

$$dx(t) = a\,\sqrt{x(t)}dt + b\,(b + 2\sqrt{x(t)})\,dJ(t)$$

where $\langle J(t)\rangle = \lambda_0 t$ and $x(0) = x_0 > 0$, $\lambda_0 \geq 0$ and a, b are real constants.

12.3 Jump-drift process

Solve the following jump-drift equation for $x(t)$, mean $\langle x(t)\rangle$ and $var[x(t)]$;

$$dx(t) = -a\,x^2(t)dt - \frac{cx^2(t)}{1 + cx(t)}\,dJ(t)$$

where $\langle J(t)\rangle = \lambda_0 t$ and $x(0) = x_0 > 0$ and a, c and λ_0 are positive constants.

12.4 Jump-diffusion process, Kramers' escape time

Consider the jump-diffusion dynamical equation (12.1) with $N(x,t) = -\partial U(x)/\partial x$ which is a force to the potential $U(x)$. Assume that $U(x)$ has a single non-degenerate minimum at $x = a$ and maximum at $x = b$ and that $U(x) \to \infty$ as $x \to -\infty$.

Show that mean exit time from interval $x \in (-\infty, b]$ satisfies the following equation with boundary conditions,

$$\mathcal{L}^{\dagger}_{KM}\tau(x) = -1, \qquad \tau(b) = 0,\;\; \tau_x(-\infty) = 0,$$

where adjoint Kramers–Moyal operator is given by,

$$\mathcal{L}^{\dagger}_{KM} = \sum_{n=1}^{\infty} \frac{1}{n!} M^{(n)}(x)\frac{\partial^n}{\partial x^n}$$

with Kramers–Moyal coefficients

$$M^{(1)}(x) = -\frac{\partial U(x)}{\partial x} + \lambda\langle\xi\rangle,\;\; M^{(2)}(x) = D(x) + \lambda\langle\xi^2\rangle,$$

and

$$M^{(j)}(x) = \lambda\langle\xi^j\rangle,\;\; j \geq 3.$$

Here λ and $\langle\xi^j\rangle$ are the jump rate and statistical moments of jump amplitude, respectively. Hint. See Problem 7.4 and [12].

12.5 Compound Poisson process

Suppose $J \equiv \{J(t) : t \geq 0\}$ be a Poisson process with jump rate λ, so that $\langle J(t)\rangle = \lambda t$ for $t \geq 0$. Let $y_1, \ldots, y_n$ identically independent distributed random variables of J. A compound Poisson process is defined as:

$$Q(t) = \sum_{i=1}^{J(t)} y_i.$$

for $t \geq 0$. The process $J(t)$ and the random sequence y_i are independent.

(a) Show that the mean and the variance of compound Poisson process and its increments are:

$$\langle Q(t)\rangle = \langle y_i\rangle\, \lambda\, t$$

$$var(Q(t)) = \langle y_i^2\rangle\, \lambda\, t$$

$$\langle (Q(s) - Q(t))\rangle = \langle y_i\rangle\, \lambda\, (s - t)$$

$$var(Q(s) - Q(t)) = \langle y_i^2\rangle\, \lambda\, (s - t).$$

(b) Verify that, if $s < t < u$, the increments $Q(u) - Q(t)$ and $Q(t) - Q(s)$ are independent.

12.6 Itô-Taylor formula for jump-diffusion processes

Let x be a diffusion process with jumps, defined as the sum of a drift term, a Brownian (Wiener) stochastic integral and a compound Poisson process [5]:

$$x(t) = x_0 + \int_0^t a(s)ds + \int_0^t b(s)dW(s) + \sum_{i=1}^{J(t)} \Delta x_i$$

where $a(s)$ and $b^2(s)$ are nonanticipating time dependent drift and diffusion coefficients. The term $\sum_{i=1}^{J(t)} \Delta x_i$ is the compound Poisson jump process and $J(t)$ is Poisson jump process with a rate $\lambda > 0$.

Prove the Itô-Taylor formula for this jump-diffusion process for any differentiable function $f(x, t)$ (first-order and second-order differentiable with respect to t and x, respectively)

$$f(x(t), t) - f(x_0, 0) = \int_0^t \left[\frac{\partial f(x(s), s)}{\partial s} + a(s)\frac{\partial f(x(s), s)}{\partial x}\right] ds$$
$$+ \frac{1}{2}\int_0^t b^2(s)\frac{\partial^2 f(x(s), s)}{\partial x^2} ds$$

$$+ \int_0^t b(s) \frac{\partial f(x(s), s)}{\partial x} dW(s)$$

$$+ \sum_{\{i \geq 1, T_i \leq t\}} [f(x_{T_i-} + \Delta x_i) - f(x_{T_i-})].$$

In differential notation

$$df(x(t), t) = \frac{\partial f(x(t), t)}{\partial t} dt + a(t) \frac{\partial f(x(t), t)}{\partial x} dt + \frac{b^2(t)}{2} \frac{\partial^2 f(x(t), t)}{\partial x^2} dt$$

$$+ b(t) \frac{\partial f(x(t), t)}{\partial x} dW(t) + [f(x_{t-} + \Delta x(t)) - f(x_{t-})].$$

References

1. M. Anvari, K. Lehnertz, M.R. Rahimi Tabar, J. Peinke, Sci. Rep. **6**, 35435 (2016)
2. M.T. Giraudo, L. Sacerdote, Biosystems **40**, 75 (1997)
3. R. Sirovich, L. Sacerdote, A.E.P. Villa, Math. Biosci. Eng. **11**, 385 (2014)
4. E. Daly, A. Porporato, Phys. Rev. E **73**, 026108 (2006)
5. R. Cont, P. Tankov, *Financial Modelling with Jump Processes* (Chapman & Hall/CRC, Boca Raton, 2004)
6. R. Stanton, A nonparametric model of term structure dynamics and the market price of interest rate risk. J. Financ. **52**, 1973 (1997)
7. M. Johannes, The statistical and economic role of jumps in continuous-time interest rate models. J. Financ. **59**, 227 (2004)
8. F.M. Bandi, T. Nguyen, On the functional estimation of jump-diffusion models. J. Econom. **116**, 293 (2003)
9. D. Madan, E. Seneta, The Variance Gamma (V.G.) model for share market returns. J. Bus. **63**, 511 (1990)
10. Y. Aït-Sahalia, J. Jacod, Testing for jumps in a discretely observed process. Ann. Stat. **37**, 184 (2009)
11. S. Lee, P.A. Mykland, Jumps in financial markets: a new nonparametric test and jump clustering. Rev. Financ. Stud. **21**, 2535 (2008)
12. Y. Xu, H. Li, H. Wang, W. Jia, X. Yue, J. Kurths, J. Appl. Mech. **84**, 091004 (2017)

Chapter 13
Two-Dimensional (Bivariate) Jump-Diffusion Processes

In this chapter we provide a generalization of jump-diffusion precesses (12.1) in two dimensions by considering a class of coupled systems that are described by a bivariate state vector $\mathbf{x}(t)$ contained in a two-dimensional state space $\{\mathbf{x}\}$. The evolution of the state vector $\mathbf{x}(t)$ is assumed to be governed by a deterministic part to which diffusion parts and jump contributions are added. Generalization of the results to higher dimensions is straightforward. We note that for N multivariate time series, by assuming the presence of two-body type of interactions between time series, the analysis will reduce to analysing $N(N-1)/2$ pairwise bivariate time series.

13.1 Bivariate Jump-Diffusion Processes

Consider bivariate (Itô) jump-diffusion processes which are given by the following stochastic dynamical equations [1, 2]:

$$\begin{pmatrix} \mathrm{d}x_1 \\ \mathrm{d}x_2 \end{pmatrix} = \underbrace{\begin{pmatrix} N_1 \\ N_2 \end{pmatrix}}_{\text{drift}} \mathrm{d}t + \underbrace{\begin{pmatrix} g_{11} & g_{12} \\ g_{21} & g_{22} \end{pmatrix}}_{\text{diffusion matrix}} \begin{pmatrix} \mathrm{d}W_1 \\ \mathrm{d}W_2 \end{pmatrix} + \underbrace{\begin{pmatrix} \xi_{11} & \xi_{12} \\ \xi_{21} & \xi_{22} \end{pmatrix} \begin{pmatrix} \mathrm{d}J_1 \\ \mathrm{d}J_2 \end{pmatrix}}_{\text{jump components}} . \tag{13.1}$$

The functions N_1, N_2 (components of drift vector $\mathbf{N} = (N_1, N_2)$) are the drift coefficients of the two components, which may depend on both (x_1, x_2) and g_{ij} are the elements of the diffusion matrix. Generally g_{ij} are function of (x_1, x_2) too. The jump amplitudes ξ_{ij} are sampled from a Gaussian distributions with zero mean $\langle \xi_{ij} \rangle = 0$ and variance $\langle (\xi_{ij})^2 \rangle = \sigma^2_{\xi_{ji}} \equiv s_{ij}$ and are assumed to be independent. The two Poisson processes $J_{1,2}$ are characterised by the jump rates λ_1 and λ_2, which may depend on (x_1, x_2). The Wiener processes $\{W_{1,2}(t), t \geq 0\}$ are assumed to be independent, and uncorrelated with the Poisson-distributed jumps $J_{1,2}$. The eight stochastic noises,

---

The original version of this chapter was revised: Belated correction has been incorporated. The correction to this chapter is available at https://doi.org/10.1007/978-3-030-18472-8_24

M. R. Rahimi Tabar, *Analysis and Data-Based Reconstruction of Complex Nonlinear Dynamical Systems*, Understanding Complex Systems,
https://doi.org/10.1007/978-3-030-18472-8_13

i.e. two Wiener processes, four jump amplitudes and two Poisson jump processes are assumed to be independent of each others.

13.2 Kramers–Moyal Coefficients for Jump-Diffusion Processes in Two Dimensions

The two-dimensional Kramers–Moyal (KM) coefficients of order $[k_1, k_2]$, are defined as,

$$\mathcal{M}^{[k_1,k_2]}(\mathbf{x}, t) = \lim_{dt \to 0} \frac{1}{dt} \left\langle dx_1^{k_1}(t) dx_2^{k_2}(t) \right\rangle |_{x_1(t)=x_1, x_2(t)=x_2} \tag{13.2}$$

with $dx_i(t) = x_i(t + dt) - x_i(t)$.

Theorem

Components of drift vector, as well as elements of diffusion matrix and jump components are given in terms of the Kramers–Moyal coefficients $\mathcal{M}^{[k_1,k_2]}(\mathbf{x}, t)$ as follows:

$$\mathcal{M}^{[1,0]}(\mathbf{x}, t) = N_1 \tag{13.3}$$

$$\mathcal{M}^{[0,1]}(\mathbf{x}, t) = N_2 \tag{13.4}$$

$$\mathcal{M}^{[1,1]}(\mathbf{x}, t) = g_{11}g_{21} + g_{12}g_{22} \tag{13.5}$$

$$\mathcal{M}^{[2,0]}(\mathbf{x}, t) = g_{11}^2 + s_{11}\lambda_1 + g_{12}^2 + s_{12}\lambda_2 \tag{13.6}$$

$$\mathcal{M}^{[0,2]}(\mathbf{x}, t) = g_{21}^2 + s_{21}\lambda_1 + g_{22}^2 + s_{22}\lambda_2 \tag{13.7}$$

$$\mathcal{M}^{[2,2]}(\mathbf{x}, t) = s_{11}s_{21}\lambda_1 + s_{12}s_{22}\lambda_2 \tag{13.8}$$

$$\mathcal{M}^{[k_1,k_2]}(\mathbf{x}, t) = \left[s_{11}^n s_{21}^m \lambda_1 + s_{12}^n s_{22}^m \lambda_2 \right] \frac{(2n)!}{2^n n!} \frac{(2m)!}{2^m m!} \tag{13.9}$$

where $k_1 = 2n,\ k_2 = 2m$, with $n, m \in \mathbb{N}^+$. We omitted the x_1, x_2- and t-dependence of all functions and parameters to enhance readability.

Proof **Kramers–Moyal coefficients $\mathcal{M}^{[1,0]}(\mathbf{x}, t)$ and $\mathcal{M}^{[0,1]}(\mathbf{x}, t)$:**
From (13.2) we have,

$$\begin{aligned}
\mathcal{M}^{[1,0]}(\mathbf{x}, t) &= \lim_{dt \to 0} \frac{1}{dt} \langle (dx_1)^1 (dx_2)^0 \rangle |_{x_1(t)=x_1, x_2(t)=x_2} \\
&= \lim_{dt \to 0} \frac{1}{\mathrm{d}t} \langle \mathrm{d}x_1 \rangle |_{x_1(t)=x_1, x_2(t)=x_2} \\
&= \lim_{dt \to 0} \frac{1}{\mathrm{d}t} \langle N_1 \mathrm{d}t + g_{11}\mathrm{d}w_1 + g_{12}\mathrm{d}w_2 + \xi_{11}\mathrm{d}J_1 + \xi_{12}\mathrm{d}J_2 \rangle |_{x_1(t)=x_1, x_2(t)=x_2} \\
&= \lim_{dt \to 0} \frac{1}{\mathrm{d}t} \left[N_1 \mathrm{d}t + g_{11}\langle \mathrm{d}w_1 \rangle + g_{12}\langle \mathrm{d}w_2 \rangle + \langle \xi_{11} \rangle \langle \mathrm{d}J_1 \rangle + \langle \xi_{12} \rangle \langle \mathrm{d}J_2 \rangle \right] \\
&= N_1
\end{aligned}$$

where we have substituted $\mathrm{d}x_1$ from the Eq. 13.1. When applying the conditional averaging to the single components in the third step, we used $\langle g_{ij}\mathrm{d}W_j\rangle = \langle g_{ij}\rangle\langle \mathrm{d}W_j\rangle = 0$, where we assumed that g_{ij} is non-anticipated function, and noted that the Wiener processes have the property that $\langle \mathrm{d}W_j\rangle = 0$. Also we used the relation $\langle \xi_{ij}\mathrm{d}J_j\rangle = \langle \xi_{ij}\rangle\langle \mathrm{d}J_j\rangle = 0$, in the last step, because of the jump amplitudes and Poisson jump processes are independent, and jump amplitudes are zero mean Gaussian processes. The derivation of $\mathcal{M}^{[0,1]}(\mathbf{x},t)$ is completely analogue to $\mathcal{M}^{[1,0]}(\mathbf{x},t)$.

Kramers–Moyal coefficient $\mathcal{M}^{[1,1]}(\mathbf{x},t)$:

From (13.2) we have,

$$
\begin{aligned}
\mathcal{M}^{[1,1]}(\mathbf{x},t) &= \lim_{dt\to 0}\frac{1}{dt}\langle \mathrm{d}x_1\mathrm{d}x_2\rangle|_{x_1(t)=x_1,x_2(t)=x_2} \\
&= \lim_{dt\to 0}\frac{1}{dt}\langle (N_1\mathrm{d}t + g_{11}\mathrm{d}w_1 + g_{12}\mathrm{d}w_2 + \xi_{11}\mathrm{d}J_1 + \xi_{12}\mathrm{d}J_2)\cdot \\
&\qquad (N_2\mathrm{d}t + g_{21}\mathrm{d}w_1 + g_{22}\mathrm{d}w_2 + \xi_{21}\mathrm{d}J_1 + \xi_{22}\mathrm{d}J_2)\rangle|_{x_1(t)=x_1,x_2(t)=x_2} \\
&= \lim_{dt\to 0}\left[N_1N_2\mathrm{d}t + g_{11}g_{21}\langle(\mathrm{d}w_1)^2\rangle\frac{1}{\mathrm{d}t} + g_{12}g_{22}\langle(\mathrm{d}w_2)^2\rangle\frac{1}{dt} + \text{h.o.t}\right] \\
&= g_{11}g_{21} + g_{12}g_{22},
\end{aligned}
$$

where $h.o.t$ means, higher order terms $\mathcal{O}(\mathrm{d}t)^{\epsilon}$ with $\epsilon > 0$. These terms are zero in the limit $dt \to 0$. Also we used the relation $\langle(\mathrm{d}w_i)^2\rangle = \mathrm{d}t$.

Kramers–Moyal coefficients $\mathcal{M}^{[2,0]}(\mathbf{x},t)$ and $\mathcal{M}^{[0,2]}(\mathbf{x},t)$:

From (13.2) we have,

$$
\begin{aligned}
\mathcal{M}^{[2,0]}(\mathbf{x},t) &= \lim_{dt\to 0}\frac{1}{dt}\langle(\mathrm{d}x_1)^2\rangle|_{x_1(t)=x_1,x_2(t)=x_2} \\
&= \lim_{dt\to 0}\frac{1}{\mathrm{d}t}\langle(N_1\mathrm{d}t + g_{11}\mathrm{d}w_1 + g_{12}\mathrm{d}w_2 + \xi_{11}\mathrm{d}J_1 + \xi_{12}\mathrm{d}J_2)^2\rangle|_{x_1(t)=x_1,x_2(t)=x_2} \\
&= \lim_{dt\to 0}\left[N_1^2\mathrm{d}t + g_{11}^2\langle(\mathrm{d}w_1)^2\rangle\frac{1}{\mathrm{d}t} + g_{12}^2\langle(\mathrm{d}w_2)^2\rangle\frac{1}{\mathrm{d}t} + \langle\xi_{11}^2\rangle\langle(\mathrm{d}J_1)^2\rangle\frac{1}{\mathrm{d}t} + \right. \\
&\qquad \left.\langle\xi_{12}^2\rangle\langle(\mathrm{d}J_2)^2\rangle\frac{1}{\mathrm{d}t} + \text{h.o.t}\right] \\
&= g_{11}^2 + g_{12}^2 + s_{11}\lambda_1 + s_{12}\lambda_2.
\end{aligned}
$$

Here, we used the relation $\langle\xi_{ij}^2\rangle = \sigma_{\xi_{ij}}^2 = s_{ij}$ as well as $\langle(\mathrm{d}J_i)^2\rangle = \lambda_i\mathrm{d}t$, for small dt. For $\mathcal{M}^{[0,2]}(\mathbf{x},t)$ the derivation is completely analogue to $\mathcal{M}^{[2,0]}(\mathbf{x},t)$.

Kramers–Moyal coefficient $\mathcal{M}^{[2,2]}(\mathbf{x},t)$:

From (13.2) we find,

$$
\begin{aligned}
\mathcal{M}^{[2,2]}(\mathbf{x},t) &= \lim_{dt\to 0}\frac{1}{\mathrm{d}t}\langle(\mathrm{d}x_1)^2(\mathrm{d}x_2)^2\rangle|_{x_1(t)=x_1,x_2(t)=x_2} \\
&= \lim_{dt\to 0}\frac{1}{\mathrm{d}t}\langle(N_1 dt + g_{11}\mathrm{d}w_1 + g_{12}\mathrm{d}w_2 + \xi_{11}\mathrm{d}J_1 + \xi_{12}\mathrm{d}J_2)^2\cdot
\end{aligned}
$$

$$
\begin{aligned}
&(N_2 dt + g_{21} dw_1 + g_{22} dw_2 + \xi_{21} dJ_1 + \xi_{22} dJ_2)^2\rangle|_{x_1(t)=x_1, x_2(t)=x_2} \\
&= \lim_{dt\to 0} \frac{1}{dt}\Big[\text{terms}(N_1, N_2, \mathcal{O}(dt^4)) + \text{terms}(g_{ij}, \mathcal{O}(dt^2)) + \text{terms}(\text{mixing } \xi_{ij}) + \\
&\quad \langle\xi_{11}^2\rangle\langle\xi_{21}^2\rangle\langle(dJ_1)^4\rangle + \langle\xi_{12}^2\rangle\langle\xi_{22}^2\rangle\langle(dJ_2)^4\rangle + \\
&\quad \langle\xi_{11}^2\rangle\langle\xi_{22}^2\rangle\langle(dJ_1)^2\rangle\langle(dJ_2)^2\rangle + \langle\xi_{12}^2\rangle\langle\xi_{21}^2\rangle\langle(dJ_1)^2\rangle\langle(dJ_2)^2\rangle\Big] \\
&= s_{11}s_{21}\lambda_1 + s_{12}s_{22}\lambda_2.
\end{aligned}
$$

Terms including dt^ϵ, with $\epsilon > 0$ in r.h.s. of above equation, again vanish as we consider the limit $dt \to 0$. We have used also the fact that $\langle\xi_{11}\xi_{12}\rangle = \langle\xi_{11}\rangle\langle\xi_{12}\rangle = 0$ and $\frac{1}{dt}[\langle(dJ_1)^2\rangle\langle(dJ_2)^2\rangle] = \frac{1}{dt}[\lambda_1 dt \lambda_2 dt] \propto dt$, which vanish in the limit $dt \to 0$.

Kramers–Moyal coefficients $\mathcal{M}^{[k_1,k_2]}(\mathbf{x}, t)$ with $k_1, k_2 > 2$ and even:

In this case we have

$$
\begin{aligned}
k_1 &= 2n, n \in \mathbb{N}^+ \\
k_2 &= 2m, m \in \mathbb{N}^+.
\end{aligned}
$$

The KM coefficients $\mathcal{M}^{[2n,2m]}(\mathbf{x}, t)$ can be written as order dt,

$$
\begin{aligned}
\mathcal{M}^{[2n,2m]}(\mathbf{x}, t) &= \lim_{dt\to 0} \frac{1}{dt}\langle(dx_1)^{2n}(dx_2)^{2m}\rangle|_{x_1(t)=x_1, x_2(t)=x_2} \\
&= \lim_{dt\to 0} \frac{1}{dt}\langle(N_1 dt + g_{11} dw_1 + g_{12} dw_2 + \xi_{11} dJ_1 + \xi_{12} dJ_2)^{2n} \cdot \\
&\quad (N_2 dt + g_{21} dw_1 + g_{22} dw_2 + \xi_{21} dJ_1 + \xi_{22} dJ_2)^{2m}\rangle|_{x_1(t)=x_1, x_2(t)=x_2} \\
&= \lim_{dt\to 0} \frac{1}{dt}\Big[\langle\xi_{11}^{2n}\rangle\langle\xi_{21}^{2m}\rangle\langle(dJ_1)^{2(n+m)}\rangle + \langle\xi_{12}^{2n}\rangle\langle\xi_{22}^{2m}\rangle\langle(dJ_2)^{2(n+m)}\rangle\Big] \\
&= \langle\xi_{11}^{2n}\rangle\langle\xi_{21}^{2m}\rangle\, \lambda_1 + \langle\xi_{12}^{2n}\rangle\langle\xi_{22}^{2m}\rangle\, \lambda_2 \\
&= \Big[s_{11}^n s_{21}^m \lambda_1 + s_{12}^n s_{22}^m \lambda_2\Big] \frac{(2n)!}{2^n n!}\frac{(2m)!}{2^m m!} .
\end{aligned}
$$

In the last step we used the Gaussian property of jump amplitudes ξ_{ij}. Therefore all KM coefficients $\mathcal{M}^{[k_1,k_1]}(\mathbf{x}, t)$ with even $k_1, k_2 > 2$ are presented.

In Chaps. 20 and 21, we describe a systematic approach to estimate the KM coefficients from univariate and bivariate time series, which enable us to estimate drift, diffusion as well as jump contributions directly from two simultaneous time series, see also [2].

Problems

13.1 Kramers–Moyal coefficients

Fill in the details in the derivation of Eq. (13.9).

13.2 Correlated Wiener processes

Assume that two Wiener processes are dependent as,

$$\langle dW_1(t)dW_2(t)\rangle = \rho dt.$$

Show that two correlated Wiener processes $dW_1(t)$ and $dW_2(t)$ can be written in terms of two independent Wiener processes dw_1 and dw_2 as,

$$dW_1 = \frac{\sqrt{1+\rho}}{\sqrt{2}}dw_1 + \frac{\sqrt{1+\rho}}{\sqrt{2}}dw_2$$
$$dW_2 = \frac{\sqrt{1+\rho}}{\sqrt{2}}dw_1 - \frac{\sqrt{1+\rho}}{\sqrt{2}}dw_2,$$

or

$$dW_1 = dw_1$$
$$dW_2 = \rho dw_1 - \sqrt{1-\rho^2}dw_2.$$

13.3 Positive correlated Poisson distributed jump processes

One simple way to correlate two independent Poisson-distributed jumps J_1 and J_2 is to use three Poisson-distributed jumps n_1, n_2 and n_3 (with jump rates λ_1, λ_2 and λ_3), so that J_1 and J_2 are defined as:

$$J_1 = n_1 + n_3, \qquad J_2 = n_2 + n_3.$$

(a) With this definitions, using the statistical moment generating functions show that changing the construction of the Poisson process J_i in this way does not change it's distribution (the mean will be $\mu_i = \lambda_i + \lambda_3$).

(b) Show that the mean, covariance and correlation coefficients of J_1 and J_2 are given by,

$$\langle dJ_i\rangle = (\lambda_i + \lambda_3)\, dt,$$

$$cov(J_1, J_2) = \langle J_1 J_2\rangle - \langle J_1\rangle\langle J_2\rangle = var(n_3) = \lambda_3$$

and

$$corr[J_1, J_2] = \frac{\lambda_3}{\sqrt{(\lambda_1+\lambda_3)(\lambda_2+\lambda_3)}}$$

where $\lambda_i > 0$ is the jump rate of process n_i. We note that using this method one will be able construct two positively correlated Poisson-distributed jumps J_1 and J_2.

13.4 Negative correlated Poisson distributed jump processes

One can use a more advanced approach for construction of negatively correlated Poisson processes (which is based on the idea of the backward simulation of the Poisson processes [3]) to generate two Poisson-distributed jump processes with negative correlations.

Follow steps presented in [3] and synthesis two Poisson jump processes with jump rates $\lambda_1 = 0.1$, $\lambda_2 = 0.2$ and correlation coefficient $corr[J_1, J_2] = -0.5$.

13.5 Correlated Poisson distributed jump and Wiener processes

(a) Suppose that Poisson jump process has jump rate λ and show that

$$\langle W(t)J(t)\rangle = \rho(t)\sqrt{\lambda}\, t$$

where $\rho(t)$ is correlation coefficient of $W(t)$ and $J(t)$.

(b) Use backward simulation of the Poisson process with jump rate $\lambda = 0.1$ and synthesis correlated jump and Wiener processes with $\rho(t) = \pm 0.2$ [3].

References

1. M. Anvari, K. Lehnertz, M.R. Rahimi Tabar, J. Peinke, Sci. Rep. **6**, 35435 (2016)
2. L.R. Gorjão, J. Heysel, K. Lehnertz, M.R. Rahimi Tabar, arXiv:1907.05371
3. A. Kreinin, Correlated poisson processes and their applications in financial modeling, Financial Signal Processing and Machine Learning, pp. 191–232 (2016)

Chapter 14
Numerical Solution of Stochastic Differential Equations: Diffusion and Jump-Diffusion Processes

Stochastic differential equations (SDE) play an important role in a range of application areas, including biology, physics, chemistry, epidemiology, mechanics, microelectronics, economics, and finance [1]. However, most SDEs, especially nonlinear SDEs, do not have analytical solutions, so that one must resort to numerical approximation schemes in order to simulate trajectories of the solutions to the given equation. The simplest effective computational method for approximation of ordinary differential equations is the Euler's method. The Euler–Maruyama method is the analogue of the Euler's method for ordinary differential equations for numerical simulation of the SDEs [2].

Another numerical scheme is the Milstein method [3], which is a Taylor method, meaning that it is derived from a truncation of the stochastic (Itô) Taylor expansion of the solution. The Milstein scheme involves the derivatives. If it happens that the derivatives do not exist, then it leads to difficulties in it implementation. In that case we need a derivative-free method. In such cases one can consider an implicit schemes that avoid the use of derivatives [2]. It can be done by replacing the derivatives there by finite difference. These methods are known as Runge–Kutta schemes [4]. In this chapter, following the exposition of Kloeden and Platen [4], we describe numerical methods for solving the SDE for diffusion and jump-diffusion processes. In addition, we provide a brief review of some computational packages in *R* and *Python* for simulating stochastic diffusion and jump-diffusion processes.

14.1 Numerical Integration of Diffusion Processes

14.1.1 Euler–Maruyama Method

Suppose that the process x under consideration satisfies the following (Itô) Langevin stochastic differential equation, i.e., to be a diffusion process,

M. R. Rahimi Tabar, *Analysis and Data-Based Reconstruction of Complex Nonlinear Dynamical Systems*, Understanding Complex Systems,
https://doi.org/10.1007/978-3-030-18472-8_14

$$dx = a(x,t)dt + b(x,t)dW(t). \tag{14.1}$$

In Euler–Maruyama scheme Eq. (14.1) has the form,

$$x_{n+1} = x_n + a(x_n, t_n)\Delta_n + b(x_n, t_n)\Delta W_n. \tag{14.2}$$

where $\Delta_n = dt_n = t_{n+1} - t_n$ is the length of the time discretization subinterval $[t_n, t_{n+1}]$ and increment of Wiener process

$$\Delta W_n = W_{t_{n+1}} - W_{t_n} \tag{14.3}$$

can be chosen from the $\mathcal{N}(0, \Delta_n)$ independent Gaussian distributed increment on $[t_n, t_{n+1}]$. Here $\mathcal{N}(0, \Delta_n)$ is a gaussian distributed random variable with zero and standard deviation Δ_n. Each random number ΔW_n is computed as

$$\Delta W_n = \Gamma_n(\Delta_n)^{1/2}$$

where uncorrelated noise Γ_n is chosen from zero mean with unit variance gaussian distribution $\mathcal{N}(0, 1)$. Each set of $\{x_0, \ldots, x_n\}$ produced by the Euler–Maruyama method is an approximate realization of the solution stochastic process $x(t)$ which depends on the random numbers Γ_n that were chosen.

For the general multi-dimensional case with $d, m \in 1, 2, \ldots$ the kth component of the Euler–Maruyama scheme has the form

$$x_{n+1}^k = x_n^k + a^k(\mathbf{x_n}, t_n)\Delta_n + \sum_{j=1}^{m} b^{k,j}(\mathbf{x_n}, t_n)\Delta W_n^j. \tag{14.4}$$

Here

$$\Delta W_n^j = W_{t_{n+1}}^j - W_{t_n}^j \tag{14.5}$$

is the $\mathcal{N}(0, \Delta_n)$ independent Gaussian distributed increment of the jth component of the m-dimensional standard Wiener process W on $[t_n, t_{n+1}]$, and $\Delta W_n^{j_1}$ and $\Delta W_n^{j_2}$ are independent for $j_1 \neq j_2$. The diffusion coefficient $\mathbf{b}(\mathbf{x_n}, t_n) = [b^{k,j}]_{k,j=1}^{d,m}$ is here a $d \times m$-matrix.

14.1.2 Milstein Scheme

The Milstein scheme is the simplest nontrivial numerical scheme for stochastic differential equations with a order of convergence one (higher than that of the Euler–Maruyama scheme). It was first derived by Milstein, who used the Itô formula to expand an integrand involving the solution in one of the error terms of the Euler–Maruyama scheme [3, 4]. For diffusion process (14.1), the Milstein scheme has in the one-dimensional case the form

$$x_{n+1} = x_n + a(x_n, t_n)\Delta_n + b(x_n, t_n)\Delta W_n + \frac{1}{2}b(x_n, t_n)b'(x_n, t_n)\{(\Delta W_n)^2 - \Delta_n\}. \tag{14.6}$$

where $b'(x, t) = \partial b/\partial x$. The additional term in the Milstein scheme, when compared with the Euler–Maruyama scheme, marks the point of divergence of stochastic numerical analysis from deterministic numerics. The accuracy of the Euler–Maruyama and Milstein schemes are of order $\Delta_n^{1/2}$ and Δ_n, respectively [4].

In the multi-dimensional case with $m, d \in 1, 2, \ldots$ the kth component of the Milstein scheme is given by

$$\begin{aligned} x_{n+1}^k &= x_n^k + a^k(\mathbf{x_n}, t_n)\Delta_n + \sum_{j=1}^{m} b^{k,j}(\mathbf{x_n}, t_n)\Delta W_n^j \\ &+ \frac{1}{2}\sum_{j_1,j_2=1}^{m}\sum_{i=1}^{d}\left[b^{i,j_1}\frac{\partial b^{k,j_2}}{\partial x^i}\right](\mathbf{x_n}, t_n)\{\Delta W_n^{j_1}\Delta W_n^{j_2} - \Delta_n\} \end{aligned} \tag{14.7}$$

for $k \in \{1, 2, \cdots, d\}$. Also there are strong schemes which avoid the use of derivatives in the same way as Runge–Kutta schemes do for ordinary differential equations in a deterministic setting. However, it must be emphasized that they cannot simply be constructed as heuristic generalizations of deterministic Runge–Kutta schemes [3, 4].

14.1.2.1 Derivation of Milstein Scheme

Consider following diffusion stochastic differential equation

$$dx = a(x, t)dt + b(x, t)dW(t).$$

and let t_n, t_{n+1}, be two consecutive points in time discretization. The Itô formula (6.35) says, that for a given function f which is two times continuously differentiable, we can write,

$$f(x_s) = f(x_n) + \int_{t_n}^{s}\left\{f'(x_u)a(x_u) + \frac{1}{2}f''(x_u)b(x_u)^2\right\}du + \int_{t_n}^{s} f(x_u)b(x_u)dW.$$

We can apply the Itô formula on the expressions $a(x_s)$ and $b(x_s)$ too and find,

$$\begin{aligned} x_{n+1} &= x_n + \int_{t_n}^{t_{n+1}}\left\{a(x_n) + \int_{t_n}^{s}\left[a'(x_u)a(x_u) + \frac{1}{2}a''(x_u)b^2(x_u)\right]du \right. \\ &\left. + \int_{t_n}^{s} a'(x_u)b(x_u)dW_u\right\}ds \end{aligned}$$

$$+\int_{t_n}^{t_{n+1}} \left\{ b(x_n) + \int_{t_n}^{s} \left[b'(x_u)a(x_u) + \frac{1}{2}b''(x_u)b^2(x_u) \right] du \right.$$
$$\left. + \int_{t_n}^{s} b'(x_u)b(x_u)dW_u \right\} dW_s.$$

By using a time discretization, the differentials dW and dt are replaced by the corresponding discrete versions ΔW and Δt. If we are seeking a method which converges strongly of order 1, we can neglect the double integrals above, which are of type $dW_s ds$ and $dsds$. We then obtain,

$$x_{n+1} \simeq x_n + \int_{t_n}^{t_{n+1}} a(x_n)ds + \int_{t_n}^{t_{n+1}} \left[b(x_n) + \int_{t_n}^{s} b'(x_u)b(x_u)dW_u \right] dW_s$$
$$\simeq x_n + a(x_n)\Delta_n + b(x_n)\Delta W_n + \int_{t_n}^{t_{n+1}} \int_{t_n}^{s} b'(x_u)b(x_u)dW_u dW_s. \tag{14.8}$$

The first three terms are well known from the Euler–Maruyama scheme. We can approximate the fourth term above by (see Problem 6.6)

$$\int_{t_n}^{t_{n+1}} \int_{t_n}^{s} b'(x_u)b(x_u)dW_u dW_s \simeq b'(x_n)b(x_n) \int_{t_n}^{t_{n+1}} \int_{t_n}^{s} dW_u dW_s$$
$$= \frac{1}{2}b'(x_n)b(x_n)\{(\Delta W_n)^2 - \Delta_n\}. \tag{14.9}$$

Substituting this expression in Eq. (14.8) we finally obtain the Milstein scheme (14.6). Derivation of Milstein scheme for multi-dimensional case is straightforward.

14.1.3 Runge–Kutta-Like Methods

A potential disadvantage of the Milstein method is that one must evaluate the first derivative of the function $b(x, t)$ that multiplies the stochastic increment. For deterministic differential equations it is the Runge–Kutta family of methods that eliminate the need to evaluate such derivatives. Similar methods can be found for stochastic equations. Here we present a first-order method of this type that was obtained by Platen, building upon Milstein's method [4, 5]. We will refer to it as the Milstein-Platen method, and it is obtained from Milstein's method above by replacing the derivative of b with an approximation that is valid to first-order. For a stochastic equation containing only a single variable x, the first-order approximation to the term bb' that appears in Milstein's method is

$$b(x, t)\frac{\partial}{\partial x}b(x, t) \approx \frac{1}{\sqrt{\Delta_n}}\left[b(q, t) - b(x, t) \right]$$

with

$$q = x + a\Delta_n + b\sqrt{\Delta_n}.$$

Substituting this into Milstein's method for a single variable, Eq. (14.6), we obtain the Milstein-Platen method for a single variable:

$$x_{n+1} = x_n + a(x_n, t_n)\Delta_n + b(x_n, t_n)\Delta W_n + \frac{1}{2\sqrt{\Delta_n}}\Big[b(q_n, t_n) - b(x_n, t_n)\Big]\{(\Delta W_n)^2 - \Delta_n\}. \tag{14.10}$$

For a vector stochastic equation with scalar noise (Eq. (14.4)), the Milstein-Platen method is

$$\sum_{m=1}^{d} b_{mj}(\mathbf{x}, t)\frac{\partial}{\partial x_m} b_{ik}(\mathbf{x}, t) \approx \frac{1}{\sqrt{\Delta_n}}\Big[b_{ij}(\mathbf{q}^{(k)}, t) - b_{ij}(\mathbf{x}, t)\Big]$$

with

$$q_i^{(k)} = x_i + a_i\Delta_n + b_{ik}\sqrt{\Delta_n}.$$

14.2 Numerical Integration of Jump-Diffusion Processes

Suppose that the process x under consideration satisfies the following Itô stochastic jump-diffusion equation

$$dx = a(x, t)dt + b(x, t)dW(t) + c(x, t)dJ \tag{14.11}$$

where $c(x, t)dJ$ is multiplicative jump process.

14.2.1 Euler–Maruyama Method

The simplest scheme is again the well-known Euler scheme, which is given by the algorithm

$$x_{n+1} = x_n + a(x_n, t_n)\Delta_n + b(x_n, t_n)\Delta W_n + \sum_{i=N(t_n)+1}^{N(t_{n+1})} c(x_n, t_n, u_i).$$

Here $N(t)$ represents the total number of Poisson jump process up to time t, which is Poisson distributed with mean λt and u_i is the ith mark of the Poisson distributed process at the ith jump time T_i. For mark-independent jump size, which means $c(x, t, u) = c(x, t)$, the Euler scheme reduces to

$$x_{n+1} = x_n + a(x_n, t_n)\Delta_n + b(x_n, t_n)\Delta W_n + c(x_n, t_n)\Delta J_n, \tag{14.12}$$

where $\Delta J_n = J_{n+1} - J_n$ follows a Poisson distribution with mean $\lambda\Delta_n$.

In the multi-dimensional (d-dimensional) case with scalar driving Wiener process, the kth component of the Euler scheme is given by

$$x_{n+1}^k = x_n^k + a^k(\mathbf{x_n}, t_n)\Delta_n + \sum_{j=1}^{m} b^{k,j}(\mathbf{x_n}, t_n)\Delta W_n^j + \sum_{i=N(t_n)+1}^{N(t_{n+1})} c^k(\mathbf{x_n}, t_n, u_i) \tag{14.13}$$

where a^k and c^k are the kth components of the drift and the jump coefficients, respectively, and $b^{k,j}$ is the component of the kth row and jth column of the diffusion matrix $\mathbf{b}$, for $k \in 1, 2, \ldots, d$, and $j \in 1, 2, \ldots, m$. For mark-independent jump size, the last term in Eq. (14.13) reduces to $c^k \Delta J_n^k$.

14.2.2 Milstein Scheme

For jump-diffusion process (14.11), the Milstein scheme has in the one-dimensional case the form [4]

$$\begin{aligned} x_{n+1} &= x_n + a(x_n, t_n)\Delta_n + b(x_n, t_n)\Delta W_n + c(x_n, t_n)\Delta J_n \\ &+ \frac{1}{2} b(x_n, t_n) b'(x_n, t_n) \left\{(\Delta W_n)^2 - \Delta_n\right\} \\ &+ b(x_n, t_n) c'(x_n, t_n) \left\{ \sum_{i=N(t_n)+1}^{N(t_{n+1})} W_{\tau_i} - \Delta J_n W_n \right\} \\ &+ \{b(x_n + c, t_n) - b(x_n, t_n)\} \left\{ \Delta J_n \Delta W_n - \left\{ \sum_{i=N(t_n)+1}^{N(t_{n+1})} W_{\tau_i} - \Delta J_n W_n \right\} \right\} \\ &+ \frac{1}{2} \{c(x_n + c, t_n) - c(x_n, t_n)\} \left\{(\Delta J_n)^2 - \Delta J_n\right\}, \end{aligned} \tag{14.14}$$

where $b' = \frac{\partial b}{\partial x}$ and $c' = \frac{\partial c}{\partial x}$.

For all the classes of methods we have presented here, there are also higher-order versions, although for strong methods these become rather complex. The interested reader can find all these methods, along with full details of their derivations, in the comprehensive book by Kloeden and Platen [4].

14.3 Stochastic Differential Equation: Packages in *R* and *Python*

In this section we provide brief review of some packages in *R* and *Python* to simulate stochastic differential equations.

14.3.1 An R Package (Langevin) for Numerical Solving and Modeling of Univariate and Bivariate "Diffusion Processes"

An R package developed by the research group Turbulence, Wind energy and Stochastic (TWiSt) at the Carl von Ossietzky University of Oldenburg, to simulate and extract the (stochastic) evolution equation underlying a set of data or measurements. Examples for the one-dimensional and two-dimensional cases are provided. For Markovian processes involving Gaussian white noise, a stochastic differential equation is derived straightforwardly from the time series and captures the full dynamical properties of the underlying diffusion process [6].

14.3.2 Simulation of Diffusion Processes, R-Package "Sim.DiffProc"

The package 'Sim.DiffProc' is for symbolic and numerical computations on scalar and multivariate systems of stochastic differential equations. It provides users with a wide range of tools to simulate, estimate, analyze, and visualize the dynamics of these systems in both forms Itô and Stratonovich. Statistical analysis with parallel Monte-Carlo and moment equations methods of stochastic differential equations are given. Enabled many researchers in different domains to use these equations to modeling practical problems in financial and actuarial modeling and other areas of application, e.g., modeling and simulate of first passage time problem [7].

14.3.3 Simulation of Diffusion Processes, R-Package "DiffusionRimp"

The DiffusionRimp package consists of various routines for analysing diffusion processes. It performs inference on a scalar diffusion model and also a bivariate diffusion model [8].

14.3.4 Simulation of Diffusion Processes, R-Package "yuima"

In the yuima package stochastic differential equations can be of very abstract type, multidimensional, driven by Wiener process or fractional Brownian motion with general Hurst parameter, with or without jumps specified as Lévy noise. The yuima package is intended to offer the basic infrastructure on which complex models and inference procedures can be built on [9].

14.3.5 Simulation of Jump-Diffusion Processes, Python-Solver "JiTCSDE"

JiTCSDE (just-in-time compilation for stochastic differential equations) is a Python implementation of the adaptive integration method for Itô SDEs proposed by Rackauckas and Nie [10], which in turn employs two Rößler-type stochastic Runge–Kutta methods [11]. It uses the difference between their results to estimate the error of the integration. These methods are intertwined to minimise the additional computational effort required to estimate the error. The step size is adapted to keep the estimated error below a user-determined threshold. If the user wishes to have jumps in addition to the Brownian noise (and thus obtain a jump-diffusion process), they can use the extension *jitcsde_jump*. It draws the waiting time between jumps as well as the value of the jump from a user-specified distribution. The jumps are positioned precisely, i.e., the dynamics is integrated up to the time of a jump, the jump is applied, and the integration is continued [12].

Problems

14.1 Euler–Maruyama scheme, diffusion processes

The simplest scheme to solve stochastic differential equations (SDE) can be obtained by using a first-order approximation, where is known as Euler–Maruyama scheme. For diffusion stochastic processes like

$$dx = a(x,t)dt + b(x,t)dW(t)$$

it has the following form

$$x_{n+1} = x_n + a(x_n, t_n)dt + b(x_n, t_n)dW_n.$$

The Wiener process $dW(t_n)$ can be approximated as follows:

$$dW_n = \Gamma_n (dt)^{1/2}$$

where the noise Γ_n is a discrete-time Gaussian white process with mean 0 and standard deviation 1. For the following cases, generate 100 realizations of $x_n = x(t_n)$ (each with $N = 10^4$ data points) with $x(t_0) = 0.001$, $dt = 0.001$ and estimate time dependent probability density function $p(x, t)$, mean $\langle x(t) \rangle$ and $\langle x^2(t) \rangle$ by averaging down the ensemble and plot these functions with time t.

(a) For $a(x, t) = b(x, t) = 1$

(b) For $a(x, t) = \mu x(t)$ and $b(x, t) = \sigma$, where $\mu = -1$ and $\sigma = 1$. This is a *Ornstein–Uhlenbeck process*.

(c) For $a(x, t) = \mu x(t)$ and $b(x, t) = \sigma x(t)$, where $\mu = 0.75$ and $\sigma = 0.3$. This is a *Black–Scholes process*.

14.2 Euler–Maruyama scheme, diffusion processes

Consider the nonlinear diffusion process

$$dx(t) = (1 + x(t))(1 + x^2(t))dt + (1 + x^2(t))dW(t).$$

(a) Use the Euler–Maruyama scheme to approximate this Langevin equation and generate 100 realizations of $x_n = x(t_n)$ (each with $N = 10^4$ data points) with $x(t_0) = x_0 = 1$, $dt = 0.001$.

(b) Using Itô calculus, verify that stochastic solution of this equation is

$$x(t) = tan\left(t + W(t) + arctan(x_0)\right).$$

Compare the integrated solution with exact stochastic solution (with same initial conditions and same realization of Wiener processes).

14.3 Euler–Maruyama scheme, diffusion process with two Wiener processes

Consider the multiplicative diffusion process

$$dx(t) = a_0 x(t)dt + b_1 x(t)dW_1(t) + b_2 x(t)dW_2(t).$$

(a) Use the Euler–Maruyama scheme to approximate this Langevin equation and generate 100 realizations of a $x_n = x(t_n)$ (each with $N = 10^4$ data points) with $x(t_0) = x_0 = 0.01$, $a_0 = 1$, $b_1 = b_2/2 = 1$ and $dt = 0.001$. Two Wiener processes are independent.

(b) Using Itô calculus, verify that exact stochastic solution of this equation is

$$x(t) = x_0 \exp\left\{\left[a_0 - \frac{1}{2}(b_1^2 + b_2^2)\right] t + b_1 W_1(t) + b_2 W_2(t)\right\}.$$

Compare the integrated solution with exact solution.

14.4 Euler–Maruyama scheme, diffusion process with two correlated Wiener processes

Consider the multiplicative diffusion process

$$dx(t) = a_0 x(t)dt + b_1 x(t)dW_1(t) + b_2 x(t)dW_2(t).$$

Here two Wiener processes are dependent as,

$$\langle dW_1(t)dW_2(t)\rangle = \rho dt.$$

Two correlated Wiener processes $dW_1(t)$ and $dW_2(t)$ can be written in terms of two independent Wiener processes dw_1 and dw_2 as (see Problem 13.2),

$$dW_1 = dw_1$$

$$dW_2 = \rho dw_1 - \sqrt{1-\rho^2}dw_2$$

where two independent Wiener processes dw_1 and dw_2 can be approximated as follows:

$$dw_i(t) = \Gamma_i(dt)^{1/2}$$

here the noises Γ_i are discrete-time Gaussian white independent processes with mean 0 and standard deviation 1.

Use the Euler–Maruyama scheme to approximate this Langevin equation and generate 100 realizations of a $x_n = x(t_n)$ (each with $N = 10^4$ data points) with $x(t_0) = x_0 = 1$, $a_0 = 3$, $b_1 = b_2/2 = 1$ and $dt = 0.001$. Compare dependence of variance $var(x(t))$ on t for trajectories with $\rho = 0.01$ and 0.9.

14.5 Time dependent drift and diffusion coefficients, The minimal market model

The minimal market model has the deterministic drift

$$a(x,t) = \alpha \exp\{\eta t\}$$

and diffusion coefficient

$$b(x,t) = \sqrt{\alpha \exp\{\eta t\} x}$$

with initial scaling parameter $\alpha > 0$ and net growth rate $\eta > 0$. Use the Euler–Maruyama scheme to approximate the corresponding Langevin equation and generate 100 realizations of $x_n = x(t_n)$ (each with $N = 10^4$ data points) with $x(t_0) = x_0 = 1$, $dt = 0.001$, $\alpha = 0.2$ and $\eta = 0.01$. Check that the generated trajectories never hit zero!.

14.6 Diffusion process in a bistable potential

Consider the motion of a particle in a bistable potential

$$dx = -\frac{\partial U(x)}{\partial x}dt + b_0\, dW(t)$$

with

$$U(x) = \frac{x^4}{4} - \frac{x^2}{2}.$$

(a) Show that deterministic dynamical system has two stable equilibria at $x = \pm 1$ and one unstable fixed point at $x = 0$.

(b) Use the Euler–Maruyama scheme to approximate this Langevin equation and generate 100 realizations of a $x_n = x(t_n)$ (each with $N = 10^4$ data points) with $x(t_0) = x_0 = 0$, $b_0 = 0.5$ and $dt = 0.01$.

(c) Show that the second statistical moment $\langle x^2(t)\rangle$ saturates for longer time.

14.7 Bivariate diffusion processes, Van-der-Pol oscillator

Use the Euler–Maruyama scheme to integrate two-dimensional diffusion process (Van-der-Pol oscillator)

$$\frac{d}{dt}x_1(t) = x_2(t)$$

$$\frac{d}{dt}x_2(t) = x_2(t)(0.3 - x_1^2(t)) - x_1 + 0.3\Gamma(t)$$

where $\Gamma(t)$ is a Gaussian white noise. Generate $x_{1,2}(t_n)$ (each with $N = 10^6$ data points) with $x_1(t_0) = x_2(t_0) = 0$ and $dt = 0.01$. Scatter-plot x_1 verses x_2 in the stationary state.

14.8 Bivariate diffusion processes, Heston model

The observed volatility for real existing price processes, such as stocks or bonds is itself a stochastic process. The following model describes this observation:

$$\frac{d}{dt}P(t) = \mu dt + \sigma(t)dW_1(t) \qquad P(0) = P_0$$

$$\frac{d}{dt}\sigma(t) = k(\theta - \sigma(t))dt + \sigma(t)\sigma_1 dW_2(t), \qquad \sigma(0) = \sigma_0$$

where θ is the average volatility, σ_1 a volatility, and k the mean reversion rate of the volatility process $\sigma(t)$. If this model is used for stock prices $S(t)$, its logarithm related to $P(t)$ as $P(t) = \ln(S(t))$. The two Brownian motions $dW_1(t)$ and $dW_2(t)$ are correlated with covariance $\langle dW_1(t)dW_2(t)\rangle = \rho dt$. This model captures the behavior of real existing prices better and its distribution of returns, i.e. $P(t)$ shows fatter tails.

For the parameters $\mu = 0.1, k = 2, \theta = 0.2, \sigma_1 = 0.5, \sigma_0 = 0.1, 0.8$ and $\rho = 0.5$, use the Euler–Maruyama scheme to integrate 100 realizations (each with $N = 10^4$ data points) of this two-dimensional diffusion process and show that variance of $P(t)$ at any time depends on σ_0.

14.9 Milstein simulation algorithm

For diffusion stochastic processes

$$dx = a(x,t)dt + b(x,t)dW(t)$$

show that Milstein scheme can be written as

$$x_{n+1} = x_n + a(x_n, t_n)dt + b(x_n, t_n)\sqrt{dt}\,\Gamma_n$$
$$-\frac{dt}{2}\, b(x_n, t_n)\, b'(x_n, t_n)\, (1 - \Gamma_n^2),$$

where Γ_n are independent samples of a zero mean gaussian distribution with unit variance.

14.10 Milstein and Euler–Maruyama schemes

For diffusion stochastic processes

$$dx = (\mu x(t) + \eta)dt + \gamma x(t)dW(t)$$

with $x(t_0) = x_0 = 1$, where μ, η and γ are real constants, write down the Euler–Maruyama and the Milstein schemes.

14.11 Colored noise driven random walk (non-Markovin random walk), first-passage time distribution

Consider following stochastic equations:

$$\frac{d}{dt}x(t) = y(t)$$
$$\frac{d}{dt}y(t) = -\frac{1}{L_0}y(t) + \frac{1}{L_0}\Gamma(t)$$

where $\Gamma(t)$ is a Gaussian white noise. For $L_0 = 1$ and $L_0 = 10$ with $dt = 0.01$,

(a) Use the Euler–Maruyama scheme to integrate two-dimensional diffusion process and numerically check that in the stationary state we have

$$\langle y(t)y(s)\rangle = \frac{1}{2L_0}\exp\left(-\frac{|t-s|}{L_0}\right).$$

(b) Generate 100 ensemble of $x(t)$ process (each with $N = 10^4$ data points), start from $x(0) = y(0) = 0$. Since the process x is a stochastic process, each realization of it reaches a given barrier $x = a$ for the first time at a different time. Therefore, it has a first-passage time probability density $f(t)$. Plot distribution $f(t)$, for $x(t)$ for when it reaches barrier $x = a = 1$ and for $L_0 = 1$ and $L_0 = 10$.

(c) In the limit $L_0 \to 0$, $y(t)$ will be Dirac delta correlated noise, therefore $x(t)$ will be the Wiener process. However for finite L_0, $y(t)$ is not Dirac delta correlated noise and is a colored noise. Compare your obtained numerical results for $L_0 = 1$ and $L_0 = 10$ with results of $f(t)$ for Wiener processes (Markovin random walk), i.e.

$$f(t) = \frac{|a|}{2\pi t^3}\exp\left(-\frac{a^2}{2t}\right).$$

(d) Estimate the right tail-exponent β of first-passage time probability density, i.e. $f(t)_{t\gg a^2} \sim t^{-\beta}$ for cases $L_0 = 1$ and $L_0 = 10$.

14.12 Simulating jump-diffusion Process, Euler–Maruyama scheme

Consider the following jump-diffusion stochastic differential equation

$$dx = \mu x(t)dt + \sigma x(t)dW(t) + (\beta - 1)x(t)dJ(t)$$

where $dJ(t)$ is Poisson jumps process with rate λ. Use the Euler–Maruyama scheme to approximate this stochastic equation and generate 100 realizations of $x_n = x(t_n)$ (each with $N = 10^4$ data points) with $x(t_0) = x_0 = 1$, $dt = 0.001$, $\mu = 0.04$, $\sigma = 0.2$, $\lambda = 0.5$ (jump rate per data point) for $\beta = 1.4$ and $\beta = 0.6$. Plot time dependent mean and variance of $x(t)$ for two cases. Ensure that in each realization the number of jumps to be $n_J \approx \lambda N$. Note that jump rate per data point is related to jump rate per unit of time as $\lambda(\text{per data point})/dt = \lambda(\text{per unit of time})$.

14.13 Simulating jump-diffusion process, Euler–Maruyama scheme

Consider the following jump-diffusion stochastic differential equation

$$dx = \mu x(t)dt + \sigma x(t)dW(t) + \xi dJ(t)$$

where $dJ(t)$ is Poisson jumps process with rate λ. The jump amplitude ξ has zero mean Gaussian distribution with variance $\sigma_\xi^2 = 1$ which can symmetrically jumps up and down, i.e. $\xi \sim \mathcal{N}(0, \sigma_\xi^2 = 1)$. Use the Euler–Maruyama scheme to approximate this stochastic equation and generate 100 realizations of a $x_n = x(t_n)$ (each with $N = 10^4$ data points) with $x(t_0) = x_0 = 0$, $dt = 0.001$, $\mu = -10$, $\sigma = 0.5$ and $\lambda = 0.5$ (jump rate per data point). Plot time dependent mean and variance of $x(t)$.

14.14 Simulating Lévy noise-driven Langevin equation

Consider the following Lévy noise-driven Langevin equation

$$dx = -10x(t)dt + dL_{\alpha}^{\gamma=1,\beta=0,\mu=0}(t)$$

where $dL(t)$ is increment of Lévy process. Use the discrete time evolution law (10.10) to approximate this stochastic equation and generate 100 realizations of a $x_n = x(t_n)$ (each with $N = 10^4$ data points) with $x(t_0) = x_0 = 0$, $dt = 0.001$, for $\alpha = 0.5$ and $\alpha = 1.5$. Plot time dependent mean and variance of $x(t)$. Use the algorithm presented in Sect. 10.6 for synthesis of Lévy noise, see also Problem 10.4.

References

1. D.J. Higham, SIAM Rev. **43**, 525 (2001)
2. E. Platen, N. Bruti-Liberati, *Numerical Solution of Stochastic Differential Equations with Jumps in Finance* (Springer, Berlin, 2010)
3. G.N. Milstein, M.V. Tretyakov, *Stochastic Numerics for Mathematical Physics* (Springer, Berlin, 2004)
4. P.E. Kloeden, E. Platen, *Numerical Solution of Stochastic Differential Equations* (Springer, Berlin, 1999)
5. K. Jacobs, *Stochastic Processes for Physicists: Understanding Noisy Systems* (Cambridge University Press, Cambridge, 2010)
6. P. Rinn, P.G. Lind, M. Wächter, J. Peinke, The Langevin approach: an R package for modeling Markov processes. J. Open Res. Softw. **4**, e34 (2016). https://doi.org/10.5334/jors.123, https://cran.r-project.org/web/packages/Langevin/index.html
7. A. Guidoum, K. Boukhetala, Sim.DiffProc: Simulation of Diffusion Processes. R package version 3.1 (2015). http://CRAN.R-project.org/package=Sim.DiffProc
8. E.A.D. Pienaar, M.M. Varughese, DiffusionRimp: Inference and Analysis for Diffusion Processes via Data Imputation and Method of Lines. R package version 0.1.0 (2015). https://CRAN.R-project.org/package=DiffusionRimp
9. A. Brouste, M. Fukasawa, H. Hino, S. Lacus, K. Kamatani, Y. Koike, H. Masuda, R. Nomura, Y. Shimuzu, M. Uchida, N. Yoshida, The YUIMA project : a computational framework for simulation and inference of stochastic differential equations. J. Stat. Soft. **57**, 1 (2014)
10. C. Rackauckas, Q. Nie, Adaptive methods for stochastic differential equations via natural embeddings and rejection sampling with memory. Discret. Cont. Dyn.-B **22**, 2731 (2017)
11. A. Rößler, Runge-Kutta methods for the strong approximation of solutions of stochastic differential equations. SIAM J. Numer. Anal. **48**, 922 (2010)
12. G. Ansmann, Efficiently and easily integrating differential equations with JiTCODE. JiTCDDE, and JiTCSDE, Chaos **28**, 043116 (2018). http://jitcsde.readthedocs.io/en/0.20/

Chapter 15
The Friedrich–Peinke Approach to Reconstruction of Dynamical Equation for Time Series: Complexity in View of Stochastic Processes

In this chapter we study stochastic properties of spatially- and temporally-disordered structures, such as turbulence and rough surfaces, or temporal fluctuations of given time series, in *scale*. Experimental observables include the field increments, such as the difference in the velocity field between two points separated by a distance r, or difference of time series in a time lag r. Therefore, the lag r can be either spatial distance or a time interval. The change of the increments' fluctuations as a function of the scale r can then be viewed as a stochastic process in a length or time scale and can, quite often, after pioneering work by Friedrich & Peinke, be mapped onto the mathematical structures of the Langevin [1] or generalised Langevin processes with jumps contributions [2]; see [3] for recent review and progresses of subject. In this chapter we discuss the stochastic analysis in scale, and provide a physical picture of the fluctuations cascade, known as cascade processes, from large to small scales. We then provide the Kramers-Moyal expansion in terms of the conditional moments of the increments. The chapter will close with the definition of "multipliers" that relate the increments' fluctuations at scale r_1 to those at scale r_2, and present the Castaing equation. Finally, multiscale correlation functions will be introduced.

15.1 Stochastic Processes in (Length or Time) Scale

15.1.1 Increments of Stochastic Processes

It is common for scale-dependent analysis of stochastic data to use the *increment* $x_r \equiv \Delta x(r,t) = x(t+r) - x(t)$ in a lag r (scale) of a data set $x(t)$ as a stochastic quantity. One can define also the *centered increment* as $x_r \equiv \Delta x(r,t) = x(t+r/2) - x(t-r/2)$, where the definitions in some cases removes spurious correlations between x_{r_1} and x_{r_2} at two scales that can be present by the common left-justified (forwarded) expression in the first definition [4]. For positive data, such as the heart interbeat or the stock price, one can define the increment as the log-returns in the lag r by,

M. R. Rahimi Tabar, *Analysis and Data-Based Reconstruction of Complex Nonlinear Dynamical Systems*, Understanding Complex Systems,
https://doi.org/10.1007/978-3-030-18472-8_15

$x_r = \ln[x(t + r)/x(t)]$. In this chapter we will use the first definition mainly, unless otherwise indicated.

15.1.2 Fractal and Multifractal Time Series: Linear and Nonlinear Processes

We begin with definition of fractal and multifractal time series and present standard method to investigate the multifractal characterization of data set using the statistical properties of increments. In the case of self-similar structures, complexity is commonly investigated by a local measure $q(r, t)$ that characterises the structure on the scale r and at the time (or position) t. The local measure is defined by, $q(r, t) \equiv |x_r| = |x(t + r) - x(t)|$, or the amplitude of the wavelet at scale r; see, for example, [5, 6]. For stationary time series the statistical moment order n, $\langle q^n(r, t)\rangle$, is independent of t, which implies $\langle q^n(r, t)\rangle = \langle q^n(r)\rangle$, and is known as the structure function of order n. For scale invariant processes, we expect

$$S_n(r) = \langle q^n(r)\rangle = C_n r^{\zeta(n)}$$

$n \geq 0$, where bracket denotes time or an ensemble average. Extension of this statistic for $n < 0$ is discussed in [7]. It is usually assumed that there is a large range of scale r values where the power law relations for the structure function apply for all the n of interest. There are general statements regarding $\zeta(n)$. First, proper normalization in $S_n(r)$ requires $\zeta(0) = 0$. Furthermore, $\zeta(n)$ is a concave function [8] and it is also nondecreasing if the increments $q(r, t)$ are bounded [9, 10]. For a special class of process one may finds $\langle q^n(r)\rangle \sim \langle q^2(r)\rangle^{n/2}$, which yields $\zeta(n) = n\zeta(2)/2$ and, therefore, one exponent $\zeta(2)$ determines the scaling of all the moments of the increments. This is called "monoscaling" (fractal or linear processes). If, however, $\zeta(n)$ is not linear in n, we are dealing with "multiscaling" (multifractal or nonlinear processes). The exponent $\zeta(2)$ is known as the roughness exponent [11]. The interested reader is referred to the seminal paper by Parisi and Frisch [8], the first to use the term multifractal. About a decade after introducing multifractality [8, 12, 13], multifractal concepts proved invaluable in nonlinear science. Their well-known applications are turbulence [14], deterministic chaos [15] and many real world time series [3, 16] (see [17] and references therein).

The simplest type of the multifractal analysis is based upon the standard partition function multifractal formalism, which has been developed for the multifractal characterization of normalized, stationary measurements [11, 18–21]. Unfortunately, this standard formalism does not yield correct results for nonstationary time series that are affected by trends, or that cannot be normalized. Thus, in the early 1990s an improved multifractal formalism was developed, the wavelet transform modulus maxima (WTMM) method [7], which is based on the wavelet analysis and involves tracing the maxima lines in the continuous wavelet transform over all the scales.

The other method, the multifractal detrended fluctuation analysis (MF-DFA) [22], is based on the identification of scaling of the qth-order moments depending on the signal length, and is a generalization of the standard DFA using only the second moment $q = 2$.

To understand the multifractality, let us study the probability distribution function (PDF) of the increments $p(q, r)$. The complete information on an increment, $q(r, t)$, is contained in the PDF $p(q, r)$ for a certain scale r. For a self-similar process (scale-invariant process) it is assumed that $q(r, t)$ exhibits scaling behavior, $q(r, t) \sim r^{\xi}$, where $\xi = \zeta(2)/2$. This means that in a certain range of r the processes

$$q(r, t) \quad , \quad \lambda^{\xi} q(\lambda r, \lambda^{\gamma} t) \tag{15.1}$$

should have the same statistics, where $\lambda > 0$ is the scaling factor. The exponent γ is known as dynamical exponent [23, 24]. More precisely, the probability distribution of the quantity q takes the form

$$p(q, r) = r^{-\xi} F(q/r^{\xi}) \ , \tag{15.2}$$

with a universal function $F(Q)$ with that property that $\int_{-\infty}^{\infty} F(Q) dQ = 1$.

Let us study the scaling behavior of the statistical moments of $q(r, t)$ as a function of the scale. Assuming that they exist, the statistical moments exhibit scaling behavior,

$$\langle q^n(r) \rangle = \int dq \quad q^n \, \frac{1}{r^{\xi}} F(q/r^{\xi}) = C_n \, r^{n\xi} = C_n r^{\zeta(n)} \tag{15.3}$$

where C_n is the amplitude of the statistical moment of order n. Such properties have been called monoscaling or fractal scaling behavior (linear processes), where the scaling exponents of the statistical moments $\langle q^n(r) \rangle$, i.e. $\zeta(n)$ vary linearly with n.

There are many experimental examples of real systems, such as turbulent fields or heart beat intervals, just to mention two, for which such a simple picture of the scaling structure is only a rough, first approximation. For instance, for turbulence in which $q(r, t)$ is taken as absolute values of the velocity increment, one can argue that *multifractal behavior* is more appropriate, the nth-order statistical moments scale according to

$$\langle q^n(r) \rangle = C_n r^{\zeta(n)} \quad , \tag{15.4}$$

where the scaling exponent $\zeta(n)$ is a nonlinear function (for nonlinear processes) of the order n,

$$\zeta(n) = n\xi_0 + n^2 \xi_1 + n^3 \xi_2 + \cdots \ . \tag{15.5}$$

Such a relation can formally be obtained by assuming that the probability distribution $p(q, r)$ has the following form

$$p(q, r) = \int d\xi \quad f(\xi, r) \, \frac{1}{r^{\xi}} F(q/r^{\xi}) \quad , \tag{15.6}$$

which is based on the assumption that in a data set regions with different scaling indices ξ exist, where $f(\xi, r)$ provides a measure (probability density) of the scaling indices ξ at scale r; see [25, 26] for a review of the recent developments on multiscale characteristics of time series.

The major shortcoming of the fractal and multifractal approaches to complexity in scale is the fact that, they only address the statistics of the measure $q(r, t)$ at a single scale r. In general though, one expects $q(r_1, t)$ and $q(r_2, t)$ to be correlated processes. The question, which we address in the following, arises as to whether there exist methods that lead to a more comprehensive characterization of the scale stochasticity by the general joint statistics for $q(r, t)$ (or x_r itself)

$$p(q_1, r_1; q_2, r_2; \ldots; q_n, r_n) \tag{15.7}$$

of the local measure q_i at multiple scales r_i [1, 3]. If the stochastic process $q(r, t)$ is a Markov process in scale (which, in practice, should be verified), the joint probability $p(q_1, r_1; q_2, r_2; \ldots; q_n, r_n)$ is determined by taking advantage of the Markovian property (in scale) in terms of the conditional probabilities, i.e.,

$$\begin{aligned} p(q_1, r_1; q_2, r_2; \ldots; q_n, r_n) = p(q_1, r_1 | q_2, r_2) \times \\ p(q_2, r_2 | q_3, r_3) \cdots p(q_{n-1}, r_{n-1} | q_n, r_n)\, p(q_n, r_n) \end{aligned} \tag{15.8}$$

where $r_1 > r_2 > \cdots > r_n$. For such Markov processes in scale we define the Kramers–Moyal (KM) coefficients and write down the KM expansion.

15.2 Intermittency and Kramers–Moyal Expansion

We now introduce the approach by Friedrich and Peinke for studying the fluctuations in scale [1]. One of the main properties of almost all the experimental data sets is their small-scale intermittency. Intuitively, an intermittent process can be interpreted as a superposition of subsets, each characterized by a given scaling exponent, with a typical increment fluctuations and is characterized by multifractal scaling behaviour. Its qualitative signature is the non-Gaussian shape of the PDF of the increments, defined by, $x_r = x(t + r) - x(t)$, for given small scales r. The stretched exponential-like tail of the PDF of x_r changes when the scale r decreases from the largest scales to the smallest ones. In this view, if the small-scale increments' PDFs are heavy tailed (with high probability of extreme events with respect to a Gaussian PDF), we define this as intermittency [5]. Studying the shape of increments' PDFs at various scales [5, 27–29] or exploring linear or nonlinear behaviour of the exponents $\zeta(n)$ [30, 31], allows quantitative evaluation of the intermittency.

Let us define the drift and diffusion coefficients or, more generally, the KM coefficients, for the *increments*, i.e., x_r of a given stochastic process or time series (see Chap. 16 for verification of Markov propertyin scale). The KM expansion will be,

$$-\frac{\partial p(x_r, r)}{\partial r} = \sum_{n=1}^{\infty}(-1)^n \frac{\partial^n}{\partial x_r{}^n}\left[D^{(n)}(x_r, r)p(x_r, r)\right]. \tag{15.9}$$

The coefficients $D^{(n)}(x_r, r)$ are the small-scale limit of the conditional KM moments, which fully characterize the statistics of the fluctuations by

$$D^{(n)}(x_r, r) = \frac{1}{n!}\lim_{r_1 \to r}\frac{1}{r_1 - r}\int (x_1 - x_r)^n p(x_1, r_1|x_r, r)dx_1.$$

The minus sign in the left side of Eq. (15.9) is due to the fact that it is common to use the scale r on a logarithmic scale, which means that $\lambda \equiv \ln(L/r)$, where L is the correlation length scale of time series, so that $r \leq L$ [16]. Generally speaking, for a lag size with $r > L$ the estimated PDF of the increments has Gaussian shape [16]. In terms of the variable λ one finds standard KM equation with positive sign.

15.2.1 Governing Equation for the Statistical Moments in Scale

Next, using the KM expansion, one is able to derive the equation for the statistical moment, such as Eq. (7.11), for the *structure functions*, i.e., $S_n(r) = \langle x_r^n \rangle$, based on which we investigate one of the most important issues in time series, namely, the scaling behavior and intermittency—which generally leads to nonlinear exponents $\zeta(n)$—of the fluctuations in various scaling region.

Using the general KM expansion, Eq. (15.9), the differential equations for the nth-order statistical moments (assuming that they exist) can be derived. Multiplying both side of Eq. (15.9) by x_r^n and integrating with respect to x_r, we obtain the governing equation for the evolution of the statistical moments of data set, given by [1, 32]:

$$\begin{aligned}
-\frac{d}{dr}\langle x_r^n \rangle &= \sum_{k=1}^{\infty}(-1)^k \int_{-\infty}^{+\infty} x_r^n \left(\frac{\partial}{\partial x_r}\right)^k D^{(k)}(x_r, r)p(x_r, r)dx_r \\
&= \sum_{k=1}^{n}\frac{n!}{(n-k)!}\int_{-\infty}^{+\infty} x_r^{n-k} D^{(k)}(x_r, r)\, p(x_r, r)dx_r \\
&= \sum_{k=1}^{n}\frac{n!}{(n-k)!}\left\langle x_r^{n-k} D^{(k)}(x_r, r)\right\rangle.
\end{aligned} \tag{15.10}$$

Here, we assumed that $p(x_r, r)$ and its derivative with respect to x_r vanish faster than polynomial functions at $x_r \to \pm\infty$. For example, for $n = 4$ we find the following equation for the fourth-order statistical moment,

$$-\frac{d}{dr}\langle x_r^4\rangle = 4\langle x_r^3\, D^{(1)}(x_r)\rangle + 12\langle x_r^2\, D^{(2)}(x_r)\rangle \\ + 24\langle x_r\, D^{(3)}(x_r)\rangle + 24\langle D^{(4)}(x_r)\rangle. \tag{15.11}$$

We note again that the KM expansion can be truncated after the second (diffusion) term, provided that the fourth-order coefficient $D^{(4)}(x_r, r)$ vanishes. If so, which is often the case (see Chap. 18 for verification of vanishing of fourth-order KM coefficient), then the KM expansion, Eq. (15.9), reduces to a Fokker–Planck evolution equation. In that case, the Fokker–Planck equation with given drift and diffusion coefficients for the PDF $p(x_r, r)$ is, in turn, used to gain information on the evolution of the shape of the PDF as a function of the scale r. We can imagine that the increments of data sets with jump discontinuities may have non-vanishing fourth-order KM coefficients and, therefore, one needs to estimate jump contributions for its increment time series. This issue is a subject of recents studies [2, 33–35].

15.3 Fokker–Planck Equation and (Multifractal) Scaling Exponents

Let us now assume that the higher-order KM coefficients ($n \geq 3$) vanish and the increment statistics of the process is describable by the Fokker–Planck (FP) equation. The FP equation for the probability density function of the increment is given by [1]

$$-\frac{\partial}{\partial r}\, p(x_r, r) = \\ \left\{ -\frac{\partial}{\partial x_r} D^{(1)}(x_r, r) + \frac{\partial^2}{\partial x_r^2} D^{(2)}(x_r, r) \right\} p(x_r, r). \tag{15.12}$$

For time series with scaling behaviour, the drift and diffusion coefficients of the increment are formulated as [1, 16]

$$D^{(1)}(x_r, r) \simeq -H\, \frac{x_r}{r} \\ D^{(2)}(x_r, r) \simeq b\, \frac{x_r^2}{r}. \tag{15.13}$$

where H and b are positive constants. The drift and diffusion coefficients are directly determined from time series; see Chaps. 16–21.

Using Eq. (15.10), we obtain the evolution of the structure functions $S_n(r) \equiv \langle (x_r)^n \rangle = \langle (x(t+r) - x(t))^n \rangle$ with r as follows,

$$-\frac{\partial}{\partial r}\langle x_r^n\rangle = n\langle x_r^{n-1} D^{(1)}(x_r, r)\rangle \\ + n(n-1)\langle x_r^{n-2} D^{(2)}(x_r, r)\rangle \tag{15.14}$$

by substituting the Eqs. (15.13) in Eq. (15.14) one finds,

$$\frac{\partial}{\partial r}\langle x_r^n \rangle = [nH - bn(n-1)]\frac{\langle x_r^n \rangle}{r}. \tag{15.15}$$

The above equation implies scaling behavior for the statistical moments of the increments, the structure function as

$$S_n(r) \equiv \langle x_r^n \rangle = \langle (x(t+r) - x(t))^n \rangle \sim r^{\zeta(n)}. \tag{15.16}$$

According to Eqs. (15.15) and (15.16), the corresponding scaling exponent are given by

$$\zeta(n) = nH - bn(n-1) \tag{15.17}$$

For mono- and multifractal processes (linear- and nonlinear processes) the exponent $\zeta(n)$ have linear and nonlinear dependence on n, respectively. We would like to point out that H is nothing but the underlying fluctuations' Hurst exponent [22, 36–39] of the time series itself. We also note that for a process with constant diffusion coefficient $D^{(2)}(x_r, r) = a$, the scaling exponents will be $\zeta(n) = nH$, and the amplitude of the statistical moments will be a function of parameter a. We should also note that exponent (15.17) are in contradiction with convexity and nondecreasing conditions for the scaling exponents of the structure functions [5] for $n \geq (H + b)/2b$. This problem can be solved by considering the higher-order KM coefficients [40, 41].

15.4 Langevin and Jump-Diffusion Modeling in Scale

For Markov processes in scale, we can generalize the time processes given by (9.1) and (12.1) to a (Itô) stochastic equation evolving in scale as

$$x_{(r+dr)} = x_r + N(x_r, r)dr + g(x_r, r)dW(r) + \xi(x_r, r)dJ(r) \tag{15.18}$$

where dW and dJ are the Wiener and Poisson jump processes, respectively. Equation (15.18) is a stochastic differential equation for x_r with r. Here, the scale r plays the role of "time" in the previous chapters. The drift and diffusion functions $N(x_r, r)$ and $g(x_r, r)$, as well as jump amplitude and jump rate are determined in terms of the KM coefficients, as described in Chap. 12. For vanishing KM coefficients $D^{(n)}(x_r, r)$ with $n \geq 3$, Eq. (15.18) will reduce to the Langevin dynamics that describes diffusive increment fluctuations in scale.

15.5 Multipliers

Let us define the multiplier $\mathcal{W}(\lambda_1, \lambda_2)$, where $\lambda_i = \ln(L/r_i)$ is a logarithmic scale, r_i is the lag of the increment $x_{r_i} = x(t + r_i) - x(t)$, and L is the correlation time scale of time series. The basic idea is that for a fixed t, the fluctuations at scales λ_1 and λ_2 are related through the cascading rule,

$$x_{\lambda_2}(t) \stackrel{\mathrm{d}}{=} \mathcal{W}(\lambda_1, \lambda_2)\, x_{\lambda_1}(t), \quad \forall \lambda_1, \lambda_2 \geq 0 \tag{15.19}$$

where $\mathcal{W}(\lambda_1, \lambda_2)$ is a random variable and $\stackrel{\mathrm{d}}{=}$ stands for equality of the distributions. In terms of scales r_i the multiplier $\mathcal{W}$ is $\mathcal{W}_r(r_1, r_2)$. Assuming that the process is homogeneous in scale leads to the conclusion that $\mathcal{W}_r(r_1, r_2)$ is a function of the ratio (r_1/r_2). The scaling $r_2 = \mu\, r_1$ is a translational symmetry in logarithmic scale, i.e., $\ln r_1 \to \ln r_1 + \ln \mu$. Iterating Eq. (15.19) forces implicitly the random variable $\mathcal{W}$ to follow a log infinitely-divisible law. Therefore, $\ln \mathcal{W}(\lambda_1, \lambda_2)$ must be an infinitely divisible random variable [42–46][1] (see problem 15.5). We note that the cascade process (15.19) is a Markov process because $\mathcal{W}$ connects the fluctuations in two succeeding scales.

Let us determine the explicit stochastic expression of the multiplier $\mathcal{W}(\lambda_1, \lambda_2)$ in Eq. (15.19) for a large class of stochastic time series in terms of the Wiener process. Consider the FP Eq. (15.12). By changing the variable $r \to \lambda = \ln(L/r)$, the PDF of the increment x_λ obeys the following equation

$$\frac{\partial}{\partial \lambda} p(x_\lambda, \lambda) = \left\{ -\frac{\partial}{\partial x_\lambda} D^{(1)}(x_\lambda, \lambda) + \frac{\partial^2}{\partial x_\lambda^2} D^{(2)}(x_\lambda, \lambda) \right\} p(x_\lambda, \lambda),$$

where $D^{(n)}(x_\lambda, \lambda) \equiv r D^{(n)}(x_r, r)$. This FP evolution equation is interpreted as if the increment x_λ is evaluated in "scale" λ (logarithmic length scale) by the (Itô) Langevin equation,

$$\frac{\partial x_\lambda}{\partial \lambda} = D^{(1)}(x_\lambda, \lambda) + \sqrt{2D^{(2)}(x_\lambda, \lambda)}\,\Gamma(\lambda) \tag{15.20}$$

---

[1]Let T be an arbitrary nonempty set. A process $x = \{x_t\}_{t \in T}$ is said to be an *infinitely divisible stochastic process* if for any $t_1, \ldots, t_n \in T$ the random vector

$$(x_{t_1}, \ldots, x_{t_n})$$

has an infinitely divisible distribution. A probability distribution p is infinitely divisible if for every $n \geq 1$ there exists a probability distribution p_n such that

$$p = \underbrace{p_n * \cdots * p_n}_{\text{n times}} .$$

Gaussian, Poisson, compound Poisson, geometric, exponential, gamma, Cauchy, etc. are the examples of infinitely divisible distributions.

where $\Gamma(\lambda)$ is a Gaussian white noise, and the diffusion term acts as a multiplicative noise.

Most of the experimental data with scaling behavior have drift and diffusion coefficients, given by [3, 16], $D^{(1)}(x_r, r) \simeq -H\, x_r/r \equiv -\alpha_1\, x_r/r$ and $D^{(2)}(x_r, r) \simeq b\, x_r^2/r \equiv \alpha_2\, x_r^2/r$, with positive constants α_1 and α_2 (with very small $D^{(n)}(x_r, r)$, for $n \geq 3$). In log scale λ, the corresponding drift and diffusion coefficients are

$$D^{(1)}(x_\lambda) = -\alpha_1 x_\lambda \qquad D^{(2)}(x_\lambda) = \alpha_2 x_\lambda^2 \quad .$$

Therefore, the resulting process can be approximated by the well-known Black–Scholes multiplicative process; see Sect. (8.3). Using the Itô prescription, one deduces that (see Eq. (8.21))

$$x_{\lambda_1}(t) = \mathcal{W}(\lambda_1, \lambda_2) x_{\lambda_2}(t). \tag{15.21}$$

The multiplier $\mathcal{W}(\lambda_1, \lambda_2)$ is then derived easily in terms of α_1 and α_2 and the Wiener process W at two logarithmic scales as [47] (see Eq. (8.21)),

$$\mathcal{W}(\lambda_1, \lambda_2) = \exp\left[(-\alpha_1 - \alpha_2)(\lambda_1 - \lambda_2) + \sqrt{2\alpha_2}\, [W(\lambda_1) - W(\lambda_2)] \right] \tag{15.22}$$

which is a positive function of scales λ_1 and λ_2.

Having the first and the second KM coefficients, one can solve the corresponding FP equation [48]. To solve the equation one needs to have the shape of the PDF at large scales. Assuming that the PDFs have Gaussian shape at large scale, $\lambda = \ln(L/r) \to 0$ (i.e. $r \to L$), then the FP equation yields the change of its shape when approaching small scales $r \to 0$ or $\lambda \to \infty$, and generally give rise to the non-Gaussian behavior [16].

15.6 Scale Dependent Solution of Fokker–Planck Equation

The FP equation when $D^{(1)}(x_\lambda) = -\alpha_1 x_\lambda$ and $D^{(2)}(x_\lambda) = \alpha_2 x_\lambda^2$, is given by;

$$\frac{\partial}{\partial \lambda} p(x_\lambda, \lambda) = \left\{ -\frac{\partial}{\partial x_\lambda}(-\alpha_1 x_\lambda) + \frac{\partial^2}{\partial x_\lambda^2}(\alpha_2 x_\lambda^2) \right\} p(x_\lambda, \lambda). \tag{15.23}$$

Using the relation $\frac{\partial}{\partial x_\lambda} x_\lambda = x_\lambda \frac{\partial}{\partial x_\lambda} + 1$, Eq. (15.23) can be written as,

$$\frac{\partial}{\partial \lambda} p(x_\lambda, \lambda) = (\alpha_1 + 2\alpha_2)\, p(x_\lambda, \lambda)$$

$$+ (\alpha_1 + 3\alpha_2)\left(x_\lambda \frac{\partial}{\partial x_\lambda} p(x_\lambda, \lambda) \right) + \alpha_2 \left(x_\lambda \frac{\partial}{\partial x_\lambda} \right)^2 p(x_\lambda, \lambda). \tag{15.24}$$

Noting the fact that the operators $(x_\lambda \frac{\partial}{\partial x_\lambda})$ and $(x_\lambda \frac{\partial}{\partial x_\lambda})^2$ are commuting allows us to write $p(x_\lambda, \lambda)$ as follows,

$$p(x_\lambda, \lambda) = e^{\gamma_0(\lambda)} e^{\gamma_1(\lambda)\left(x_\lambda \frac{\partial}{\partial x_\lambda}\right)} e^{\gamma(\lambda)\left(x_\lambda \frac{\partial}{\partial x_\lambda}\right)^2} p(x_\lambda, \lambda = 0), \tag{15.25}$$

which is also valid for λ-dependent α_1 and α_2 and here, $\gamma_0(\lambda) = \int_0^\lambda (\alpha_1(\lambda') + 2\alpha_2(\lambda'))d\lambda'$, $\gamma_1(\lambda) = \int_0^\lambda (\alpha_1(\lambda') + 3\alpha_2(\lambda'))d\lambda'$, and $\gamma(\lambda) = \int_0^\lambda \alpha_2(\lambda')d\lambda'$. Also $p(x_\lambda, \lambda = 0)$ is the PDF of incement at scale L.

Changing the variable to $\ln(x_\lambda) = v$ and defining $H(\gamma, \lambda) = e^{\gamma(\partial_v)^2} p(e^v, 0)$, one can show that H satisfies the following equation:

$$\frac{\partial H}{\partial \gamma} = \frac{\partial^2 H}{\partial v^2}. \tag{15.26}$$

The solution of Eq. (15.26) is given by,

$$H(v, \gamma) = \int_{-\infty}^{+\infty} \frac{e^{-\frac{s^2}{4\gamma(\lambda)}}}{\sqrt{4\pi\gamma(\lambda)}} p(e^{v-s}, 0) ds. \tag{15.27}$$

Operation of the operator $e^{\gamma_1(\lambda)(x_\lambda \frac{\partial}{\partial x_\lambda})}$ on H will act as a shift in v with value $\gamma_1(\lambda)$, so that,

$$e^{\gamma_1 \partial_v} H(v, \gamma) = \int_{-\infty}^{+\infty} \frac{e^{-\frac{s^2}{4\gamma(\lambda)}}}{\sqrt{4\pi\gamma(\lambda)}} p\left(e^{v-s+\gamma_1(\lambda)}, 0\right) ds. \tag{15.28}$$

Therefore, the solution of the FP equation when $D^{(1)}(x_\lambda) = -\alpha_1 x_\lambda$ and $D^{(2)}(x_\lambda) = \alpha_2 x_\lambda^2$, is given by [40, 47, 49, 50];

$$p(x_\lambda, \lambda) = \frac{e^{\gamma_0(\lambda)}}{\sqrt{4\pi\gamma(\lambda)}} \int_{-\infty}^{+\infty} e^{-\frac{s^2}{4\gamma(\lambda)}} \phi\left(x_\lambda e^{\gamma_1(\lambda)-s}\right) ds. \tag{15.29}$$

One can consider the Gaussian distribution, $\phi(x) \cong e^{-mx^2}$ in the scale $r = L$, which is a reasonable choice [40, 47]. Equation (15.29) enables us to determine the PDF of the increments at scale λ (r) in terms of the PDF in scale $\lambda = 0$ (r = L), where the width of kernel is controlled via $\gamma(\lambda)$, which in turn is a function of the coefficient α_2. For a constant α_2 one finds $\gamma(\lambda) = \alpha_2\lambda$.

Starting with a Gaussian distribution of the increments x_r at scale L and with the drift and diffusion coefficients $D^{(1)}(x_\lambda) = -x_\lambda$ and $D^{(2)}(x_\lambda) = x_\lambda^2$, we numerically calculated the PDFs for various length scales whose plot is presented in Fig. 15.1. These graphs were obtained numerically from the integral representation of the PDF [Eq. (15.29)] from the FP equation. The curves correspond with the scales $L/r \in \{1.5, 2, 5, 10, 20\}$.

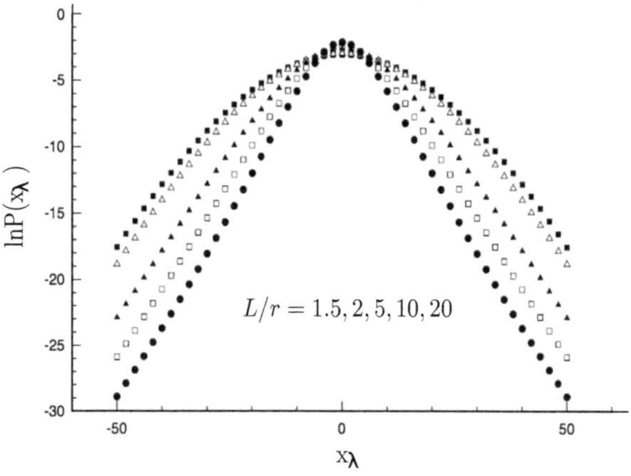

Fig. 15.1 Schematic view of the logarithm of PDF in terms of different length scales. These graphs are numerically obtained from the integral representation of PDF from the Fokker–Planck equation. The plots from top to bottom are correspond with the scales $L/r = 1.5, 2, 5, 10, 20$. Source from [40]

15.7 The Castaing Equation

A model for the description of the increments' PDFs was introduced by Castaing [51–53]. Within the multifractal framework introduced in Sec. 15.3, for each scale the increments' fluctuations in lag r ($x_r = x(t + r) - x(t)$) have non-homogeneous scaling properties such as, for example, the scaling exponent ζ in Eq. (15.4) that has different values in different points of time series. A multifractal process may be interpreted as a superposition of subsets, each characterized by a given scaling exponent, and with a typical increment fluctuations. Each region can then be reasonably assumed to have the same distribution $p_0(x_r, \sigma_r)$ of the field fluctuations, with variable width σ_r (that can be related to the local scaling indices) and weight $L(\sigma_r)$ (that depends on the fraction of points characterized by the same statistics). The PDF of the Castaing model consists of continuous superposition of such distributions, each contributing to the statistics with its appropriate weight. The latter is introduced through the distribution function of the widths σ_r. This leads, for each scale r, to the convolution [51]:

$$p(x_r, r) = \int_0^{\infty} L_\lambda(\sigma_r)\; p_0(x_r, \sigma_r)\; d\; \ln \sigma_r\,. \tag{15.30}$$

Based on empirical large scale PDF shape, a Gaussian distribution for $p_0(x_r, \sigma_r)$ is normally used.

In some time series, such as those for turbulence, the third statistical moment of the increments is not zero, which means that the PDFs of the fluctuations must be skewed. Thus, in order to account for this, a skewness parameter a_s must also be included, so that [51]

$$p_0(x_r, \sigma_r, a_s) = \frac{1}{\sqrt{2\pi}\sigma} \exp\left[-\frac{x_r^2}{2\sigma^2}\left(1 + a_s \frac{x_r/\sigma}{(1 + x_r^2/\sigma^2)^{1/2}}\right)\right]\,. \tag{15.31}$$

The PDF of the variances $L_\lambda(\sigma_r)$ needs theoretical prescription. A log-normal ansatz has been often used, as conjectured in the framework of the multifractal cascade [51–53]

$$L_\lambda(\sigma_r) = \frac{1}{\lambda_r\sqrt{2\pi}} \exp\left[\frac{\ln^2 \sigma_r/\sigma_0}{2\lambda_r^2}\right]. \tag{15.32}$$

Such a choice is justified by assuming that the cascade process is the result of multifractal fragmentation, giving rise to a multiplicative hierarchy of sub-structures (see, for example, problem 15.7 below). By assuming a random distribution of the multipliers, the central-limit theorem suggests a log-normal distribution for $L_\lambda(\sigma_r)$. In Eq. (15.32), for $\lambda_r^2 \to 0$ the log-normal PDF is a δ function, so that the convolution (15.30) gives a Gaussian of width σ_0, the most probable value of σ_r [54]. As λ_r^2 increases, the convolution includes more and more values of σ_r, and the PDF's tails are enhanced. Therefore, the scaling of the parameter λ_r^2 controls the shape of the PDF's tails, and describes the deviation from the distribution in large scale.

For a time series with zero skewness, Eq. (15.30) has the following explicit expression,

$$P(x_r, r) = \frac{1}{2\pi\lambda_r} \int_0^\infty \frac{d\sigma}{\sigma^2} \exp\left[-\frac{x_r^2}{2\sigma^2}\right] \exp\left[-\frac{\ln^2(\sigma/\sigma_0)}{2\lambda_r^2}\right] \tag{15.33}$$

In this equation the PDF is considered as a continuous superposition of Gaussian distributions with different standard deviations, which are weighted by a log-normal distribution function. The shape of the resulting PDF is determined by the two parameters λ_r and σ_0. Here, σ_0 fixes the median of the log-normal function, whereas λ_r^2 determines mainly the shape of the distribution and, accordingly, is called the shape parameter. The limit $\lambda_r \to 0$ represents Gaussian distributions, whereas for positive λ_r non-Gaussian (intermittent) distributions are obtained. Having the values of $\langle x_r^2 \rangle$ enables us to eliminate σ_0 from Eq. (15.33). By multiplying Eq. (15.33) in x_r^2 and integrating from $-\infty$ to ∞ we find [55]

$$\sigma_0^2 = \langle x_r^2 \rangle \exp\left[-2\lambda_r^2\right]. \tag{15.34}$$

Therefore, we have only one unknown shape parameter (the Castaing parameter) λ_r^2 in Eq. (15.33). The dependence of the Castaing parameter λ_r^2 on the scale r is an interesting quantity for characterisation of small-scale intermittent behaviour of time series [56, 57]. The standard procedure to estimate λ_r^2 is to use optimisation, such as, for example, the χ^2 method, in order to determine the most accurate parameter that fits the estimated $p(x_r, r)$ from data [58].

Another method of estimation of λ_r^2 is based on the flatness of the increments' statistics. Define the flatness or kurtosis as $f(r) = S_4(r)/S_2(r)^2$ (where $S_{2k} = \langle [x(t+r) - x(t)]^{2k} \rangle$), which is a measure of non-Gaussianity and quantifies the deviations of the ratio of the fourth-order statistical moment and its Gaussian values, i.e. three times the squared second-order moment of a stochastic variable. It can be shown that the Castaing parameter λ_r^2 can be determined from the flatness of the increments at scale r by [59]

$$\lambda_r^2 = \ln[f(r)/3].$$

The third method of estimating λ_r^2 is in terms of the statistical moments of the absolute values of the increments x_r, i.e., $\langle |x_r|^q \rangle$. For normalised increments, i.e., $\langle x_r^2 \rangle = 1$, from Eq. (15.34) one finds, $\sigma_0^2 = \exp\left[-2\lambda_r^2\right]$. Therefore, we find [58]

$$\langle |x_r|^q \rangle = \int_{-\infty}^{\infty} |x_r|^q \; p(x_r, r)\, dx_r = \frac{2^{\frac{q}{2}} \Gamma\left(\frac{q+1}{2}\right)}{\sqrt{\pi}} \exp\left\{\frac{q(q-2)\lambda_r^2}{2}\right\} \tag{15.35}$$

where $\Gamma(\cdots)$ is the gamma function, and $q > -1$. From Eq. (15.35) we obtain an estimate for λ_r^2, $\hat{\lambda}_{r_q}^2$ the Castaing parameter

$$\hat{\lambda}_{r_q}^2 = \frac{2}{q(q-2)}\left[\ln\left(\frac{\sqrt{\pi}\langle |x_r|^q \rangle}{2^{\frac{q}{2}}}\right) - \ln\Gamma\left(\frac{q+1}{2}\right)\right] \tag{15.36}$$

where $q \neq 0, 2$. In practice, $\langle |x_r|^q \rangle$ is calculated from a zero-mean unit-variance x_r series. We expect that for arbitrary values of q one finds a constant $\hat{\lambda}_{r_q}^2$ (independent of q). In [58] it is shown that one can numerically compare the root mean-squared errors (RMSE) of $\hat{\lambda}_{r_q}^2$ for various values of q and select the optimal one. The optimum values of q depends on the number of available data points. For example, for $n = 10^6$ data points all values of $q \in (-1, 1)$ (excluding $q = 0$) lead to a constant $\hat{\lambda}_{r_q}^2$. Therefore, it will be safe to choose, for example, $q = 0.5$ and estimate $\hat{\lambda}_{r_q}^2$ from Eq. (15.36) [60].

Any non-vanishing value of $\hat{\lambda}_{r_q}^2$ will be a measure of non-Gaussian fluctuations in scale r. Any estimate of ${\lambda_r}^2$ from any of the aforementioned methods will be function of the scale r. For a complex intermittent time series, as one moves down the cascade process from the large to small scales, one expects the statistics to increasingly deviate from Gaussianity. Note that, a non-Gaussian PDF with fat tails on small scales indicates an increased probability of the occurrence of short-time extreme fluctuations. Therefore, we expect ${\lambda_r}^2$ be a decreasing function of r.

Precise relation between the Castaing equation, Eq. (15.33), and the solution of FP equation Eq. (15.29), in the Friedrich–Peinke approach (which is a data-driven method based on estimation of the KM coefficients from time series) is not known. Assuming, however, that both distributions have finite second- and fourth-order structure functions with scaling behaviour like Eq. (15.4) and have same flatnesses, using the relation $\lambda_r^2 = \ln[f(r)/3]$ gives us the Castaing parameter (one can also choose to have the same statistical moments of the absolute value of the increments of order q),

$$\lambda_r^2 = a_0 + [\zeta(4) - 2\zeta(2)]\ln(r) = a_0 - 8\alpha_2 \ln(r) \tag{15.37}$$

where $\alpha_2 \geq 0$ is in the diffusion coefficient, $D^{(2)}(x_r) = \alpha_2 x_r^2/r$. This gives a logarithmic dependent λ_r^2 as function of scale r. Here, we used the structure functions in the FP approach that are given by, $S_n(r) \equiv \langle x_r^n \rangle = \langle [x(t+r) - x(t)]^n \rangle = C_n r^{\zeta(n)}$, with the scaling exponents, $\zeta(n) = n\alpha_1 - \alpha_2 n(n-1)$ [Eq. (15.17)] and C_n is the amplitude of structure function. In Eq. (15.37) the non-universal parameter a_0 is $\ln(C_4/3C_2^2)$. Logarithmic dependence (with minus factor) of the Castaing parameters on the scale r has been observed in analysing time series of turbulence and other complex time series [16].

At the level of FP equation, the Castaing and Friedrich–Peinke approaches predict quadratic nonlinearity for the scaling exponents $\zeta(n)$. However, having the KM coefficients, such as $D^{(n)}(x_r) = \alpha_n x_r^n/r$, as predicted by some authors [34, 35, 40, 41], the Friedrich–Peinke approach provide functionality of $\zeta(n)$ for any order n. To determine $P(x_r, r)$, one should solve the KM equation.

Finally, we remark that, although Eq. (15.33) is equivalent to that for a log-normal cascade model, originally introduced to study fully-developed turbulence [51–53, 61, 62], it also describes approximately the non-Gaussian PDFs observed in a wide range of other phenomena, such as the foreign exchange markets [56, 63–66], heartbeat interval fluctuations [67], and seismic time series [57, 68].

In practice, for given time series, one can estimate the scale-dependence of the Castaing parameter λ_r^2, which provide a measure to study non-Gaussianity of the fluctuation statistics in different scales r. Moreover, we can fix the scale r at a certain r_* a priori, and by moving with a given window size estimate $\lambda_{r_*}^2(t)$ and determine the local temporal dependance of the Castaing parameter.

Analysis of the fluctuations of the increments of the financial index dynamics and seismic time series for the Earth's velocity reveals a pronounced change of the shapes of the PDF of the series increments, henceforth $\lambda_{r_*}^2(t)$, close to an extreme event, such as the black Monday in October 1987 and earthquake [57, 63, 68], as shown in Fig. 15.2. Although every increasing $\lambda_{r_*}^2(t)$ does not necessarily mean an extreme event will occur [57, 63], but if the same characteristics can be observed in other time series, the approach may be applicable to quantitative risk evaluation in complex systems; see also [69].

15.8 Multiscale Correlation Functions

In [70–72] it is proposed that it would be more natural for scale-invariant series (such as flow velocity in turbulence, order parameter fluctuations in critical phenomena, etc.) to look at the correlations in the fluctuations of a time series increments at various scales:

$$\mathcal{F}_n(t|r_1, r_2, \ldots, r_n) = \langle x_{r_1}(t) x_{r_2}(t) \cdots x_{r_n}(t) \rangle. \tag{15.38}$$

All the scales r_i are within the region $r_i \leq L$, where L is correlation time (or length) scale and $x_{r_i} = x(t + r_i) - x(t)$. Writing $\mathcal{F}_n(t|r_1, r_2, \ldots, r_n)$ in terms of the structure

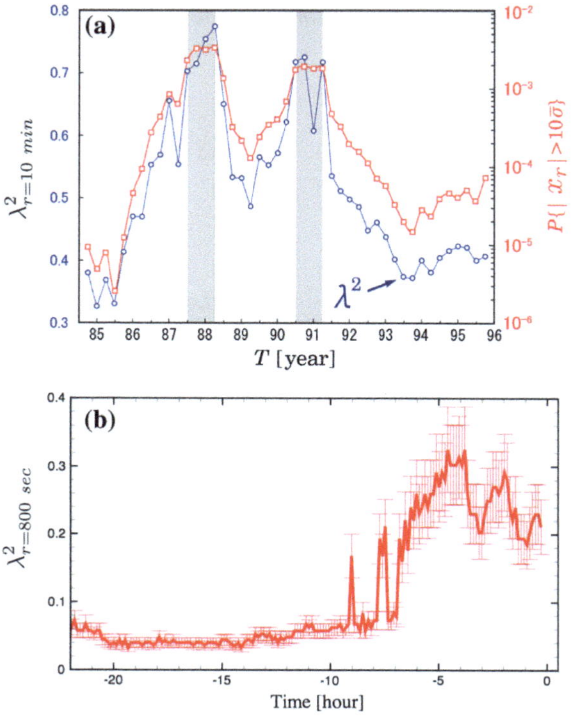

Fig. 15.2 **Temporal variations of Castaing parameter**. **a** (circles) The temporal dependence of the $\lambda^2_{r=10\ min}$ over a one-year time span (window size) (5×10^4 with sample rate 2min) of S&P 500 index time series $Z(t)$ over the period 1984–1995. Increment data are log-return $x_r = \log Z(t+r) - \log Z(t)$, and scale $r = 10$min has been chosen to estimate time dependent Castaing parameter. (squares) The probability of a large change, greater than $10\bar{\sigma}$, where $\bar{\sigma}$ is the average of standard deviation σ of log-returns over the period 1984–1995, and the probability is numerically estimated from the value of $\lambda^2_{r=10\ min}$. The first gray region contains black Monday in October 1987 and the second gray region contains data in the years 1990 and 1991, (Iraq's attack on Kuwait, which began in August 1990, and the Persian Gulf War 1991). Source from [63]. **b** The temporal dependence of the $\lambda^2_{r=800\ sec}$ estimated in each window with size 1 h (data sampled with frequency 100 Hz). Data are Earth's vertical velocity $V_z(t)$, collected by broadband stations. Increment data are $x_r = V_z(t+r) - V_z(t)$, and scale $r = 800$ sec has been chosen to estimate time dependent Castaing parameter. Data are belong to May 21, 2003, Mw = 6.8 event in Boumerdes (Algeria), detected in Ibiza (Balearic Islands). The earthquake has been occurred at time $t = 0$. Similar results has been obtained for many earthquake in several regions around the world. Source from [57, 68]

function of each increment is known as the *fusion rules* [70–74]. The fusion rules describe the asymptotic properties of the n-point correlation functions, when some of the coordinates approach one, say p of coordinates [75–77] . The fusion rules have been tested experimentally *only for turbulence data*, and good agreement between experiment and theory has been observed [47, 49, 78, 79]. For given $p < n$ pairs of the coordinates of a time series increments merge with the typical separations $r_i \sim r$

for $i \leq p$, while the remaining separation is on the order of R, such that R belongs to the region $r_i \leq L$. The fused multiscale correlation is defined by

$$\mathcal{F}_{p+q}(r,R) \equiv \left\langle [x(t+r)-x(t)]^p[x(t+R)-x(t)]^q \right\rangle$$
$$\equiv \left\langle [x_r]^p[x_R]^q \right\rangle \tag{15.39}$$

where $0 \ll r \ll R \ll L$. Here, we show that the multiscale correlation functions for a wide class of processes can be written in terms of structure functions as [16, 79]

$$\mathcal{F}_{p+q}(r,R) \sim S_p(r)S_{p+q}(R)/S_p(R). \tag{15.40}$$

where $S_k(l) = \langle x_l{}^k \rangle$. We recall that for the Wiener processes (or any process with independent increments), increment independence implies that, $\mathcal{F}_{p+q}(r,R)|_{Wiener} \sim S_p(r)S_q(R)$.

Proof

Suppose that one is able to determine all the KM coefficients and, therefore, the solution of Eq. (15.9) for $p(x_r, r)$ can obviously be written as a scale-ordered exponential [47, 80],

$$p(x_r,r) = \mathcal{T}\left(e_+^{\int_{r_0}^{r} dr' \mathcal{L}_{KM}(x_{r'},r')}\right) p(x_0,r_0) \tag{15.41}$$

where $p(x_0, r_0)$ is the PDF of x_{r_0} at scale r_0 and $\mathcal{L}_{KM}$ is the KM evolution operator. Using the properties of the scale-order $\mathcal{T}$ exponentials, the conditional probability density will satisfy the Chapman–Kolmogorov equation [47]. Similarly, the conditional PDF of the increments in two scales is written as a scalar-ordered operator as

$$p(x_1,\lambda_1|x_2,\lambda_2) = \mathcal{T}\left(e_+^{(\int_{\lambda_2}^{\lambda_1} d\lambda \mathcal{L}_{KM}(x_1,\lambda))}\right)\delta(x_1-x_2)$$

where we define $\lambda_1 = \ln(L/r)$ and $\lambda_2 = \ln(L/R)$ [47]. Explicitly, we examine the behavior of $\mathcal{F}_{p+q}(\lambda_1, \lambda_2)$ defined in Eq. (15.39)

$$\mathcal{F}_{p+q}(\lambda_1,\lambda_2) = \langle x^p(\lambda_1)x^q(\lambda_2)\rangle$$
$$= \int dx_1 dx_2 p(x_1,\lambda_1|x_2,\lambda_2)p(x_2,\lambda_2)x_1^p x_2^q$$
$$= \int dx_1 dx_2 p(x_2,\lambda_2)x_1^p x_2^q \left(e^{-(\lambda_1-\lambda_2)\mathcal{L}_{KM}(x_1)}\delta(x_1-x_2)\right)$$

$$
\begin{aligned}
&= \int dx_1 dx_2 x_2^q \delta(x_1 - x_2) p(x_2, \lambda_2) \left(e^{-(\lambda_1-\lambda_2)\mathcal{L}_{KM}^{\dagger}(x_1)} x_1^p\right) \\
&= \int dx_1 dx_2 x_2^q \delta(x_1 - x_2) p(x_2, \lambda_2) \left(e^{-(\lambda_1-\lambda_2)\zeta(p)} x_1^p\right) \\
&= \int dx_2 x_2^{q+q} p(x_2, \lambda_2) \left(\frac{r}{R}\right)^{\zeta(p)} \\
&\sim \frac{S_p(r)}{S_p(R)} S_{p+q}(R).
\end{aligned} \tag{15.42}
$$

Here, $\mathcal{L}_{KM}^{\dagger}$ is adjoint KM operator [80], defined as

$$
\mathcal{L}_{KM}^{\dagger}(x_1, r_1) = \sum_{n=1}^{\infty} D^{(n)}(x_1, r_1)(\partial/\partial x_1)^n.
$$

In Eq. (15.42) we also used the relation $\mathcal{L}_{KM}^{\dagger}(x_1)x_1^p = \zeta(p)x_1^p$ [34, 35, 47].[2]

The limitations of Eq. (15.42) has been discussed in [79], where they checked the validity of the cascade assumption, i.e. Eq. (15.19) in the context of turbulence. It was found that whenever a simple scaling ansatz based on uncorrelated multiplicative processes [79] for turbulence is not prevented by symmetry arguments, the multiscale correlations are in good agreement with the fusion rule prediction, even if strong corrections due to subleading terms are seen for small-scale separation $r/R \sim \mathcal{O}(1)$.

Problems

15.1 Increment statistics

Show that the joint $(n+1)$-point probability distribution (PDF) of the process x can be rewritten without loss of information as a joint $(n+1)$-point PDF of n increments and the statistics of the reference value x_n,

$$
\begin{aligned}
& p(x_0, t_0; \ldots; x_n, t_n) dx_0 \cdots dx_n \\
= \;& p(q_0, r_0; q_1, r_1; \ldots, ; q_{n-1}, r_{n-1}; x_n, t_n) dq_0 \cdots dq_{n-1} dx_n \\
= \;& p(q_0, r_0; q_1, r_1; \ldots, ; q_{n-1}, r_{n-1} | x_n, t_n) p(x_n, t_n) dq_0 \cdots dq_{n-1} dx_n
\end{aligned}
$$

---

[2] We assumed that structure functions, $\langle x_1^p \rangle$, have scaling behaviour with exponents $\zeta(p)$, so that $\frac{\partial \langle x_1^p \rangle}{\partial \lambda} = \zeta(p)\langle x_1^p \rangle$. For example, by assuming that the nth KM coefficient behaves as $D^{(n)}(x_r) \sim x_r^n/r$, we apply the adjoint KM operator $\mathcal{L}_{KM}^{\dagger}$ to x_r^p.

where $q_i := q(t_i, r_i) = x(t_n) - x(t_i) = x(t_n) - x(t_n - r_i)$ and $r_i := t_n - t_i$, with $t_n > t_{n-1} > \cdots > t_0$. See [3] for possible consequences of dependence of the increments' PDF on the value of the time series at point t_n.

15.2 Multiplier

Fill in the details in the derivation of Eq. (15.22).

15.3 Scale-dependent Fokker–Planck equation

Fill in the details in the derivation of Eqs. (15.25), (15.27), and (15.29).

15.4 Statistical moments of the absolute values of the increments

Fill in the details in the derivation of Eq. (15.35).

15.5 Monoscaling (fractality): Linear processes

For the following distributions of the increments x_r

(a) zero-mean Gaussian distribution function with the variance $\sigma^2 = \langle x_r^2 \rangle$ and

(b) *one-parameter* normalizable distribution with finite statistical moments

$$p(x_r) = A \exp\left[-\frac{|x_r|^3}{\alpha^3}\right]$$

where A is a normalization constant and $\alpha > 0$, show that $\langle |x_r|^n \rangle \sim (\langle |x_r|^2 \rangle)^{n/2}$. This means that both distributions have nth-order structure functions with exponent $\zeta(n) = n\zeta(2)/2$ and, thus, they describe monoscaling (linear) processes. Therefore monoscaling doesn't necessarily mean a Gaussian distribution for increments.

15.6 Cascade processes

Assume that there exists some random variable z independent of x_1 and x_2, such that

$$x_2 \stackrel{\mathrm{d}}{=} z\, x_1.$$

where $\stackrel{\mathrm{d}}{=}$ stands for the equality of distributions.

(a) Denoting by $p_z(z)$ the probability density function of z, show that

$$p(x_2|x_1) = \frac{1}{|x_1|} p_z\left(\frac{x_2}{x_1}\right).$$

(b) If x_1 and x_2 have zero mean, unit variance independent Gaussian random variables, show $z = x_2/x_1$ has the standard Cauchy distribution, i.e., $p(z) = 1/\left(\pi(z^2+1)\right)$ with $-\infty < z < \infty$.

15.7 Cascade processes

Assume that there exists positive independent and identically distributed variables z_m that relate the fluctuations of stochastic quantities at two scales r_m and r_{m-1} as (see, for example, [81])

$$x_{r_m} = z_m \; x_{r_{m-1}} \quad .$$

(a) Show that by iterating the cascade process for x_{r_m},

$$\log\left(x_{r_m}/x_{r_0}\right) = \sum_{i=1}^{m} \log z_i .$$

(b) By applying the central-limit theorem, show that $p(x_{r_m})$ has a log-normal distribution

$$p(x_{r_m})dx_{r_m} = \frac{1}{\Lambda\sqrt{2\pi}} \exp\left(-\frac{\ln(x_{r_m}/x_{r_0})}{2\Lambda^2}\right)$$

where $\Lambda^2 = \langle(\log(x_{r_m}/x_{r_0}))^2\rangle = m\langle(\log(z))^2\rangle$ (for large m).

15.8 Cascade processes

Suppose that the increments on scale sl are related to scale l obey the following cascade rule

$$x_{sl} \stackrel{\mathrm{d}}{=} \mathcal{W}_s \; x_l .$$

where $\stackrel{\mathrm{d}}{=}$ stands for equality of distributions. Here, $\mathcal{W}_s$ is a positive random factor independent of x_l, which depends only on the scale ratio s.

Consider a coarse scale L such that $l = sL$. Denote $G_s(u)$ as the PDF of $\ln(\mathcal{W}_s)$ and use the cascade rule to show that the PDF of the increments on scale l can be written in terms of the PDF on scale L as (Castaing equation),

$$p_l(x_l, l) = \int G_s(u) \, e^{-u} \; p_L(e^{-u}x_l, L)du \quad .$$

15.9 Cascade processes

One of the simplest candidates for increment fluctuations in cascade processes is, $x_r(t) = \zeta_r(t)\exp[\omega_r(t)]$, where ζ_r and $\omega_r(t)$ are independent Gaussian variables with zero mean and variances σ_ζ^2 and σ_ω^2.

(a) Show that the probability density function of x_r is given by following Castaing equation [63]

$$p(x_r, r) = \int F_r\left(\frac{x_r}{\sigma}\right)\frac{1}{\sigma}G_r(\ln\sigma)d\ln\sigma \; ,$$

where $F_r(\zeta_r)$ and $G_r(\omega_r)$ are both Gaussian with zero mean and variances σ_r^2 and λ_r^2, respectively. Here, $G_r(\ln \sigma) = (\sqrt{2\pi}\lambda_r)^{-1} \exp(-\ln^2 \sigma / 2\lambda_r^2)$.

(b) Show that $p(x_r, r)$ converges to a Gaussian distribution when $\lambda_r \to 0$.

References

1. R. Friedrich, J. Peinke, Phys. Rev. Lett. **78**, 863 (1997)
2. M. Anvari, K. Lehnertz, M.R. Rahimi Tabar, J. Peinke, Sci. Rep. **6**, 35435 (2016)
3. J. Peinke, M.R. Rahimi Tabar, M. Wächter, Annu. Rev. Condens. Matter Phys. **10**, (2019)
4. M. Wächter, A. Kouzmitchev, J. Peinke, Phys. Rev E **70**, 055103(R) (2004)
5. U. Frisch, *Turbulence: the legacy of AN Kolmogorov*. (Cambridge University Press, Cambridge, England, 1995)
6. R. Lopes, N. Betrouni, Fractal and multifractal analysis: a review. Med. Image Anal. **13**, 634 (2009)
7. J.F. Muzy, E. Bacry, A. Arneodo, Phys. Rev. Lett. **67**, 3515 (1991)
8. G. Parisi, U. Frisch, A multifractal model of intermittency. Turbulence and Predictability in Geophysical Fluid Dynamics, vol. 84, ed. by M. Ghil, R. Benzi, G. Parisi (North Holland, 1985)
9. U. Frisch, From global scaling, á la Kolmogorov, to local multifractal in fully developed turbulence, Proc. Roy. Soc. London, ser. A**434**, 89 (1991)
10. A. Marshak, A. Davis, R. Cahalan, W. Wiscombe, Bounded cascade models as non-stationary multifractals. Phys. Rev. E **49**, 55 (1994)
11. A.L. Barabási, T. Vicsek, Phys. Rev. A **44**, 2730 (1991)
12. P. Grassberger, Generalized dimensions of strange attractors. Phys. Lett. A **97**, 227 (1983)
13. H.G.E. Hentschel, I. Procaccia, The infinite number of generalized dimensions of fractals and strange attractors. Phys. D **8**, 435 (1983)
14. K.R. Sreenivasan, Fractals and multifractals in fluid turbulence. Annu. Rev. Fluid. Mech. **23**, 539 (1991)
15. E. Ott, *Chaos in Dynamical Systems* (Cambridge University Press, 1993)
16. R. Friedrich, J. Peinke, M. Sahimi, M.R. Rahimi Tabar, Phys. Rep. **506**, 87 (2011)
17. G. Valenza, H. Wendt, K. Kiyono, J. Hayano, E. Watanabe, Y. Yamamoto, P. Abry, R. Barbieri, in IEEE Transactions on Biomedical Engineering, (2018), https://doi.org/10.1109/TBME.2018.2797158.
18. J. Feder, *Fractals* (Plenum Press, New York, 1988)
19. H.O. Peitgen, H. Jürgens, D. Saupe, *Chaos and Fractals* (Springer-Verlag, New York, 1992)
20. E. Bacry, J. Delour, J.F. Muzy, Phys. Rev. E **64**, 026103 (2001)
21. M.R. Rahimi Tabar, Disordered Systems and Logarithmic Conformal Field Theory. Int. J. Mod. Phys. A **18**, 4703 (2003)
22. E. Koscielny-Bunde, A. Bunde, S. Havlin, H.E. Roman, Y. Goldreich, H.J. Schellnhuber, Phys. Rev. Lett. **81**, 729 (1998)
23. A.-L. Barabási, H.E. Stanley, *Fractal Concepts in Surface Growth* (Cambridge University Press, New York, 1995)
24. M. Marsili, A. Maritan, F. Toigo, J.R. Banavar, Rev. Mod. Phys. **68**, 963 (1996)
25. F. Schmitt, Y. Huang, *Stochastic Analysis of Scaling Time Series: From Turbulence Theory to Applications* (Cambridge University Press, Cambridge, 2016)
26. A. Jiang, W. Xie, W. Zhou, D. Sornette, Multifractal analysis of financial markets, arXiv:1805.04750
27. B. Castaing, Y. Gagne, E. Hopfinger, Phys. D **46**, 177 (1990)
28. R. Benzi, L. Biferale, G. Paladin, A. Vulpiani, M. Vergassola, Phys. Rev. Lett. **67**, 2299 (1991)
29. U. Frisch, D. Sornette, J. Phys. I France **7**, 1155 (1997)

30. C. Meneveau, K.R. Sreenivasan, Phys. Rev. Lett. **59**, 1424 (1987)
31. Z.-S. She, E. Lévêque, Phys. Rev. Lett. **72**, 336 (1994)
32. R. Friedrich, T. Galla, A. Naert, J. Peinke, Th. Schimmel, Disordered structures analysed by the theory of Markov processes, in a Perspective Look at Nonlinear Media, volume 503 of Lecture Notes in Physics, ed. J. Parisi, St. C. Müller, W. Zimmermann (Springer Verlag, Berlin, 1998), pp. 313–326
33. J. Friedrich, R. Grauer, Generalized description of intermittency in turbulence via stochastic methods, (2016), arXiv:1610.04432
34. J. Friedrich, Closure of the Lundgren-Monin-Novikov hierarchy in turbulence via a Markov property of velocity increments in scale, Doctoral thesis, Ruhr-Universität Bochum, Universitätsbibliothek (2017)
35. J. Friedrich, G. Margazoglou, L. Biferale, R. Grauer, Phys. Rev. E **98**, 023104 (2018)
36. H.E. Hurst, R.P. Black, Y.M. Simaika, *Long-term storage: An experimental study* (Constable, London, 1965)
37. F. Shayeganfar, S. Jabbarifarouji, M.S. Movahed, G.R. Jafari, M.R. Rahimi Tabar, Phys. Rev. E **80**, 061126 (2009)
38. A. Eke, P. Herman, L. Kocsis, L.R. Kozak, Physiol. Meas. **23**R1, (2002)
39. E. Koscielny-Bunde, H.E. Roman, A. Bunde, S. Havlin, H.J. Schellnhuber, Phil. Mag. B **77**, 1331 (1998)
40. J. Davoudi, M.R. Rahimi Tabar, Phys. Rev. Lett. **82**, 1680 (1999)
41. V. Yakhot, Phys. Rev. E **57**, 1737 (1998)
42. B. Dubrulle, Phys. Rev. Lett. **73**, 959 (1994)
43. Z.-S. She, E.C. Waymire, Phys. Rev. Lett. **74**, 262 (1995)
44. B. Dubrulle, Eur. Phys. J. B **13**, 1 (2000)
45. P. Chainais, R. Riedi P. Abry, On non-scale-invariant infinitely divisible cascades, in IEEE Transactions on Information Theory, **51**, 1063 (2005)
46. P. Chainais, Infinitely divisible cascades to model the statistics of natural images, in IEEE Transactions on Pattern Analysis and Machine Intelligence **29**, 12, 2105 (2007)
47. J. Davoudi, M.R. Rahimi Tabar, Phys. Rev. E **61**, 6563 (2000)
48. V. Holubec, K. Kroy, S. Steffenoni, Phys. Rev. E **99**, 032117 (2019)
49. M.R. Rahimi Tabar, S. Rouhani, Phys. Lett. A **224**, 331 (1997)
50. A.A. Donkov, A.D. Donkov, E.I. Grancharova, Int. J. Mod. Phys. A **12**, 165 (1997)
51. B. Castaing, Y. Gagne, E.J. Hopfinger, Phys. D **46**, 177 (1990)
52. B. Castaing, Phys. D **73**, 31 (1994)
53. B. Castaing, J. Phys. II France **6**, 105 (1996)
54. L. Sorriso-Valvo, R. Marino, L. Lijoi, S. Perri, V. Carbone, Astrophys. J. **807**, 86 (2015)
55. C. Beck, Europhys. Lett. **64**, 151 (2003)
56. S. Ghashghaie, W. Breymann, J. Peinke, P. Talkner, Y. Dodge, Nature **381**, 767 (1996)
57. P. Manshour, S. Saberi, M. Sahimi, J. Peinke, A.F. Pacheco, M.R. Rahimi Tabar, Phys. Rev. Lett. **102**, 014101 (2009)
58. K. Kiyono, Phys. Rev. E **79**, 031129 (2009)
59. C. Beck, Phys. D **193**, 195 (2004)
60. A. Leövey, T. Lux, Phys. Rev. E **85**, 046114 (2012)
61. B. Jouault, P. Lipa, M. Greiner, Phys. Rev. E **59**, 2451 (1999)
62. J. Cleve, M. Greiner, B.R. Pearson, K.R. Sreenivasan, Phys. Rev. E **69**, 066316 (2004)
63. K. Kiyono, Z.R. Struzik, Y. Yamamoto, Phys. Rev. Lett. **96**, 068701 (2006)
64. H.E. Stanley, V. Plerou, Quant. Fin. **1**, 563 (2001)
65. E. Bacry, J. Delour, J.F. Muzy, Phys. Rev. E **64**, 026103 (2001)
66. G.R. Jafari, M.S. Movahed, P. Noroozzadeh, A. Bahraminasab, M. Sahimi, F. Ghasemi, M.R. Rahimi Tabar, Int. J. Mod. Phys. C **18**, 1689 (2007)
67. K. Kiyono, Z.R. Struzik, N. Aoyagi, F. Togo, Y. Yamamoto, Phys. Rev. Lett. **95**, 058101 (2005)
68. P. Manshour, F. Ghasemi, T. Matsumoto, G. Gómez, M. Sahimi, J. Peinke, A.F. Pacheco, M.R. Rahimi Tabar, Phys. Rev. E **82**, 036105 (2010)

69. T. Rings, M. Mazarei, A. Akhshi, C. Geier, M.R.R. Tabar, K. Lehnertz, Traceability and dynamical resistance of precursor of extreme events. Sci. Rep. **9**, 1744 (2019)
70. G. Eyink, Phys. Lett. A **172**, 355 (1993)
71. G. Eyink, Phys. Rev. E **48**, 1823 (1993)
72. V.S. L'vov, I. Procaccia, Phys. Fluids **8**, 2565 (1996)
73. V.S. L'vov, I. Procaccia, Phys. Rev. Lett. **76**, 2898 (1996)
74. V.S. L'vov, I. Procaccia, Phys.Rev. E **54**, 6268 (1996)
75. M. Khorrami, A. Aghamohammadi, M.R. Rahimi Tabar, Phys. Lett. B **419**, 179 (1998)
76. A. Shafiekhani, M.R. Rahimi Tabar, Int. J. Mod. Phys. A **12**, 3723 (1997)
77. M.R. Rahimi Tabar, Int. J. Mod. Phys. A **18**, 4703 (2003)
78. A.L. Fairhall, B. Druhva, V.S. L'vov, I. Procaccia, K.R. Sreenivasan, Phys. Rev. Lett. **79**, 3174 (1997)
79. R. Benzi, L. Biferale, F. Toschi, Phys. Rev. Lett. **80**, 3244 (1998)
80. H. Risken, *The Fokker–Planck Equation* (Springer, Berlin, 1989)
81. A. Morales, M. Wächter, J. Peinke, Wind. Energy **15**, 391 (2012)

Chapter 16
How to Set Up Stochastic Equations for Real World Processes: Markov–Einstein Time Scale

In Chaps. 16–21 we address a central question in the field of complex systems: *Given a fluctuating (in time or space), sequentially uni- or multi-variant measured set of experimental data (even noisy data), how should one analyse the data non-parametrically, assess their underlying trends, discover the characteristics of the fluctuations, including diffusion and jump parts, and construct stochastic evolution equation for the data?*

16.1 From Time Series to Stochastic Dynamical Equation

In recent years, significant progress has been made in addressing the question *how to reconstruct a stochastic dynamical equation from time series?*, for a class of stochastic processes that can be modeled by Langevin or jump-diffusion equations, including additive as well as multiplicative diffusive fluctuations and jump contribution. Important results have emerged from the analysis of temporal data for such diverse fields as neuroscience, cardiology, finance, economy, surface science, turbulence, seismic time series and epileptic brain dynamics, to name but a few [1]. Furthermore, it has been recognized that, as we have seen in Chap. 15, similar approach can be applied to the data that depend on a length or time scale (lag), such as velocity increments in fully-developed turbulent flow [2], or height increments that characterize rough surfaces [3], etc.

In Chaps. 16–21, we propose a *nonlinear method* for the analysis of the time series, which enables us to reconstruct a stochastic dynamical equation, describing its fluctuations. The main advantage of the method is that all the functions and parameters of the dynamics are determined directly (non-parametrically) from the measured series. The development of such methods has been stimulated by research on *turbulent flows* and *neuroscience* [2, 4].

The original version of this chapter was revised: Belated correction has been incorporated. The correction to this chapter is available at https://doi.org/10.1007/978-3-030-18472-8_24

M. R. Rahimi Tabar, *Analysis and Data-Based Reconstruction of Complex Nonlinear Dynamical Systems*, Understanding Complex Systems,
https://doi.org/10.1007/978-3-030-18472-8_16

A basic ingredient of the approach to the analysis of fluctuating data is the presence of a Markovian property, which can be detected in real systems above a certain time or length scale. This scale is referred to as the *Markov–Einstein* (ME) time scale, and has turned out to be a useful characteristic of complex systems.

A complete stochastic analysis of experimental data, which is generated by the interplay of deterministic dynamics and dynamical noise, has to address the following issues:

- Identification of the order parameters, for instance state variable of a system or its increment in given lag (scale).
- Verifying the Markovianity.
- Extracting the deterministic dynamics.
- Evaluating the properties of the fluctuations, including diffusion and jump contributions.
- Reconstructing the data.
- Checking of self-consistency.

In this Chapter, we address in detail the following issues: (i) Estimation of Markov–Einstein (ME) time or length scale for given measured time series (ii) Estimation of Kramers–Moyal coefficients from data (iii) Reconstruction of stochastic evolution equations from data in terms of the Langevin or jump-diffusion equations. Some important technical details will be given in Chaps. 17–21.

16.2 Markov–Einstein Time (Length) Scale

In this section we introduce the Markov–Einstein time (length) scale of a data set, which is the basic characteristic time (length) scale of the proposed reconstruction method.

In some systems such as hydrodynamic instabilities in the laminar flow regime, such as the Rayleigh-Bénard convection or the Taylor-Couette experiment, thermal fluctuations are usually small and can be neglected [5]. The equations for the order parameters can be derived from the basic equations that characterize the system under consideration close to instabilities [6].

In contrast, in biological systems the impact of noise has been shown to be of great importance and the basic equations are not available. For such cases the order parameter concept yields a top-down approach to complexity [7].

In the top-down approach the analysis of the experimental time series becomes a central issue. Methods of analysis of nonlinear time series (see, e.g., [8]) have been widely applied to analyze complex systems. However, such methods aim at the understanding of deterministic systems and can only be successful if the stochastic forces are weak. Thus, it appears that these methods must be extended in order to include stochastic contributions. The ultimate goal of analysis of nonlinear time series, applied to deterministic systems, is to extract directly the underlying nonlinear dynamical system from the measured time series in the form of a system of stochastic differential equations [8]. The role played by dynamic fluctuations has not been fully appreciated. Instead, noise has usually been considered as a random variable that is additively superimposed on a trajectory generated by a deterministic dynamical system. In this view noise has been usually considered as extrinsic or measurement noise.

For an extension of the methods for the analysis of nonlinear time series to continuous processes or processes with jump discontinuity, one needs to estimate the *Markov–Einstein time (length) scale* from the time series and write down time series–based stochastic differential equations that describe the observed stochastic behavior. In recent years, it has become evident that such an approach is feasible [1]. In fact, noise may help in the estimation of the deterministic ingredients of the dynamics. Due to the dynamical noise the system explores a larger part of the phase space and, thus, measurements of the series yield considerably more information about the dynamics as compared to the purely deterministic case, where the trajectories rapidly converge to attractors, hence providing only limited information [8].

In Chaps. 2–15, we provided complete mathematical tools which enable us to write down the Langevin or jump-diffusion stochastic equations for given time series. Let us start with continuous stochastic Markov processes. As we have seen in Chap. 9, a complex system with continuous stochastic behaviour, can be described in terms of the state variables $x_i(t)$ that obey a set of (Itô) stochastic differential equations of the form

$$\frac{dx_i}{dt} = N_i[x_1, \dots, x_n] + \sum_j g_{ij}(x_1, \dots, x_n)\Gamma_j(t), \tag{16.1}$$

where $N_i[x_1, \dots, x_n]$ is the i-th component of drift vector $\mathbf{D}^{(1)}(\mathbf{x}(t_j), t_j)$ and $\Gamma_j(t)$ are independent zero-mean gaussian white noises. Although the state vector $\mathbf{x}(t)$ of the complex system under consideration is high dimensional, its long time behavior is entirely governed by the dynamics of typically a few order parameters [6, 7, 9, 10],

$$\mathbf{q}(t) = \mathbf{Q}(x_1, \dots, x_n). \tag{16.2}$$

In what follows by $\mathbf{x}(t)$, we mean the multivariate time series itself or order parameter $\mathbf{q}(t)$.

The modeling of stochastic processes with Langevin equation, i.e. Eq. (16.1) or with jump-diffusion processes, the main assumption is the Markov property of measured time series. In general, a given dynamic process with a degree of stochasticity may have a finite Markov–Einstein (ME) time t_M - the minimum time interval over which the data can be considered as a Markov process. Therefore one should first estimate t_M for measured time series.

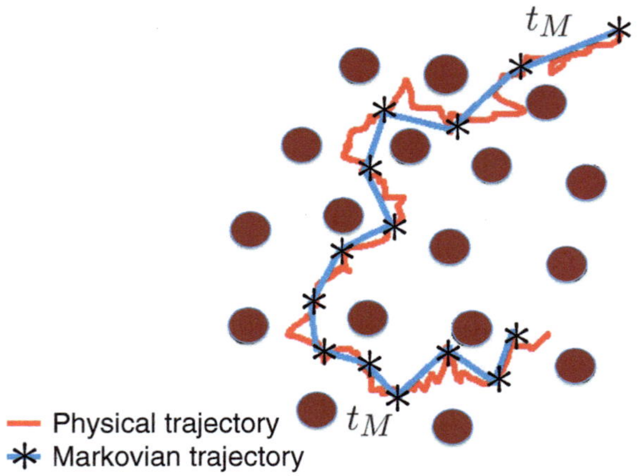

Fig. 16.1 Schematic representation of a physical versus coarse-grained trajectory. The physical trajectory can be in general non-Markov process due to interaction with environment. However one can imagine a coarse-grained trajectory with time step t_M, where resulting process be a Markov trajectory. The time scale t_M is known as the *Markov–Einstein* time scale and one can consider it as constant blackout interval during eyeblinks to watch a real non-Markov trajectory

16.3 Evaluating Markovian Properties

It is a difficult task, in principle, to decide by inspection whether some experimental data possess Markovian properties. There are at least two reasons for this fact. The main point to emphasize is that such properties are usually violated in small time increments $|t - t'| < t_M$, as was pointed out by Einstein [11]. In fact the dynamical noise ($\Gamma(t)$) sources become correlated over small time differences. If we consider Gaussian noise sources, one usually observes an exponential decay of the correlations

$$\langle \Gamma_i(t)\Gamma_j(t')\rangle = \delta_{ij}\frac{1}{t_M}\exp(-|t - t'|/t_M). \tag{16.3}$$

Markovian properties may be expected to hold only for time increments, $|t - t'| > t_M$. Second, microstructure (or measurement) noise can spoil Markovian properties [12]. This steams from the fact that microstructure noise has large impacts on the statistics of measurements on small scales and due to its white noise type behaviour in such scales spoils Markov properties of an underlying process, see also Chap. 17.

Thus, estimation of the Markovian time scale t_M is a necessary step for stochastic data analysis. Several methods have been proposed for testing and detecting the Markov properties. In Fig. 16.1, schematic representation of a physical versus coarse-grained trajectory is given. The physical trajectory can be in general non-Markov process (due to its interactions with the environment around it). However one can imagine a coarse-grained trajectory with time step t_M, where resulting process be a Markov trajectory.

16.4 Methods for Estimation of Markov–Einstein Time or Length Scale

In general, an experimental data set is either stationary or nonstationary. Establishing or rejecting the stationarity of the set can be done in a straightforward manner, by measuring, e.g., the numerical stability (constancy) of the average and variance of the stochastic process $x(t)$ in a moving time window. Such time series are known as *second-order stationary*. In practice if we wish to test whether a given time series is second-order stationary, then can examine it by methods described in [13, 14]. Also one can use the Wavelet transform to study the stationarity of time series [15, 16]. As an alternative method, detrended fluctuation analysis (DFA) of given time series provides a scaling exponent α, for which for $\alpha \geq 1$ and $\alpha < 1$ the corresponding time series are non-stationary and stationary, respectively [17–19].

Non-stationary behaviors can be trends, cycles, time dependent stochastic behavior, etc. In principle for non-stationary processes one should study the ensemble averages. However in real world data, such as stock market index or pattern of cosmic background radiation have unique trajectory. In order to receive consistent, reliable results, the non-stationary data needs to be transformed into stationary data. For given time series with "drifts" it is quite often that one can investigate the "derivative" of the data set, $D(t) \equiv x(t+1) - x(t)$, or its log-returns defined as, $R(t) \equiv \ln[x(t+1)/x(t)]$ (if the data are positive numbers, such as the heart interbeat and the stock price). Frequently, the derivative or returns turn out to be stationary, since a drifting mean value that renders the time series nonstationary is subtracted.

Now we check whether the data follow a Markov process and, if so, estimate the ME time (length) scale t_M. The ME time (length) scale is the minimum time interval (length scale) over which the data can be considered as a Markov process. Here, we present three different methods of estimating the ME time scale.

Method 1: In the first method, the ME time scale is estimated using the least square test. The exact mathematical definition of the Markov process is given by, $p(x_k, t_k|x_{k-1}, t_{k-1}; \cdots; x_1, t_1; x_0, t_0) = p(x_k, t_k|x_{k-1}, t_{k-1})$. Intuitively, the physical interpretation of a Markov process is that, it is a process with almost, no memory; it "forgets" its past. In other words, only the most nearby condition of data, say (x_{k-1}, t_{k-1}), is relevant to the probability of finding the system at a particular state x_k at time t_k. Thus, the ability to predict the value of $x(t)$ at time t is not enhanced by knowing its values in steps prior to the the most recent one. Therefore, an important simplification for a Markov process is that, a conditional multivariate joint PDF can be written in terms of the products of the simple two-point conditional PDFs as,

$$p(x_k, t_k; x_{k-1}, t_{k-1}; \cdots; x_1, t_1; x_0, t_0)$$

$$= \prod_{i=1}^{k} p(x_i, t_i|x_{i-1}, t_{i-1})\, p(x_0, t_0),$$

where we have used the definition of Markov process, i.e.,

$$p(x_k, t_k|x_{k-1}, t_{k-1}; \cdots ; x_1, t_1; x_0, t_0) = p(x_k, t_k|x_{k-1}, t_{k-1}). \tag{16.4}$$

Testing Eq. (16.4) for large values of k is beyond the present computational capability. However, for $k = 3$, i.e., for three points, the relation,

$$p(x_3, t_3|x_2, t_2; x_1, t_1) = p(x_3, t_3|x_2, t_2), \tag{16.5}$$

should hold for any value of t_2 in the interval $t_1 < t_2 < t_3$.

A process is Markov with a ME time scale t_M, if Eq. (16.5) is satisfied for a *certain* time separation or interval. To estimate the ME time scale, we use a fundamental theory of probability that allows us to write any three point PDF in terms of the conditional probability functions as,

$$\begin{aligned} &p(x_3, t_3; x_2, t_2; x_1, t_1) \\ &= p(x_3, t_3|x_2, t_2; x_1, t_1)p(x_2, t_2; x_1, t_1). \end{aligned} \tag{16.6}$$

Using the properties of Markov processes and substituting Eq. (16.6), we obtain,

$$\begin{aligned} &p_{\text{Markov}}(x_3, t_3; x_2, t_2; x_1, t_1) \\ &= p(x_3, t_3|x_2, t_2)p(x_2, t_2; x_1, t_1). \end{aligned} \tag{16.7}$$

We then compute the deviations of p_{Markov} from that given by Eq. (16.6), using the least squares method:

$$\begin{aligned} \chi^2 = \int dx_3 dx_2 dx_1 \times \\ \frac{[p(x_3, t_3; x_2, t_2; x_1, t_1) - p_{\text{Markov}}(x_3, t_3; x_2, t_2; x_1, t_1)]^2}{\sigma^2 + \sigma^2_{\text{Markov}}} \end{aligned} \tag{16.8}$$

where $\sigma^2 + \sigma^2_{\text{Markov}}$ is the sum of the errors of the terms in the numerator. For stationary time series, one then takes, $t_1 = 0$ and $t_2 = t_3/2$ and plots the reduced χ^2, $\chi^2_\nu = \frac{\chi^2}{\mathcal{N}}$ ($\mathcal{N}$ is the number of degrees of freedom), as a function of the time scale $t_3 = 2, 4, \ldots$ (in units of data points). By definition, the ME time scale t_M is that value of t_2 at which χ^2_ν is minimum [20, 21].

Method 2: The second method is based on the direct evaluation of the definition of Markovian properties (16.5), considering the higher-order conditional probability distributions:

$$\begin{aligned} p(x_3, t_3|x_2, t_2; x_1, t_1) &= \frac{p(x_3, t_3; x_2, t_2; x_1, t_1)}{p(x_2, t_2; x_1, t_1)} \\ &= p(x_3, t_3|x_2, t_2). \end{aligned} \tag{16.9}$$

This procedure is feasible if used for large data sets. Due to the different conditioning, both probabilities are typically based on different numbers of events. As an appropriate method to show the similarity of (16.9), the Wilcoxon test (for one dimensional time series) has been proposed [22, 23], see Appendix 1. In principle, higher-order conditional probability distributions should be considered in a similar way, which becomes numerically intractable. However, the validity of relation (16.9) is a strong indication that the data set possesses Markovian properties.

Method 3: A refined method can be based on validation of the Chapman–Kolmogorov equation, i.e., by considering several intermediate times [1]. This method is based on a comparison between the conditional PDF

$$p(x_3, t_3|x_1, t_1) \tag{16.10}$$

taken from the experiment and the one calculated by the Chapman–Kolmogorov equation

$$\tilde{p}(x_3, t_3|x_1, t_1) = \int dx_2 p(x_3, t_3|x_2, t_2) p(x_2, t_2|x_1, t_1) \tag{16.11}$$

where t_2 is an intermediate time $t_1 < t_2 < t_3$. We can compare the resulting two distributions in terms of, for instance, a χ^2-test [20, 24, 25]

$$\chi^2(t_1, t_2, t_3) = \int dx_3 dx_1 \times \frac{[p(x_3, t_3|x_1, t_1) - \tilde{p}(x_3, t_3|x_1, t_1)]^2}{p(x_3, t_3|x_1, t_1) + \tilde{p}(x_3, t_3|x_1, t_1)} \tag{16.12}$$

which easier to implement than (16.8). For stationary processes, one can set again $t_1 = 0$, $t_2 = t_3/2$ and check $\chi^2(t_1 = 0, t_2 = t_3/2, t_3)$ with varying $t_3 = 2, 4, \ldots$. By definition, the Markov–Einstein time scale t_M is that value of $t_2 = t_3/2$ at which $\chi^2(t_1 = 0, t_2 = t_3/2, t_3)$ will be minimum.

Verifying Chapman–Kolmogorov equation for multivariate time series will provide an overall Markov–Einstein time scale t_M of system, which in general could be a complex function of individual Markov–Einstein time scales of the univariate series.

Finally, we note that *Markov–Einstein* time scale is independent of correlation time scale of process. For instance, Wiener process has infinite correlation time scale (it has power-law correlation function and incremental moments) and is Markov process with unit time step.

16.5 Estimation of Drift and Diffusion Coefficients from Time Series

We now describe an operational approach, which allows one to estimate the drift vector and the diffusion matrix from data. We consider the case in which the averages are taken with respect to an ensemble of experimental realizations of the stochastic process under consideration, in order to include nonstationary processes in the discussions. Replacing the ensemble averages by time averages for statistically stationary processes is straightforward. For a *unique* non-stationary trajectory, one can use the methods presented in Chap. 17, to estimate time dependent drift vector and the diffusion matrix from data.

The procedure is as follows:

- The data is represented in a d-dimensional phase space.
- The phase space is partitioned into a set of finite, but small d-dimensional volume elements.
- For each bin (denoted by α), located at point $\mathbf{x}_\alpha$ of the partition we consider the quantity (see Eq. 9.9),

$$\mathbf{x}(t_j+\tau) = \mathbf{x}(t_j) + \mathbf{D}^{(1)}(\mathbf{x}(t_j), t_j)\tau + \sqrt{\tau}\mathbf{g}(\mathbf{x}(t_j), t_j)\boldsymbol{\eta}(t_j) \tag{16.13}$$

where $\mathbf{D}^{(1)}(\mathbf{x}(t_j), t_j)$ is drift vector and $\mathbf{g}(\mathbf{x}(t_j), t_j)$ is related to the diffusion matrix, see below. Since we consider time-dependent processes, this must be done for each time step t_j separately.

16.5.1 Estimation of the Drift Vector

The drift vector assigned to the bin located at $\mathbf{x}_\alpha$ is determined as the small τ-limit of

$$\mathbf{D}^{(1)}(\mathbf{x}_\alpha, t) = \lim_{\tau\to 0} \frac{1}{\tau}\mathbf{K}^{(1)}(\mathbf{x}_\alpha, t, \tau) \tag{16.14}$$

of the conditional moment

$$\mathbf{K}^{(1)}(\mathbf{x}_\alpha, t_j, \tau) = \frac{1}{N_\alpha}\sum_{\mathbf{x}(t_j)\in\alpha}[\mathbf{x}(t_j+\tau) - \mathbf{x}(t_j)]. \tag{16.15}$$

The sum is over all the N_α points contained in the bin α.

Proof The drift vector assigned to the bin α located at $\mathbf{x}_\alpha$ is approximated by the conditional averaged value (from Eq. (16.13))

$$\mathbf{K}^{(1)}(\mathbf{x}_\alpha, t_j, \tau) = \tau\frac{1}{N_\alpha}\sum_{\mathbf{x}(t_j)\in\alpha}\mathbf{D}^{(1)}(\mathbf{x}_j, t_j) + \sqrt{\tau}\frac{1}{N_\alpha}\sum_{\mathbf{x}(t_j)\in\alpha}\mathbf{g}(\mathbf{x}(t_j), t_j)\,\boldsymbol{\eta}(t_j). \tag{16.16}$$

Assuming that $\mathbf{D}^{(1)}(\mathbf{x}, t)$ and $\mathbf{g}(\mathbf{x}, t)$ do not vary significantly within the bin, the second contribution drops out since

$$\frac{1}{N_\alpha}\sum_{\mathbf{x}_j\in\alpha}\boldsymbol{\eta}(t_j) \to 0. \tag{16.17}$$

16.5.2 Estimation of the Diffusion Matrix

The diffusion matrix for stationary process is estimated by the small τ-limit of (see Chap. 9)

$$\mathbf{D}^{(2)}(\mathbf{x}_\alpha, t) = \mathbf{D}^{(2)}(\mathbf{x}_\alpha) = \lim_{\tau\to 0} \frac{1}{\tau}\mathbf{K}^{(2)}(\mathbf{x}_\alpha, \tau) \tag{16.18}$$

of the second conditional moment

$$\mathbf{K}^{(2)}(\mathbf{x}_\alpha, \tau) = \frac{1}{N_\alpha}\sum_j \sum_{\mathbf{x}(t_j)\in\alpha} \left\{[\mathbf{x}(t_j+\tau) - \mathbf{x}(t_j)] - \tau\mathbf{D}^{(1)}(\mathbf{x}_j, t_j)\right\}^2. \tag{16.19}$$

Proof We consider the quantity,

$$\mathbf{K}^{(2)}(\mathbf{x}_\alpha, t_k, \tau) = \tau\frac{1}{N_\alpha}\sum_j \sum_{\mathbf{x}_k\in\alpha} \sum_{\mathbf{x}_j\in\alpha} \mathbf{g}(\mathbf{x}(t_k), t_k)\boldsymbol{\eta}(t_k)\mathbf{g}(\mathbf{x}(t_j), t_j)\boldsymbol{\eta}(t_j). \tag{16.20}$$

If the bin size is small compared to the scale in which the matrix $\mathbf{g}(\mathbf{x}, t)$ varies significantly, we can replace $\mathbf{g}(\mathbf{x}(t_l), t_l)$ by $\mathbf{g}(\mathbf{x}_\alpha, t_l) = \mathbf{g}(\mathbf{x}_\alpha)$ such that

$$\begin{aligned}\mathbf{K}^{(2)}(\mathbf{x}_\alpha, t_k, \tau) &= \tau\sum_j \mathbf{g}(\mathbf{x}_\alpha, t_k)\left[\frac{1}{N_\alpha}\sum_{\mathbf{x}_k\in\alpha}\sum_{\mathbf{x}_j\in\alpha} \boldsymbol{\eta}(t_k)\boldsymbol{\eta}^T(t_j)\right]\mathbf{g}^T(\mathbf{x}_\alpha, t_j)\\ &= \tau\mathbf{g}(\mathbf{x}_\alpha, t_k)\mathbf{g}^T(\mathbf{x}_\alpha, t_k) = \tau\mathbf{g}(\mathbf{x}_\alpha)\mathbf{g}^T(\mathbf{x}_\alpha)\\ &= \mathbf{K}^{(2)}(\mathbf{x}_\alpha, \tau).\end{aligned} \tag{16.21}$$

Here, we have used the assumption of the statistical independence of the fluctuations

$$\frac{1}{N_\alpha}\sum_{\mathbf{x}(t_j)\in\alpha}\sum_{\mathbf{x}(t_k)\in\alpha} \boldsymbol{\eta}(t_k)\boldsymbol{\eta}^T(t_j) = \delta_{kj}. \tag{16.22}$$

In Sect. 20.1.2, we describe methods for estimation of the elements of matrix $\mathbf{g}(\mathbf{x}, t)$ in terms of diffusion matrix $\mathbf{D}^{(2)}(\mathbf{x}_\alpha, t)$. For time dependent $\mathbf{D}^{(1)}(\mathbf{x}_\alpha, t)$ and $\mathbf{D}^{(2)}(\mathbf{x}_\alpha, t)$ the summation in Eqs. (16.15) and (16.19) will be ensemble averaging at time t_j. For time–homogeneous diffusion processes drift vector and diffusion matrix are not explicitly time dependent.

16.5.3 Higher Order Kramers–Moyal Coefficients

In a similar way one may estimate higher conditional moments $K^{(n)}(\mathbf{x}_\alpha, t, \tau)$, with $n > 2$ that, in the small time limit converge to the Kramers–Moyal coefficients [4,

26–28]. The estimation of such quantities allows one to estimate the jump rate and jump amplitude in the jump-diffusion modeling, see Chap. 20 for details.

16.5.4 Estimation of Drift and Diffusion Coefficients from Sparcely-Sampled Time Series

The results of an analysis of data sets can be reconsidered self-consistently. This fact can be used to extend the procedure to data sets with insufficient amount of data, or sparcely-sampled time series, for which the estimation of the conditional moments $K^{(1)}(\mathbf{x}, t, \tau)$, $K^{(2)}(\mathbf{x}, t, \tau)$ and the subsequent limiting procedure $\tau \to 0$ cannot be performed accurately [29].

In this case, one may proceed as follows. In the first step one obtains a zeroth-order approximation of the drift vector $\mathbf{D}^{(1)}(\mathbf{x})$ and the diffusion matrix $\mathbf{D}^{(2)}(\mathbf{x})$. Based on the preformed estimates as the second step, a suitable ansatz for the drift vector and the diffusion matrix that contain a set of free parameters $\boldsymbol{\sigma}$ are assumed as follows,

$$\mathbf{D}^{(1)}(\mathbf{x}, \boldsymbol{\sigma}), \mathbf{D}^{(2)}(\mathbf{x}, \boldsymbol{\sigma}) \tag{16.23}$$

that defines a class of the Langevin equations. From each Langevin equation one can estimate the n-point joint probability distribution

$$p(\mathbf{x}_n, t_n; \cdots; \mathbf{x}_1, t_1; \boldsymbol{\sigma}). \tag{16.24}$$

The joint probability distribution can be compared with the experimental one, $p(\mathbf{x}_n, t_n; \cdots; \mathbf{x}_1, t_1; \text{exp.})$. The best representative of the class of the Langevin equations for the reconstruction of experimental data is then obtained by minimizing a suitably defined distance between the two distributions:

$$\text{Distance}\{p(\mathbf{x}_n, t_n; \cdots; \mathbf{x}_1, t_1; \boldsymbol{\sigma}) - p(\mathbf{x}_n, t_n; \cdots; \mathbf{x}_1, t_1; \text{exp.})\} = \text{Min.} \tag{16.25}$$

A reasonable choice is the so-called Kullback–Leibler distance between two distributions, defined as

$$K = \int dx_1 \cdots dx_n p(\mathbf{x}_n, t_n; \cdots; \mathbf{x}_1, t_1, \text{exp.}) \ln \frac{p(\mathbf{x}_n, t_n; \cdots; \mathbf{x}_1, t_1; \text{exp.})}{p(\mathbf{x}_n, t_n; \cdots; \mathbf{x}_1, t_1; \boldsymbol{\sigma})}. \tag{16.26}$$

It was shown how the iteration procedure can be obtained from the maximum likelihood arguments. For more details, we refer the reader to [12, 30, 31]. A technical question concerns the determination of the minimum, [32] used the limited-memory Broyden–Fletcher–Goldfarb–Shanno algorithm for constraint problems, in order to determine the solution of the optimization problem. Estimation of higher order Kramers–Moyal coefficients in short and nonstationary data sets has been investigated in [25, 33–37], see Chap. 17.

16.6 Deriving an Effective Stochastic Equation

Up to now, we evaluated the Markovian property and estimated the Markov–Einstein time scale of given time series. Now for Markov processes we can write down Kramers–Moyal (KM) expansion for dynamics of probability distribution function. To derive an effective stochastic diffusion equation for time series one should verify the Pawula theorem, which states that the Kramers–Moyal expansion of the probability distribution can be truncated after the second (diffusive) term, provided that the fourth-order Kramers–Moyal coefficient $D^{(4)}(x)$ vanishes. For vanishing fourth-order KM coefficient one can construct the Langevin equation with the estimated drift function and diffusion coefficient.

In practice for given time series all the KM coefficients could be nonzero. In this situation, comparison of the fourth-order KM coefficients with second-order one, can justifies keeping only the two first KM coefficients. In Chap. 18, we show that one can use the criterion $K^{(4)}(x,\tau) \approx 3(K^{(2)}(x,\tau))^2$ in small limit τ to judge whether the fourth-order KM moment is ignorable or not.

In this case, the KM expansion reduces to a Fokker-Planck equation. We note that the Fokker-Planck equation is in turn equivalent to the Langevin equation (using for instance, Itô's interpretation of stochastic integrals). Ignorable KM coefficients of order higher than two, particularly the fourth-order coefficient $D^{(4)}(x)$, will guarantee that time series is statistically continuous.

Non-vanishing higher-order (> 2) KM coefficients, however, have been observed in various systems [1, 38–42]. A priori, it is not evident if such observations are due to the finiteness of the respective sampling intervals or whether the measured time series do not belong to the class of continuous diffusion processes and contain discontinuous, abrupt changes or jumps. In Chap. 19, we show how one can use the information about higher-order KM moments, to check whether for a given, even noisy time series the underlying process has a continuous or discontinuous trajectory. The latter were shown recently to have pronounced contributions to higher-order KM coefficients [4] and to account for the non-Gaussian behaviour of increment statistics of empirical time series [43]. In order to improve modeling of such processes, one can argue that if all the conditional moments of KM coefficients of order larger than two are non-vanishing, jump events play a significant role in the underlying stochastic process. In this situation we can build a dynamical equation (for instance, jump-diffusion equation) that is able to generate trajectories with jumps.

16.7 Self-consistency

For continuous stochastic process with estimated drift vector and the characteristics of the noise sources from data, it is straightforward to synthetically generate data sets by iterating the corresponding stochastic evolution equations. Subsequently, their statistical properties can be compared with the properties of the actual data. This yields a self-consistent check of the results [1, 44].

Consider the case with vanishing KM coefficients of order higher than two, particularly the fourth-order coefficient $D^{(4)}(x)$. In this case we can write a Langevin equation with estimated drift a diffusion coefficients. As a check for *self-consistency* one should verify if the stochastic force in Langevin equation fulfils the conditions of a δ-correlated Gaussian white noise. Suppose for one dimensional time series one estimates the drift and diffusion coefficients $D^{(1)}(x)$ and $D^{(2)}(x)$ and substitute them in Langevin equation and solves it with respect to $\eta(t)$:

$$\eta(t) = \frac{x(t+\tau) - x(t) - D^{(1)}(x)\tau}{\sqrt{2D^{(2)}(x)\tau}} . \tag{16.27}$$

Taking τ as the time-step of the observed time-series and substituting in $x(t+\tau)$ and $x(t)$ successive values of that series one construct a series for $\eta(t)$ which should be uncorrelated normally distributed.

Another check, for instance, in case of stationary continuous one-dimensional systems allows for the following treatment due to the fact that the stationary probability distribution function (PDF), which is assumed to exist, can be determined analytically:

$$p(x) = \frac{N}{D^{(2)}(x)} e^{\int^{x} dx' \frac{D^{(1)}(x')}{D^{(2)}(x')}} \tag{16.28}$$

where N is normalization constant. As a consequence, we have the relationship

$$D^{(1)}(x) = D^{(2)}(x)\frac{d}{dx}\ln p(x) + \frac{d}{dx}D^{(2)}(x). \tag{16.29}$$

Since $p(x)$ can be determined from the time series an estimate in terms of a polynomial ansatz for the diffusion term suffices. In fact, one may use the ansatz $D^{(2)}(x) = Q + ax^2 + bx^4 + \dots$, which helps in lowering the number of parameters σ to be estimated by the above procedure of minimization the Kullback–Leibler information (16.26). For such classes of stochastic processes the above procedure can be reduced considerably by the fact that only few free parameters for the parametrization of drift and diffusion terms have to be introduced. As a consequence the minimization procedure of the Kullback–Leibler information is greatly simplified.

For jump-diffusion modeling the *self-consistency* check can be done via study of increment statistics

$$x_r(t) = x(t+r) - x(t)$$

to have the same distribution as the original series for a fixed r spanning from the time-step of the original series up to two or more orders of magnitude larger. This *self-consistency* check is also applicable to Langevin modeling as well. Another possible *self-consistency* check which can be used for both Langevin and jump-diffusion modeling is to compare the conditional short time propagator $p(x, t+\tau|x_0, t)$ of original time series with the synthetic ones, as suggested in [45]. Finally, we remark that it would be interesting to extend the tools discussed in this subsection, specially by introducing new self-consistency check to the class of stochastic processes with jump discontinuity, for instance by comparing the record statistics [46], level crossing statistics [47, 48], Fano-factor of jump events [49], etc. of original and synthetic time series from constructed dynamical equations.

Problems

16.1 Drift vector and diffusion matrix

Fill in the details in the derivation of Eqs. (16.14) and (16.21).

16.2 Markovian embedding of a non-Markov process

Suppose we have a time series with Markov–Einstein time scale $t_M = n\tau$, where n and τ are integer number and sampling interval, respectively. Argue that this non-Morkov process can be transformed to a Markov process in n-dimensions which is known as the Markovian embedding of a non-Markov process.

16.3 Estimation of Markov–Einstein time scale t_M

Consider following stochastic equations:

$$\frac{d}{dt}x(t) = -x(t) + y(t)$$

$$\frac{d}{dt}y(t) = -\frac{1}{L_0}y(t) + \frac{1}{L_0}\Gamma(t)$$

where $\Gamma(t)$ is a Gaussian white noise, for $L_0 = 1$ and $L_0 = 10$ with $dt = 0.001$.

(a) Use the Euler-Maruyama scheme to integrate the coupled two-dimensional diffusion process and by checking the χ^2-test (16.12) for Chapman–Kolmogorov equation or Wilcoxon test (Appendix 1), in the stationary state estimate the Markov–Einstein time scale of the process x (compare estimated t_M with L_0).

(b) Show that the correlation time scale of y is $\simeq L_0$ (see Problem 14.11).

References

1. R. Friedrich, J. Peinke, M. Sahimi, M.R. Rahimi Tabar, Phys. Rep. **506**, 87 (2011)
2. R. Friedrich, J. Peinke, Phys. Rev. Lett. **78**, 863 (1997)
3. G.R. Jafari, S.M. Fazeli, F. Ghasemi, S.M. Vaez Allaei, M.R. Rahimi Tabar, A. Iraji Zad, G. Kavei, Phys. Rev. Lett. **91**, 226101 (2003)
4. M. Anvari, K. Lehnertz, M.R. Rahimi Tabar, J. Peinke, Sci. Rep. **6**, 35435 (2016)
5. J. Argyris, G. Faust, M. Haase, R. Friedrich, *An Exploration of Dynamical Systems and Chaos* (Springer, New York, 2015)

6. H. Haken, *Information and Self-Organization: A Macroscopic Approach to Complex Systems* (Springer, Berlin, 2000)
7. H. Haken, *Synergetics: Introduction and Advanced Topics* (Springer, Berlin, 2004)
8. H. Kantz, T. Schreiber, *Nonlinear Time Series Analysis* (Cambridge University Press, Cambridge, 2003)
9. H. Haken, *Synergetics: An Introduction* (Springer, Berlin, 1983)
10. H. Haken, *Advanced Synergetics* (Springer, Berlin, 1987)
11. A. Einstein, Über die von der molekularkinetischen Theorie der Wärme geforderte Bewegung von in ruhenden Flüssigkeiten suspendierten Teilchen. Ann. Phys. **17**, 549 (1905)
12. D. Kleinhans, R. Friedrich, M. Wächter, J. Peinke, Phys. Rev. E **76**, 041109 (2007)
13. M.B. Priestley, R.T. Subba, A test for non-stationarity of time-series. J. R. Stat. Soc., Ser. B **31**, 140 (1969)
14. G.P. Nason, A test for second-order stationarity and approximate confidence intervals for localized autocovariances for locally stationary time series. J. R. Stat. Soc., Ser. B **75**, 879 (2013)
15. G.P. Nason, R. von Sachs, G. Kroisandt, Wavelet processes and adaptive estimation of the evolutionary wavelet spectrum. J. R. Stat. Soc., Ser. B **B62**, 271 (2000)
16. R. von Sachs, M.H. Neumann, A wavelet-based test for stationarity. J. Time Ser. Anal. **21**, 597–613 (2000)
17. J.W. Kantelhardt, S.A. Zschiegner, E. Koscielny-Bunde, S. Havlin, A. Bunde, H.E. Stanley, Phys. A **316**, 87 (2002)
18. M.S. Movahed, G.R. Jafari, F. Ghasemi, S. Rahvar, M.R. Rahimi Tabar, J. Stat. Mech. **P02003** (2006)
19. M.S. Movahed, F. Ghasemi, S. Rahvar, M.R. Rahimi Tabar, Phys. Rev. E **84**, 039902 (2011)
20. F. Ghasemi, J. Peinke, M. Sahimi, M.R. Rahimi Tabar, Euro. Phys. J. B **47**, 411 (2005)
21. F. Ghasemi, A. Bahraminasab, M.S. Movahed, K.R. Sreenivasan, S. Rahvar, M.R. Rahimi Tabar, J. Stat. Mech. **P11008** (2006)
22. F. Wilcoxon, Biometrics **1**, 80 (1945)
23. C. Renner, J. Peinke, R. Friedrich, J. Fluid Mech. **433**, 383 (2001)
24. W.H. Press, S.A. Teukolsky, W. Vetterling, B.P. Flannery, *Numerical Recipes in C: The Art of Scientific Computing*, 2nd edn. (Cambridge University Press, Cambridge, 1994)
25. A.M. Van Mourik, A. Daffertshofer, P.J. Beek, Biol. Cybern. **94**, 233 (2006)
26. J. Gottschall, J. Peinke, New J. of Phys. **10**, 083034 (2008)
27. J. Gottschall, J. Peinke, V. Lippens, V. Nagel, Phys. Lett. A **373**, 811 (2008)
28. M. Wächter, A. Kouzmitchev, J. Peinke, Phys. Rev. E **70**, 055103(R) (2004)
29. M. Ragwitz, H. Kantz, Phys. Rev. Lett. **87**, 254501 (2001)
30. D. Kleinhans, R. Friedrich, A. Nawroth, J. Peinke, Phys. Lett. A **346**, 42 (2005)
31. D. Kleinhans, R. Friedrich, Phys. Lett. A **368**, 194 (2007)
32. A. Nawroth, J. Peinke, D. Kleinhans, R. Friedrich, Phys. Rev. E **76**, 056102 (2007)
33. A.M. Van Mourik, A. Daffertshofer, P.J. Beek, Phys. Lett. A **351**, 13 (2006)
34. R. Riera, C. Anteneodo, J. Stat. Mech. **P04020**, (2010)
35. S.J. Lade, Phys. Lett. A **373**, 3705 (2009)
36. C. Anteneodo, S.M. Queirós, S.M. Duarte, Phys. Rev. E **82**, 041122 (2010)
37. C. Honisch, R. Friedrich, Phys. Rev. E **83**, 066701 (2011)
38. M. Tutkun, L. Mydlarski, New J. of Phys. **6**, 49 (2004)
39. J. Prusseit, K. Lehnertz, Phys. Rev. Lett. **98**, 138103 (2007)
40. G. Lim, S. Kim, E. Scalas, K. Kim, K.-H. Chang, Analysis of price fluctuations in futures exchange markets. Phys. A **387**, 2823 (2008)
41. M. Petelczyc, J.J. Żebrowski, R. Baranowski, Kramers-Moyal coefficients in the analysis and modeling of heart rate variability. Phys. Rev. E **80**, 031127 (2009)
42. M. Petelczyc, J.J. Żebrowski, E. Orlowska-Baranowska, A fixed mass method for the Kramers-Moyal expansion - application to time series with outliers. Chaos **25**, 033115 (2015)
43. M. Anvari, G. Lohmann, M. Wächter, P. Milan, E. Lorenz, D. Heinemann, M.R. Rahimi Tabar, J. Peinke, New J. Phys. **18**, 063027 (2016)

44. J. Gradisek, S. Siegert, R. Friedrich, I. Grabec, J. Sound Vib. **252**(3), 545 (2002)
45. F. Ghasemi, M. Sahimi, J. Peinke, R. Friedrich, G.R. Jafari, M.R. Rahimi Tabar, Phys. Rev. E **75**, R060102 (2007)
46. J. Franke, G. Wergen, J. Krug, Phys. Rev. Lett. **108**, 064101 (2012)
47. F. Shahbazi, S. Sobhanian, M.R. Rahimi Tabar, S. Khorram, G.R. Frootan, H. Zahed, J Phys. A **36**, 2517 (2003)
48. G.R. Jafari, M.S. Movahed, S.M. Fazeli, M.R.R. Tabar, S.F. Masoudi, J. Stat. Mech. **P06008**, (2006)
49. M.R. Rahimi Tabar, F. Nikakhtar, S. Baghram, S. Rahvar, R.K. Sheth, K. Lehnertz, M. Sahimi, Phys. Rev. E (2019)

Chapter 17
The Kramers–Moyal Coefficients of Non-stationary Time Series and in the Presence of Microstructure (Measurement) Noise

Most real world time series have transient behaviours and are *non-stationary*. They exhibit different type of non-stationarities, such as trends, cycles, random-walking, and generally exhibit strong intermittency. Therefore local stochastic characteristics of time series, such as the drift and diffusion coefficients, as well as the jump rate and jump amplitude, will provide very important information for understanding and quantifying "real time" variability of time series. For diffusive processes the systems have a longer memory and a higher correlation time scale and, therefore, one expects the stochastic features of dynamics to change slowly. In contrast, a rapid change of dynamics with jumps will cause strong ramp events (abrupt changes) in small time scales. Beside nonstationarity, in real world data, such as stock market index, pattern of cosmic background radiation, genomic data, etc., there is unique trajectory. Therefore, we require analysing techniques to estimate local stochastic behaviour of time series, which will be applicable to stationary and nonstationary time series and those with unique trajectory. This chapter contains technical aspects of the approach for *real-time* estimating of the Kramers–Moyal (KM) coefficients, and the drift and diffusion coefficients, as well as jump contributions for time series. We present the Kernel method (Nadaraya-Watson estimator) to estimate the time-dependent KM coefficients, which can be used in analysing stationary and nonstationary time series. We will also provide the details of estimating the KM coefficients in the presence of microstructure (measurement) noise, and show how the statistical properties of the noise can be determined from vanishing τ limit behaviour of the KM conditional moments.

The original version of this chapter was revised: Belated correction has been incorporated. The correction to this chapter is available at https://doi.org/10.1007/978-3-030-18472-8_24

M. R. Rahimi Tabar, *Analysis and Data-Based Reconstruction of Complex Nonlinear Dynamical Systems*, Understanding Complex Systems,
https://doi.org/10.1007/978-3-030-18472-8_17

17.1 The Kramers–Moyal Coefficients for Non-stationary Time Series: Nadaraya-Watson Estimator

The most of measured time series are non-stationary and this means the Kramers–Moyal (KM) coefficients could be explicitly time dependent. The Nadaraya-Watson (NW) estimator allows us to determine "local" KM coefficients, which means that one can estimate such conditional averaging in each step of time. The main ingredient of NW method is to estimate the conditional averaging in KM coefficients, using a kernel. We assume the reader is familiar with kernel estimations methods, otherwise we provide a short review in Appendix 2.

Let us start with a conditional averaging $\langle Y|X = x\rangle$, where can be written in terms of the joint PDF $p(x, y)$ as,

$$\langle Y|X = x\rangle = \int yp(y|x)dy = \frac{1}{p(x)}\int yp(x, y)dy = \frac{\int yp(x, y)dy}{\int p(x, y)dy}. \quad (17.1)$$

The joint PDF $p(x, y)$ in the numerator and denominator of Eq. (17.1) can be estimated using kernels $k_x(x)$ and $k_y(y)$ estimators as (Appendix 2)

$$p(x, y) = \frac{1}{nh_xh_y}\sum_{i=1}^{n} k_x\left(\frac{x - x_i}{h_x}\right) k_y\left(\frac{y - y_i}{h_y}\right) = \frac{1}{n}\sum_{i=1}^{n} k_x(x - x_i)k_y(y - y_i)$$

where h_x and h_y are the bandwidths of the kernels $k_x(x)$ and $k_y(y)$, respectively. Using expression for $p(x, y)$ one can show that $\int yp(x, y)dy = \frac{1}{n}\int y \sum_{i=1}^{n} k_x(x-x_i)$ $k_y(y - y_i)dy$. Here n is the number of data points and kernel $k_u(u - u_i)$ is a localized function around $u \simeq u_i$. Noting to the fact that $\int yk_y(y - y_i)dy = y_i$ (see Appendix 2), then we have,

$$\int yp(x, y)dy = \frac{1}{n}\sum_{i=1}^{n} k_x(x - x_i)\, y_i. \quad (17.2)$$

In the denominator of Eq. (17.1) also we can write $p(x)$ as $p(x) = \frac{1}{n}\sum_{i=1}^{n} k_x(x - x_i)$. Putting all together we find the Nadaraya-Watson estimator for conditional averaging as,

$$\langle Y|X = x\rangle = \frac{\sum_{i=1}^{n} k_x(x - x_i)y_i}{\sum_{i=1}^{n} k_x(x - x_i)}. \quad (17.3)$$

This kernel regression estimator was first proposed by Nadaraya (1964) and Watson (1964) [1, 2]. The KM conditional moments are in the form of $\langle Y|X = x\rangle$, where $Y \equiv (x(t + \tau) - x(t))^j$ for $j \in \{1, 2, \cdots\}$. Therefore the j-th order KM conditional moment can be written as,

$$\frac{1}{\tau}\langle (x_{(i+1)\tau} - x_{i\tau})^j |_{x_i=x}\rangle = \frac{\sum_{i=1}^{n} k\left(\frac{x_{i\tau}-x}{h}\right)(x_{(i+1)\tau} - x_{i\tau})^j}{\sum_{i=1}^{n} k\left(\frac{x_{i\tau}-x}{h}\right)\tau}$$

where $h \equiv h_x$ is the band-width of kernel $k(u)$ which has the property $\int u^2 k(u)du < \infty$. In general, kernel k is any smooth function satisfying,

$$k(x) \geq 0, \qquad \int k(x)dx = 1, \qquad \int xk(x)dx = 0. \tag{17.4}$$

The most common kernels are Gaussian kernel $k(x) = \frac{1}{\sqrt{2\pi}}\exp(-x^2/2)$ and the Epanechnikov kernel $k(x) = \frac{3}{4}(1 - x^2)$, $x \in [-1, 1]$ and 0 otherwise.

The Epanechnikov kernel can be employed easily because it leads to continuous and (at least piecewise) differentiable estimates and is numerically cheaper than other kernels that have this property [3–5].

In *Nadaraya-Watson estimator*, the bandwidth selection remains the main problem for practitioners. There are several methods for selecting the bandwidth that have been introduced in kernel regression. Given the need of automatic selection of data-driven bandwidth for applications, in [6] a review is provided to explain and compare different methods. About 20 different selection methods have been described, implemented and compared in an extensive simulation study in [6], see also [7–9] for bandwidth selection in nonparametric kernel estimation. For bandwidth selection of multi-dimensional kernel estimation, see [10]. The problem of *bandwidth selection* is a research topic in statistics.[1]

[1] To have an accurate estimation for marginal probability density, if one needs n data points with bin size δ_1, for two point joint or conditional probability density will need approximately order $10n$ data points. This increases the bin size to $\delta_2 = 2.15\delta_1$. The bin size for marginal density will be $\delta_1 \leq 2.6(IQ)n^{-1/3}$ (found by Terrell), where (IQ) is the inter-quartile range of data set. The inter-quartile range is given by $(IQ) = Q_3 - Q_1$. To determine the values of Q_1 and Q_3, find the median of whole data set, separate data to "above" and "below" the median, find median for each new data set, name the median of "above" and "below" sets as Q_3 and Q_1, respectively. Typical bin numbers for normalised time series with 10^6 data points, will be $M_1 \simeq 121$ and $M_2 \simeq 55$ bins for estimation of marginal and conditional PDFs. In practice we use an odd number of bins. In Kramers–Moyal coefficients we need to estimate conditional PDF $p(x', t + \tau|x, t)$ therefore can use *an approximate* bin size δ_2 in the analysis. For more detailed analysis use the method proposed by Knuth to obtain optimal bin numbers for joint probability distribution.

For stationary time series with n data points, (1) estimate the Kramers–Moyal coefficients via binning of state variable and using the Nadaraya-Watson estimator with different bandwidth h. (2) Select bandwidth h^* in Nadaraya-Watson estimator, in such a way that the estimated KM coefficients from two approaches converge. This gives bandwidth h^*, which is cross-validated with binning method. (3) Now apply kernel estimator with obtained bandwidth to original data and compute time dependent KM coefficients.

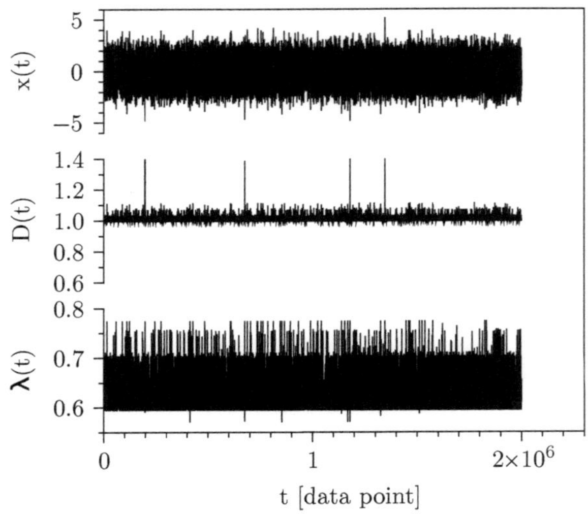

Fig. 17.1 Estimated local jump rate (bottom), local diffusion coefficient (middle), and time series (top) for a normalized synthetic Ornstein–Uhlenbeck process with jumps. To estimate the conditional moments $M^{(j)}$, we used the Nadaraya-Watson estimator with gaussian kernel with band-width $h \simeq 0.1$. The drift and diffusion coefficients as well as jump amplitude and jump rate are given in the Sect. (21.2.1)

17.1.1 Time Dependent Kramers–Moyal Coefficients

As an example, we consider an Ornstein–Uhlenbeck process in the presence of jumps, so that $D^{(1)}(x) = -x$, $D(x) = 1$ with unit jump amplitude $\sigma_\xi^2(x) = 1$ and constant jump rate $\lambda = 0.6$, see Sect. 21.2.1. We generate a synthetic time series $x(t_n)$ with $n = 10^6$ data points by a numerical simulation of the corresponding dynamical Eq. (17.7) in Euler-Maruyama scheme, using a sampling interval $\tau = 0.01$. The number of jumps will be $n_j \simeq \lambda n$. The KM coefficients M^j, for $j = 1, 2, 4$, and 6 are then estimated using the Nadaraya-Watson estimator with gaussian kernel. Therefore using the relations (12.2) and (12.8) one can find local drift, diffusion and jump properties from estimated KM coefficients.

A sample of stationary time series as well as local diffusion coefficient and jump rate has been illustrated in Fig. 17.1. In Fig. 17.2 same plot is given for recorded intracranial electroencephalographic (iEEG) non-stationary time series which includes a seizure. Strong intermittent behaviour is present in (iEEG) time series and is visible in Fig. 17.2.

17.2 The Kramers–Moyal Coefficients in the Presence of Microstructure (Measurement) Noise

In the recent years high frequency data are available in many area of research. However in high frequency time series generally microstructure noise are present and the direct application of many techniques, such as reconstruction of Langevin dynamics or jump-diffusion modeling run into problem. Generally, the term 'microstructure noise' refers to any superimposed uncorrelated noise that is present in time series

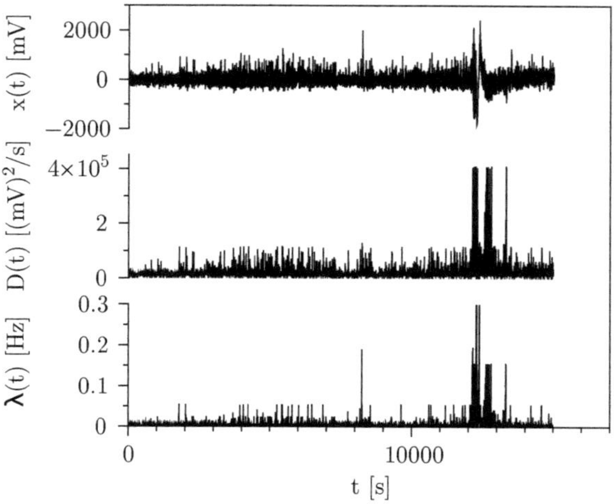

Fig. 17.2 Estimated local jump rate (bottom), local diffusion coefficient (middle), and time series (top) for recorded intracranial electroencephalographic (iEEG) non-stationary time series with sample rate 200 Hz, which includes a seizure at $t \simeq 12500\,s$. To estimate the conditional moments $M^{(j)}$, $j \in \{1, 2, 4, 6\}$, we used the Nadaraya-Watson estimator with gaussian kernel and bandwidth $h \simeq 0.1$

of complex system; it might be generated by the complex system itself or to be measurement noise.

We now consider the case that a time series $x(t)$ is contaminated with some noise $\eta(t)$, which is not assimilated by the stochastic process, leading to $y(t) = x(t) + \eta(t)$ [11–14]. The noise $\eta(t)$ is supposed to be uncorrelated with $x(t)$, and we assume that it has zero mean and finite statistical moments $\langle \eta^n \rangle$ for $n = 2, \cdots$. To reconstruct the unknown dynamics $x(t)$ from zero mean measurement $y(t)$, it is essential to quantify $\eta(t)$ and its influence on the estimation of KM coefficients.

Using the definition of the KM conditional moments $K^{(j)}(x, \tau) = \langle (x(t + \tau) - x(t))^j|_{x(t)=x} \rangle$ and relation $y(t) = x(t) + \eta(t)$, we can derive the following expressions for the first- and second-order conditional moments of $y(t)$ [14]:

$$
\begin{aligned}
K^{(1)}(y, \tau) &= K^{(1)}(x, \tau) + \gamma_1(y), \\
K^{(2)}(y, \tau) &= K^{(2)}(x, \tau) + \gamma_2(y),
\end{aligned}
\tag{17.5}
$$

where τ independent terms $\gamma_{1,2}(y)$ can be found as follows. Let us decompose the expression $K^{(1)}(y, \tau)$ into two parts. The one part contains all those cases where y is very close to x and thus $\eta = 0$. The second part covers all other states where y differs from x, i.e. $\eta \neq 0$. We thus obtain the following expression:

$$
\begin{aligned}
K^{(1)}(y, \tau) \quad &= \quad \langle y(t + \tau) - y(t)|_{y=x} \rangle + \langle y(t + \tau) - y(t)|_{\eta(t)=y-x} \rangle \\
&= \quad \langle x(t + \tau) - x(t)|_{x=y} \rangle + \langle \underbrace{x(t + \tau) - x(t)}_{\text{mean}\,=\,0}|_{\eta(t)=y-x} \rangle \\
&\quad + \langle \underbrace{\eta(t + \tau)}_{=0} - \underbrace{\eta(t)}_{=0}|_{x=y} \rangle + \langle \underbrace{\eta(t + \tau) - \eta(t)}_{\text{mean}\,=\,0}|_{\eta(t)=y-x} \rangle
\end{aligned}
$$

$$
\begin{aligned}
&= \langle x(t+\tau) - x(t)|_{x=y}\rangle - \langle \eta(t)|_{\eta(t)=y-x}\rangle \\
&= K^{(1)}(x=y,\tau) + \langle x(t) - y\rangle \\
&= K^{(1)}(x=y,\tau) + \gamma_1(y). \qquad (17.6)
\end{aligned}
$$

Here we used the stationarity of process x and zero mean property of the noise $\eta(t)$. With zero mean $y(t)$, the process $x(t)$ also will be zero mean. Equation (17.6) shows that the additional offset comes from the expectation value of the difference between y and $x(t)$. For an underlying linear drift it might be expected that smaller and larger values compensate each other because the measurement noise is distributed symmetrically around the process $x(t)$. In this case $\langle x(t) - y\rangle$ is expected to be zero in Eq. (17.6). A similar analysis gives us

$$
\gamma_2(y) = \langle (x(t) - y)^2\rangle. \qquad (17.7)
$$

where in weak noise limit ($\sqrt{\langle \eta^2(t)\rangle} \leq 0.04\sqrt{D^{(2)}(x)}$) one finds $\gamma_2(y) \simeq 2\langle \eta^2\rangle$ [14, 15].

The finite and τ-independent terms on the r.h.s. of Eq. (17.5) cause a strong overestimation of KM coefficients $M^{(m)}(y) = \lim_{\tau\to 0} \frac{1}{\tau} K^{(m)}(y,\tau)$ for small τ and thus of the functions of interest in the Langevin and the jump-diffusion equation. Equation (17.5) states that there are extra τ-independent terms contributing to the conditional moments and these constants are related to different statistical moments of the noise, at least in the weak noise limit. The latter can easily be estimated by dividing the even-order conditional moments for $y(t)$ by τ (because this leads to a divergence of $K^{(m)}(y,\tau)/\tau$), and averaged higher-order statistical moments of η can then be derived from the knowledge of the lower-order ones [16].

17.2.1 *Real-World Data with Microstructure Noise*

In [14], for DM-US-Dollar exchange rate (from the one-year period October 1992 to September 1993) and climate North Atlantic oscillation (NAO), it is shown that these data sets have strong microstructure noise. In Fig. 17.3, the effect of measurement noise on $K^{(1)}$ is shown for these two real world examples, in which a strong offset at $\tau = 0$ causes a divergence of $K^{(1)}/\tau$. This means that for analysed data $\gamma_1(y)$ is not zero.

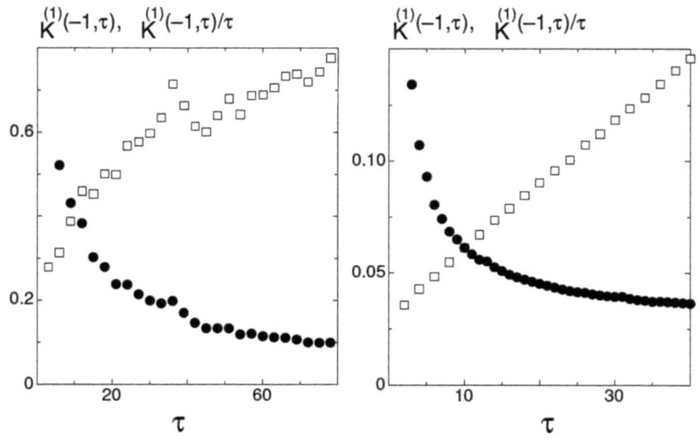

Fig. 17.3 Both plots show measured $K^{(1)}$ (squares) and $K^{(1)}/\tau$ (circles) as a function of lag τ. The left plot refers to the North Atlantic Oscillation (NAO) index [17], the right one to an increment time series of financial exchange rates [18]. Units are arbitrary for clarity of presentation. For both systems a significant offset of the first conditional moment is observed causing a divergence of $K^{(1)}/\tau$, which means that they possesses non vanishing $\gamma_1(y)$. *Source* from [14]

17.2.2 Real-World Data without Microstructure Noise

As an example for a data set that has not microstructure noise, we present results for stochastic behavior of epileptic brain dynamics by analysing intracranial electroencephalographic (iEEG) recording from epilepsy patient [19]. To check whether the data set has microstructure noise or not, we checked whether the limit $\tau \to 0$ exists for different KM moments. As shown in Fig. 17.4, this limit exists and KM coefficients have not $1/\tau$ behaviour in the limit $\tau \to 0$. We therefore can safely assume that there is not microstructure noise contribution in estimation of KM coefficients.

Problems

17.1 Nadaraya-Watson Estimator, Ornstein–Uhlenbeck Process with Jumps

Use Nadaraya-Watson estimator (use a Gaussian kernel with bandwidth $h \simeq 0.1$) to estimate time dependent Kramers–Moyal coefficients and reproduce the results presented in Fig. 17.1. The drift and diffusion coefficients as well as jump amplitude and jump rate are given in the Sects. 17.1.1 and 21.2.1. Use the Euler-Maruyama scheme to integrate related jump-diffusion dynamical equation.

17.2 Local Drift and Diffusion of the Minimal Market Model

The minimal market model is given by a (Itô) Langevin dynamics with time dependent drift $a(x, t)$ and multiplicative $b(x, t)$ as,

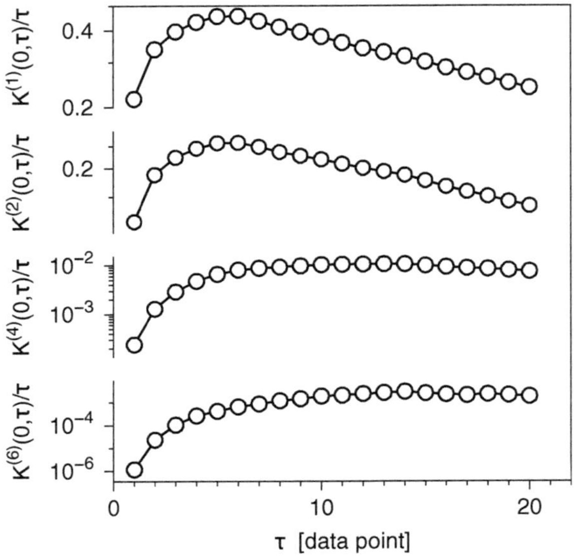

Fig. 17.4 Estimated $K^{(1)}(0,\tau)/\tau$, $K^{(2)}(0,\tau)/\tau$, $K^{(4)}(0,\tau)/\tau$ and $K^{(6)}(0,\tau)/\tau$ for an iEEG recording (2×10^6 data points) from within the epileptic focus [19]. The lag τ is in the units of the data points

$$a(x,t) = \alpha \exp\{\eta t\}, \qquad b(x,t) = \sqrt{\alpha \exp\{\eta t\} x}$$

with scaling parameter $\alpha > 0$ and growth rate $\eta > 0$.

(a) Use the Euler-Maruyama scheme to approximate this Langevin equation and generate 10^5 data points of $x(t_n)$ with $x(t_0) = x_0 = 1$, $dt = 0.001$, $\alpha = 0.2$ and $\eta = 0.01$.

(b) Use the Nadaraya-Watson estimator with Gaussian kernel (with different bandwidth $h \in \{0.01, 0.1, 0.6\}$) to estimate time dependence of the functions a and b. The first and second order Kramers–Moyal coefficients are given by $D^{(1)}(x,t) = a(x,t)$ and $b(x,t) = \sqrt{2D^{(2)}(x,t)}$.

17.3 Microstructure Noise

Derive the third- and fourth-order ($j = 3, 4$) Kramers–Moyal conditional moments $K^{(j)}(y,\tau)$ in terms of $K^{(j)}(x,\tau)$ and statistical moments of noise $\eta(t)$ in the *weak noise limit*. Assume that the noise to be uncorrelated with $x(t)$, that it has finite statistical moments $\langle \eta^n \rangle$ for $n = 2, \cdots$.

17.4 Microstructure Noise

Consider a Ornstein–Uhlenbeck process $x(t)$ with drift and diffusion coefficients as $D^{(1)}(x,t) = -10x$ and $D^{(2)}(x,t) = 1$, respectively, and synthesis a time series with 10^6 data points (with $x(t=0) = 0$ and $dt = 0.01$).

Now generate a new time series with microstructure noise as,

$$y(t) = x(t) + \sigma \zeta(t)$$

where $\sigma = 0.01$ and $\zeta(t)$ is uncorrelated zero mean, unit variance Gaussian noise $\zeta \sim \mathcal{N}(0, \sigma_\zeta^2 = 1)$.

(a) Estimate second order Kramers–Moyal conditional moment for process $y(t)$, and using the method presented in Sect. 17.2, determine σ.

(b) Estimate drift and diffusion coefficients of original process $x(t)$.

(c) Find the explicit expressions for $\gamma_{1,2}(y)$ in terms of y, σ and variance of x [14].

References

1. E.A. Nadaraya, On estimating regression, theory of probability and its applications. Theory Probab. Appl. **9**, 141 (1964)
2. G.S. Watson, Smooth regression analysis. Sankhya: Indian J. Stat., Ser. A **26**, 359 (1964)
3. H.J. Bierens, Kernel estimators of regression functions, in *Advances in Econometrics, Fifth World Congress of the Econometric Society*, ed. by T.F. Bewley, vol. I of Econometric Society Monographs, chap. 3, pp. 99-144. (Cambridge University Press, Cambridge, 1987)
4. B.W. Silverman, *Density Estimation* (Chapman and Hall, London, 1986)
5. C. Honisch, *Analysis of Complex Systems: From Stochastic Time Series to Pattern Formation in Microscopic Fluidic Films* (Dissertation, University of Münster) (Westfalen, 2014)
6. M. Köhler, A. Schindler, S. Sperlich, A review and comparison of bandwidth selection methods for kernel regression. Int. Stat. Rev. **82**, 243 (2014)
7. J. Gao, I. Gijbels, Bandwidth selection in nonparametric kernel testing. J. Amer. Statist. Assoc. **103**, 1584–1594 (2008)
8. G.R. Terrell, The maximal smoothing principle in density estimation. J. Am. Stat. Assoc. **85**, 470 (1990)
9. S. Ghosh, *Kernel Smoothing: Principles, Methods And Applications* (Wiley, New York, 2017)
10. K.H. Knuth, Optimal data-based binning for histograms, ArXiv Physics e-prints (May, 2006) [physics/0605197]
11. H. Kantz, T. Schreiber, *Nonlinear Time Series Analysis* (Cambridge University Press, Cambridge, 2003)
12. E.J. Kostelich, T. Schreiber, Phys. Rev. E **48**, 1752 (1993)
13. J.P.M. Heald, J. Stark, Phys. Rev. Lett. **84**, 2366 (2000)
14. F. Böttcher, J. Peinke, D. Kleinhans, R. Friedrich, P.G. Lind, M. Haase, Phys. Rev. Lett. **97**, 090603 (2006)
15. M. Siefert, A. Kittel, R. Friedrich, J. Peinke, Europhys. Lett. **61**, 466 (2003)
16. K. Lehnertz, L. Zabawa, M. Reza Rahimi Tabar, New J. Phys. **20**, 113043 (2018)
17. P.G. Lind, A. Mora, J.A.C. Gallas, M. Haase, Int. J. Bifurc. Chaos **17**, 3461 (2007)
18. C. Renner, J. Peinke, R. Friedrich, Phys. A **298**, 499 (2001)
19. M. Anvari, K. Lehnertz, M.R. Rahimi Tabar, J. Peinke, Sci. Rep. **6**, 35435 (2016)

Chapter 18
Influence of Finite Time Step in Estimating of the Kramers–Moyal Coefficients

Data sampled at discrete times appear as successions of discontinuous jump events, even if the underlying trajectory is continuous. In this chapter we study finite sampling τ expansion of the Kramers-Moyal conditional moments for the Langevin and jump-diffusion dynamics. Using the expansion for the Langevin dynamics, we introduce a criterion to validate the method numerically, namely, the Pawula theorem, to judge whether the fourth-order KM moment tends to zero. The criterion is a relation between the fourth- and second-order KM conditional moments for small time lag τ [1].

18.1 Diffusion Processes

It is common to model a given time series via a Langevin equation, which generally includes state-dependent drift and multiplicative diffusion coefficients. In this modeling, it is assumed that the underlying process has continuous trajectories and the resulting dynamics is diffusion process. In this section, we demonstrate that sampling at discrete times not only influences the aforementioned coefficients but also causes non-vanishing higher-order ones.

Statistically continuous processes are commonly modeled by a Langevin equation as

$$dx(t) = a(x,t)\tau + b(x,t)dW(t), \tag{18.1}$$

where $\{W(t), t \geq 0\}$ is a scalar Wiener (Brownian) motion, and $a(x,t)$ and $b^2(x,t)/2$ denote the state-dependent deterministic drift and the diffusion coefficients. For this type of process and using the conditional probability distribution, $x(t)$ satisfies the Lindeberg's continuity condition, given some $\delta > 0$ (see Chap. 4),

M. R. Rahimi Tabar, *Analysis and Data-Based Reconstruction of Complex Nonlinear Dynamical Systems*, Understanding Complex Systems,
https://doi.org/10.1007/978-3-030-18472-8_18

$$C(t) = \lim_{\tau \to 0^+} \frac{Prob[|\Delta x(t)| > \delta \quad |_{x(t)=x}]}{\tau} = 0, \tag{18.2}$$

where $\Delta x(t) = x(t + \tau) - x(t)$.

When analyzing empirical data, one might want to ensure that the corresponding time series can indeed be modeled by Eq. (18.1). For this purpose, one should check if the fourth-order KM coefficient $M^{(4)}(x, t)$ (or $D^{(4)}(x, t)$) vanishes. If this was the case, then—according to the Pawula theorem (see Chap. 3)—all higher-order (>2) coefficients will vanish too ($M^{(4)} = 0 \to M^{(3)} = M^{(5)} = M^{(6)} = \cdots = 0$), and the probability density of the underlying process can be described by a Fokker–Planck equation.

18.2 The Kramers–Moyal Conditional Moments for the Langevin Equation

In general, for a given time series, the non-vanishing of higher-order (>2) KM coefficients can be related for example to the existence of jumps in the time series. In this case, the KM coefficients provide an upper limit for the continuity condition (Eq. (18.2)): $C(t) \leq \frac{M^{(m)}(x,t)}{\delta^m}$, which holds for any $m \geq 3$ (provided the first- and second-order coefficients are non-vanishing), see Chap. 11. Therefore, any vanishing higher-order KM coefficients, particularly the fourth-order one ($M^{(4)}(x, t)$), guarantee that the underlying process is statistically continuous.

Here we demonstrate that a finite time interval τ not only influences the first- and second-order KM coefficients [1–10] but also causes non-vanishing higher-order (>2) ones.

Let us begin by deriving of conditional moments of the Langevin equation for different orders of the time interval τ. From Eq. (18.1) it is evident that only two Kramers–Moyal coefficients, namely $M_{\mathrm{d}}^{(1)}(x, t) = a(x, t)$ and $M_{\mathrm{d}}^{(2)}(x, t) = b^2(x, t)$ (the subscript "d" denotes diffusion and $M_{\mathrm{d}}^{(n)}(x, t) = \lim_{\tau \to 0} \frac{1}{\tau} K_{\mathrm{d}}^{(n)}(x, t, \tau)$) are non-vanishing. Here $K_{\mathrm{d}}^{(n)}(x, t, \tau)$ is the KM conditional moment order n.

Therefore conditional probability distribution of the process x satisfy the truncated Kramers–Moyal differential equation (Fokker–Planck equation) as,

$$\frac{\partial p(x, t|x', t')}{\partial t} = \mathcal{L}_{FP}(x)\; p(x, t|x', t') \tag{18.3}$$

with initial condition $p(x, t|x', t) = \delta(x - x')$ and $\mathcal{L}_{FP}$ is given by,

$$\mathcal{L}_{FP} = \sum_{n=1}^{2} \frac{1}{n!} \left(-\frac{\partial}{\partial x} \right)^{n} M_{\mathrm{d}}^{(n)}(x, t) \quad . \tag{18.4}$$

The formal solution of (18.3) reads

$$p(x, t + \tau | x', t) = \exp\{\tau \, \mathcal{L}_{FP}\} \, \delta(x - x') \quad . \tag{18.5}$$

The conditional moments $K_{\mathrm{d}}^{(n)}(x, t, \tau)$ with finite τ can be written as,

$$\begin{aligned} K_{\mathrm{d}}^{(n)}(x_i, t, \tau) &= \int_{-\infty}^{\infty} (x - x_i)^n \exp\{\tau \, \mathcal{L}_{FP}\} \, \delta(x - x_i) dx \\ &= \exp\{\tau \, \mathcal{L}_{FP}^{\dagger}\}(x - x_i)^n|_{x=x_i} \end{aligned} \tag{18.6}$$

where $\mathcal{L}_{FP}^{\dagger}$ is the adjoint operator of $\mathcal{L}_{FP}$ and is given by,

$$\mathcal{L}_{FP}^{\dagger} = \sum_{n=1}^{2} \frac{1}{n!} M_{\mathrm{d}}^{(n)}(x, t) \left(\frac{\partial}{\partial x} \right)^{n} \quad . \tag{18.7}$$

An explicit expression of $\mathcal{L}_{FP}^{\dagger}$ will be,

$$\mathcal{L}_{FP}^{\dagger} = a \frac{\partial}{\partial x} + \frac{1}{2!} b^2 \frac{\partial^2}{\partial x^2} \quad . \tag{18.8}$$

Equation (18.6), for $K_{\mathrm{d}}^{(n)}(x_i, t, \tau)$ can be written as,

$$K_{\mathrm{d}}^{(n)}(x_i, t, \tau) = \left(1 + \tau \mathcal{L}_{FP}^{\dagger} + \frac{1}{2} \tau^2 \mathcal{L}_{FP}^{\dagger} \mathcal{L}_{FP}^{\dagger} + \cdots \right) (x - x_i)^n|_{x=x_i} \quad . \tag{18.9}$$

Considering terms up to the order of the first non-vanishing power in $\mathcal{O}(\tau^2)$, one can find the following expressions for the conditional moments $K_{\mathrm{d}}^{(m)}(x, \tau)$, for $m \in \{1, 2, 3, 4\}$ and $\mathcal{O}(\tau^3)$ for $m \in \{5, 6\}$ (in the following, we omit the x- and t-dependence of a and b to enhance readability) [1],

$$\begin{aligned} K_{\mathrm{d}}^{(1)}(x, \tau) &= a\tau + \frac{1}{2} \left[aa' + \frac{1}{2} b^2 a'' \right] \tau^2 + \mathcal{O}(\tau^3), \\ K_{\mathrm{d}}^{(2)}(x, \tau) &= b^2 \tau \\ &\quad + \frac{1}{2} \left[2a(a + bb') + b^2(2a' + b'^2 + bb'') \right] \tau^2 \\ &\quad + \mathcal{O}(\tau)^3, \end{aligned}$$

$$K_{\mathrm{d}}^{(3)}(x,\tau) = 3b^2(a+bb')\tau^2 + \mathcal{O}(\tau)^3,$$

$$K_{\mathrm{d}}^{(4)}(x,\tau) = 3b^4\tau^2 + \mathcal{O}(\tau)^3,$$

$$K_{\mathrm{d}}^{(5)}(x,\tau) = 15b^4(a+2bb')\tau^3 + \mathcal{O}(\tau)^4$$

$$K_{\mathrm{d}}^{(6)}(x,\tau) = 15b^6\tau^3 + \mathcal{O}(\tau)^4. \tag{18.10}$$

where a' and b' denote derivatives with respect to state variable x. Dividing by τ and in the limit $\tau \to 0$, the first terms on the r.h.s. of the first two equations will give the drift and diffusion coefficients. For finite τ all conditional moments possess some correction terms. For d-dimensional estimation of drift vector and diffusion matrix with finite τ see [11] for details.

We can check the validity of expansions in Eq. (18.10) by reconstructing stochastic processes with known drift and diffusion coefficients from synthetic time se-

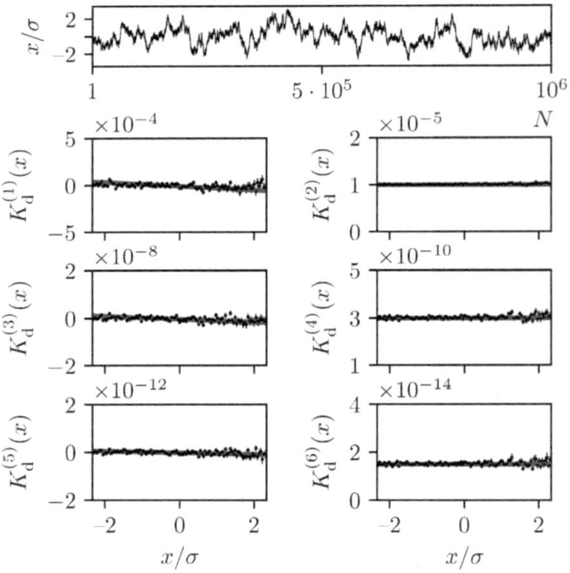

Fig. 18.1 Conditional moments $K_{\mathrm{d}}^{(1)}(x,\mathrm{d}t)$ to $K_{\mathrm{d}}^{(6)}(x,\mathrm{d}t)$ estimated from exemplary time series of a continuous linear diffusion process (Eq. 18.1), generated with $a(x) = -px$ and $b(x) = b_0$ ($p = 10$, $b_0 = 1$) and with a time interval $\mathrm{d}t = 10^{-5}$. Time series consisted of $N = 3 \times 10^6$ data points. We observed extreme events up to about $\pm 3\sigma$ for these time series (σ denotes the standard deviation). Error bars indicate the standard error of the mean (SEM) in each bin. Theoretically expected values are shown as red lines, and in the range $\pm 2\sigma$ more than two-thirds of these values lie within the SEM interval of estimated conditional moments. Deviations in the tails are due to low statistics. Source from [1]

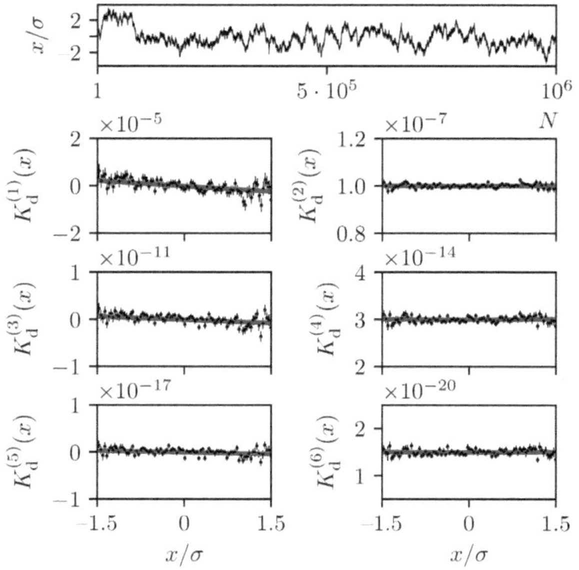

Fig. 18.2 Same as Fig. 18.1 but for exemplary time series of continuous nonlinear diffusion process, generated with $a(x) = -px$ and $b(x) = b_0 + b_1x^2$ ($p = 5$, $b_0 = 0.1$, $b_1 = 0.05$) and with a time interval $\mathrm{d}t = 10^{-5}$. Time series consisted of $N = 3 \times 10^6$ data points. We observed extreme events up to $\pm 4\sigma$ for these time series. Theoretically expected values are shown as red lines, and in the range $\pm 1.5\sigma$ more than two-thirds of these values lie within the SEM interval of estimated conditional moments. Deviations in the tails are due to low statistics. Source from [1]

ries sampled with time intervals τ spanning three orders of magnitude (with $\tau \in \{10^{-6}, 2 \times 10^{-6}, 4 \times 10^{-6}, 6 \times 10^{-6}, \ldots, 10^{-3}\}$).

Conditional moments $K_d^{(1)}(x, \tau)$ to $K_d^{(6)}(x, \tau)$ estimated from time series of continuous (linear and nonlinear) diffusive processes (Eq. (18.1)). Linear diffusive process generated with $a(x, t) = -px$ and $b(x, t) = b_0$ ($p = 10$, $b_0 = 1$) as state-dependent deterministic drift and constant diffusion coefficients, and nonlinear diffusive process generated with $a(x, t) = -px$ and $b(x, t) = b_0 + b_1x^2$ ($p = 5$, $b_0 = 0.1$, $b_1 = 0.05$), integrated with time interval $\tau = 10^{-5}$ in Euler–Maruyama scheme. For both linear and nonlinear continuous process described by the Langevin equation (Eq. (18.1)), one finds a very good agreement between estimated conditional moments and the respective theoretical predictions (see Figs. 18.1 and 18.2), demonstrating the validity of the approach for such systems. As expected, we find the second-order conditional moment to depend linearly on τ while the fourth-order one scales with τ^2 (see Fig. 18.3). These findings demonstrate that—even for diffusive continuous processes—non-vanishing higher-order conditional moments can originate from a finite time interval τ.

Now using the KM conditional moments fourth- and second order for small τ, we find [1],

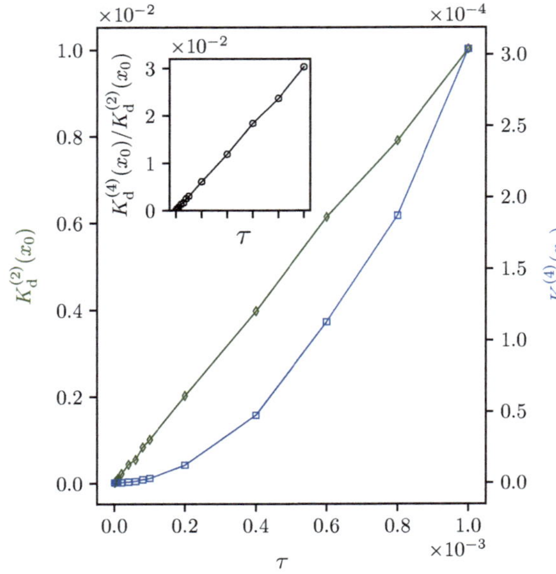

Fig. 18.3 Dependencies of conditional moments $K_{\mathrm{d}}^{(2)}(x_0, \tau)$ and $K_{\mathrm{d}}^{(4)}(x_0, \tau)$ for some $x = x_0$ on time intervals τ for a continuous nonlinear diffusive processes (we obtained similar findings a continuous linear diffusive process). As expected, we find the second-order conditional moment to depend linearly on τ while the fourth-order one scales with τ^2. The conditional moments are estimated from normalised time series, so that be able compare two conditional moments. We chose x_0 around the mean of the respective time series. Source from [1]

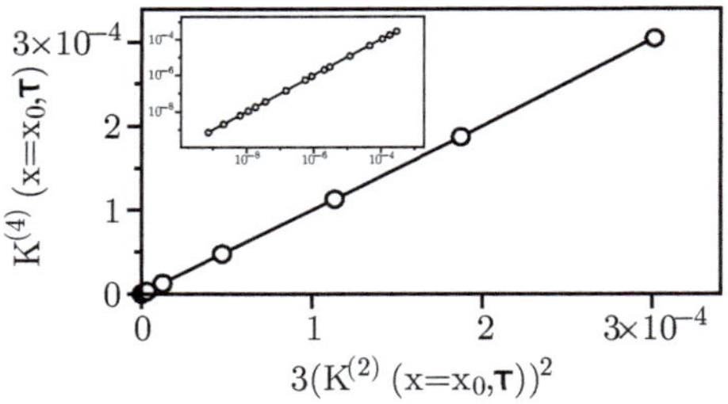

Fig. 18.4 Dependencies of conditional moments $K_{\mathrm{d}}^{(4)}(x_0, \tau)$ and $(K_{\mathrm{d}}^{(2)}(x_0, \tau))^2$ for some $x = x_0$ with different time intervals τ ($\tau \in \{10^{-6}, 2 \times 10^{-6}, 4 \times 10^{-6}, 6 \times 10^{-6}, \cdots, 10^{-3}\}$) for a continuous nonlinear diffusive processes. We find $K_{\mathrm{d}}^{(4)}(x_0, \tau) \simeq 3(K_{\mathrm{d}}^{(2)}(x_0, \tau))^2$. Similar findings has been obtained for continuous linear diffusive process. We chose x_0 around the mean of the respective time series. Source from [1]

$$K^{(4)}(x, \tau) \simeq 3(K^{(2)}(x, \tau))^2 \tag{18.11}$$

which confirms Wick's theorem and follows from the fact that the short-time propagator of the Langevin dynamics (Eq. (18.1)) is a Gaussian distribution. Therefore one can use this criterion as a possible numerical verification of Pawula theorem. No such short-time propagator, however, is known for general jump-diffusion processes. In Fig. 18.3, we plot $K^{(4)}(x,\tau)$ verses $(K^{(2)}(x,\tau))^2$ for aforesaid nonlinear diffusion process. We obtained similar findings for continuous linear diffusive process. Figure 18.3 shows clearly a linear behaviour.

Therefore when analyzing empirical data, the relation between fourth- and second-order conditional moment, i.e. $K^{(4)}(x,\tau) \simeq 3(K^{(2)}(x,\tau))^2$ can be used to judge whether the fourth-order moment tends to zero.[1] We note that a normalization of the time series leaves this relationship between the fourth- and second-order conditional moments unaffected (Fig. 18.4).

18.3 Conditional Moments of the Jump-Diffusion Equation for Different Orders of the Time Interval

Let us consider the jump-diffusion equation, that is capable of generating a trajectory with jump discontinuity [12],

$$dx(t) = a(x,t)\tau + b(x,t)\,dW(t) + \xi\,dJ(t), \tag{18.12}$$

where $\{W(t), t \geq 0\}$ is a scalar Wiener process, $a(x,t)$ and $b(x,t)$ are again the state-dependent deterministic drift and the multiplicative diffusion, and $J(t)$ is a time-homogeneous Poisson jump process. Jumps have rate $\lambda(x)$ and size ξ, which we assume to be Gaussian distributed with zero mean and variance σ_ξ^2 (or to follow any symmetric distribution with finite statistical moments). The jump rate $\lambda(x)$ and statistical moments of ξ can be state dependent. For infinitesimal τ, it was shown in Sect. 12.1, that [12],

$$\begin{aligned} K_{\rm j}^{(1)}(x,\tau) &= a\tau \\ K_{\rm j}^{(2)}(x,\tau) &= \left[b^2 + \langle\xi^2\rangle\lambda(x)\right]\tau \\ K_{\rm j}^{(2m)}(x,\tau) &= \langle\xi^{2m}\rangle\lambda(x)\tau, \quad \text{for } 2m > 2 \end{aligned} \tag{18.13}$$

(the subscript "j" denotes jumpy). In the presence of jump events with rate $\lambda(x)$, we thus expect to find, for instance for the fourth-order conditional moment, a term of $\mathcal{O}(\tau)$ (instead of $\mathcal{O}(\tau^2)$ dependence for diffusion processes) if we estimate KM conditional moments using Eq. 18.13.

Now we derive the conditional moments of the jump-diffusion equation (Eq. (18.12)) for different orders of the time interval τ. From Eq. (18.13) it is ev-

[1] This criterion has been proposed at first by Lina Zabawa and Klaus Lehnertz in numerical simulations of Langevin dynamics and later analytically supported in [1].

ident that all of Kramers–Moyal coefficients are not vanishing (except odd order KM coefficients with $n > 1$). Therefore conditional probability distribution of the process x in jump-diffusion equation satisfies the Kramers–Moyal differential equation as [1],

$$\frac{\partial p(x,t|x',t')}{\partial t} = \mathcal{L}_{KM}\ p(x,t|x',t') \tag{18.14}$$

with initial condition $p(x,t|x',t) = \delta(x-x')$ and $\mathcal{L}_{KM}$ is given by,

$$\mathcal{L}_{KM} = \sum_{n=1}^{\infty} \frac{1}{n!} \left(-\frac{\partial}{\partial x}\right)^n M_{\mathrm{j}}^{(n)}(x,t) \quad . \tag{18.15}$$

The formal solution of (18.14) reads

$$p(x,t+\tau|x',t) = \exp\{\tau\ \mathcal{L}_{KM}\}\ \delta(x-x') \quad . \tag{18.16}$$

The conditional moments $K_{\mathrm{j}}^{(n)}(x,t,\tau)$ with finite τ can be written as,

$$\begin{aligned} K_{\mathrm{j}}^{(n)}(x_i,t,\tau) &= \int_{-\infty}^{\infty} (x-x_i)^n \exp\{\tau\ \mathcal{L}_{KM}\}\ \delta(x-x_i) dx \\ &= \exp\{\tau\ \mathcal{L}_{KM}^{\dagger}\}(x-x_i)^n|_{x=x_i} \end{aligned} \tag{18.17}$$

where $\mathcal{L}_{KM}^{\dagger}$ is the adjoint operator of $\mathcal{L}_{KM}$ and is given by,

$$\mathcal{L}_{KM}^{\dagger} = \sum_{n=1}^{\infty} \frac{1}{n!}\ M_{\mathrm{j}}^{(n)}(x,t) \left(\frac{\partial}{\partial x}\right)^n . \tag{18.18}$$

where $M_{\mathrm{j}}^{(n)}(x,t) = \lim_{\tau\to 0} \frac{1}{\tau} K_{\mathrm{j}}^{(n)}(x,t,\tau)$ and $K_{\mathrm{j}}^{(n)}(x,t,\tau)$ are given by Eq. (18.17). Explicit expansion of $\mathcal{L}_{KM}^{\dagger}$ will be,

$$\begin{aligned} \mathcal{L}_{KM}^{\dagger} = {} & \underbrace{a}_{\mathrm{A}} \frac{\partial}{\partial x} + \underbrace{\frac{[b^2 + \langle\xi^2\rangle\lambda(x)]}{2!}}_{\mathrm{B}} \frac{\partial^2}{\partial x^2} \\ & + \underbrace{\frac{\langle\xi^4\rangle\lambda(x)}{4!}}_{\mathrm{C}} \frac{\partial^4}{\partial x^4} + \underbrace{\frac{\langle\xi^6\rangle\lambda(x)}{6!}}_{\mathrm{D}} \frac{\partial^6}{\partial x^6} \\ & + \underbrace{\frac{\langle\xi^8\rangle\lambda(x)}{8!}}_{\mathrm{E}} \frac{\partial^8}{\partial x^8} + \cdots \quad . \end{aligned} \tag{18.19}$$

Operation of $\mathcal{L}^{\dagger}_{KM}$ in Eq. (18.17) gives [1],

$$
\begin{aligned}
K_{\rm j}^{(1)}(x,\tau) &= A\tau \\
&+ \frac{1}{2}\Big[AA^{\{1\}} + BA^{\{2\}} + CA^{\{4\}} + DA^{\{6\}} + \mathcal{O}(\delta)\ \Big]\tau^2 + \mathcal{O}(\tau^3), \\
K_{\rm j}^{(2)}(x,\tau) &= 2B\tau \\
&+ \frac{1}{2}\Big[2A^2 + 2AB^{\{1\}} + 4BA^{\{1\}} + 2BB^{\{2\}} + 8CA^{\{3\}} + 2CB^{\{4\}} \\
&+ 12DA^{\{5\}} + 2DB^{\{6\}} + \mathcal{O}(\delta)\ \Big]\tau^2 + \mathcal{O}(\tau^3), \\
K_{\rm j}^{(3)}(x,\tau) &= \frac{1}{2}\Big[12AB + 12BB^{\{1\}} + 36CA^{\{2\}} + 24CB^{\{3\}} + 90DA^{\{4\}} \\
&+ 36DB^{\{5\}} + \mathcal{O}(\delta)\ \Big]\tau^2 + \mathcal{O}(\tau^3), \\
K_{\rm j}^{(4)}(x,\tau) &= 4!C\tau \\
&+ \frac{1}{2}\Big[4!AC^{\{1\}} + 4!B^2 + 4!BC^{\{2\}} + 4\times 4!CA^{\{1\}} + (4!)^2/(2!2!)CB^{\{2\}} \\
&+ 4!CC^{\{4\}} + 4\times 6!/3!DA^{\{3\}} + 6!/2!DB^{\{4\}} + 4!DC^{\{6\}} + \mathcal{O}(\delta)\ \Big]\tau^2 + \mathcal{O}(\tau^3), \\
K_{\rm j}^{(5)}(x,\tau) &= \frac{1}{2}\Big[2\times 5!AC + 2\times 5!BC^{\{1\}} + 5\times (4!)^2/3!CB^{\{1\}} + 4\times 5!CC^{\{3\}} \\
&+ 5\times 6!/2!DA^{\{2\}} + 20\times 6!/3!DB^{\{3\}} + 6!DC^{\{5\}} + 5\times 8!/4!EA^{\{4\}} \\
&+ 20\times 8!/5!EB^{\{5\}} + 8\times 5!EC^{\{7\}} + \mathcal{O}(\delta)\ \Big]\tau^2 + \mathcal{O}(\tau^3), \\
K_{\rm j}^{(6)}(x,\tau) &= 6!D\tau \\
&+ \frac{1}{2}\Big[6!AD^{\{1\}} + 2\times 6!BC + 6!BD^{\{2\}} + 6\times 6!CC^{\{2\}} + 6!CD^{\{4\}} \\
&+ 6\times 6!DA^{\{1\}} + (6!)^2/(4!2!)DB^{\{2\}} + (6!)^2/(4!2!)DC^{\{4\}} + 6!DD^{\{6\}} \\
&+ 6\times 8!/3!EA^{\{3\}} + 30\times 8!/4!EB^{\{4\}} + 8!/2!EC^{\{6\}} + 6!ED^{\{8\}} \\
&+ \mathcal{O}(\delta)\ \Big]\tau^2 + \mathcal{O}(\tau^3),
\end{aligned}
\tag{18.20}
$$

where the superscript $\{i\}$ denotes the i-th derivative with respect to x. The term $\mathcal{O}(\delta)$ comprises higher order-derivatives of A, B, etc. We kept the terms up to third order derivative of A and fourth order of B.

We note that jump events induce terms of order $\mathcal{O}(\tau)$ in the conditional moments of even orders. In addition, jump properties $\lambda(x)$ and σ_ξ^2 induce terms of order $\mathcal{O}(\tau^2)$ in all conditional moments.

We check the validity of expansions in Eq. (18.20) for conditional moments of orders $m \in \{1, 2, 3, 4, 5, 6\}$ by reconstructing a jump-diffusion process with known drift and diffusion coefficients as well as known properties of jumps from synthetic time series sampled with time intervals $\tau = 10^{-6}$ (in Euler–Maruyama scheme). Jump-diffusion process generated with $a(x,t) = x - x^3$ and $b(x,t) = x^2$ and with jump rate $\lambda = 0.33$ and size $\sigma_\xi^2 = 1$. For our examples, the fourth-order derivative of A and the fifth-order derivative of B are vanishing. In addition, all derivatives of terms C, D, etc. in Eq. (18.20) are zero. We find again a very good agreement between estimated conditional moments and the respective theoretical predictions (see Fig. 18.5), demonstrating the validity of expansion (18.20).

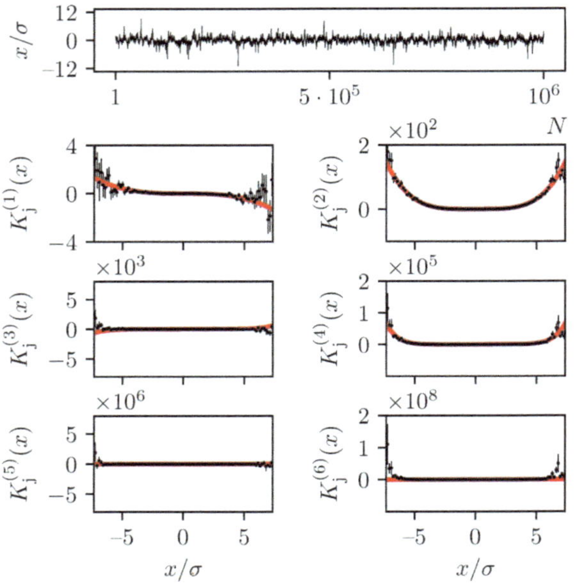

Fig. 18.5 Conditional moments $K_j^{(1)}(x, dt)$ to $K_j^{(6)}(x, dt)$ estimated from exemplary time series of a discontinuous jump-diffusion process (Eq. (18.12)) generated with $a(x) = x - x^3$ and $b(x) = x^2$, with jump rate $\lambda = 0.3$ and size $\sigma_\xi^2 = 1$ and with a time interval $dt = 10^{-6}$. Time series consisted of $N = 3 \times 10^6$ data points. We observed extreme events up to about $\pm 10\sigma$. Error bars indicate the standard error of the mean (SEM) in each bin. Theoretically expected values are shown as red lines, and in the range $\pm 5\sigma$ more than two-thirds of these values lie within the SEM interval of estimated conditional moments. Deviations in the tails are due to low statistics. Source from [1]

It is worth mentioning that using the slope of $K_j^{(4)}(x, \tau)$ for small τ we can estimate also $\langle \xi^4 \rangle \lambda(x)$. Similar analysis can be done for $K_j^{(6)}(x, \tau)$, and in this way we can estimate $\langle \xi^6 \rangle \lambda(x)$. With Gaussian distributed jump amplitudes, it follows that $\langle \xi^{2n} \rangle = \frac{(2n)!}{2^n n!} \langle \xi^2 \rangle^n$ and $\langle \xi^{2n-1} \rangle = 0$, for $n = 1, 2, \cdots$, where $\sigma_\xi^2 = \langle \xi^2 \rangle$. Therefore, small τ limit both $K_j^{(4)}(x, \tau)$ and $K_j^{(6)}(x, \tau)$, and these gives us $3\sigma_\xi^4 \lambda(x)$ and $30\sigma_\xi^6 \lambda(x)$. Now we can estimate both, jump amplitude σ_ξ^2 and jump rate $\lambda(x)$ from $K_j^{(4)}(x, \tau)$ and $K_j^{(6)}(x, \tau)$. Once the jump characteristics are identified, $K_j^{(2)}(x, \tau)$ identifies the diffusion coefficient $b(x, t)$ and $K_j^{(1)}(x, \tau)$ provides an estimate for the drift coefficient $a(x, t)$.

Finally we note that for jump-diffusion processes and for small τ the quantity $K^{(4)}(x, \tau)/3(K^{(2)}(x, \tau))^2$ diverges as $1/\tau$

$$\frac{K^{(4)}(x, \tau)}{3(K^{(2)}(x, \tau))^2} \simeq \frac{\langle \xi^4 \rangle \lambda(x) \tau}{3[b^2 + \langle \xi^2 \rangle \lambda(x)]^2 \tau^2} \sim 1/\tau, \tag{18.21}$$

instead of approaching unity as observed for continuous diffusion processes.

18.4 The Kramers–Moyal Coefficients in Vanishing Time Interval Limit

In the previous sections, we checked the validity of expansions in Eqs. (18.10) and (18.20), by reconstructing stochastic processes with known drift and diffusion coefficients as well as with known jump properties from synthetic time series sampled with time intervals τ. In practice, we are in the opposite position, so that from time series with given sampling interval τ, we are interested in estimating Kramers–Moyal conditional moments, and then should search for the "true" (in the zero limit of τ) drift and diffusion coefficients, for instance in Langevin modeling.

In general, up to know there is not concrete technique to find $a(x, t)$ and $b(x, t)$ (without making some presumption) from Eq. (18.10) and $a(x, t)$, $b(x, t)$, $\lambda(x)$ and statistical moments of jump amplitude from Eq. (18.20), by knowing the Kramers–Moyal conditional moments $K^{(n)}(x, \tau)$. In what follows we present some available methods to compute or estimate "true" drift and diffusion coefficients in the vanishing limit of τ, for diffusion processes.

18.4.1 "Apparent" and "True" (in Vanishing Time Interval Limit) Drift and Diffusion Coefficients in Diffusion Processes

Let us start with the Langevin dynamics and from Eq. (18.6) we have

$$K_{\mathrm{d}}^{(n)}(x, \tau) = \exp\{\tau\, \mathcal{L}_{FP}^{\dagger}(x')\}(x' - x)^n|_{x'=x} \tag{18.22}$$

where $\mathcal{L}_{FP}^{\dagger}$ is the adjoint operator of $\mathcal{L}_{FP}$ and is given by

$$\mathcal{L}_{FP}^{\dagger} = D^{(1)}(x, t)\frac{\partial}{\partial x} + D^{(2)}(x, t)\frac{\partial^2}{\partial x^2} \quad .$$

here $D^{(1)}(x, t) = a(x, t)$ and $D^{(2)}(x, t) = b^2(x, t)/2$ are the drift and diffusion coefficients, respectively. The operator $\exp\{\tau\, \mathcal{L}_{FP}^{\dagger}\}$ can be understood as the propagator of the partial differential equation [7, 13],

$$\frac{\partial W_{n,x}(x', t)}{\partial t} = \mathcal{L}_{FP}^{\dagger}(x')\; W_{n,x}(x', t) \tag{18.23}$$

which acts on the initial condition

$$W_{n,x}(x', 0) = (x' - x)^n \quad . \tag{18.24}$$

Therefore, the KM conditional moments are given by

$$K_{\mathrm{d}}^{(n)}(x,\tau) = W_{n,x}(x' = x, t = \tau) \quad . \tag{18.25}$$

Advantage of this interpretation over the series expansion (18.10) lies in the fact that the KM conditional moments can be calculated exactly or at least up to numerical accuracy. On the contrary, the usage of the series expansion is connected to the problem that one does not know a priori how many terms are necessary.

18.4.1.1 Example

For simple drift and diffusion coefficients, Eq. (18.23) can be solved analytically [13]. For a linear drift $D^{(1)}(x) = -\gamma x$ and quadratic diffusion $D^{(2)}(x) = \alpha + \beta x^2$ where $\gamma, \alpha, \beta > 0$ ($\beta \geq \gamma$), one finds for $n = 1$[2]

$$W_{1,x}(x',t) = x' e^{-\gamma t} - x$$

which leads to "apparent" (with finite time interval) drift coefficient

$$D_{\tau}^{(1)}(x) = \frac{1}{\tau} W_{1,x}(x' = x, t = \tau) = -\frac{x}{\tau}\left(1 - e^{-\gamma\tau}\right) = -\gamma_{app}(\tau)\, x \leq -\gamma\, x \tag{18.26}$$

where apparent spring constant γ_{app} is equal $\frac{1}{\tau}\left(1 - e^{-\gamma\tau}\right)$. The expression for $D_{\tau}^{(1)}(x)$, shows that with finite τ, one underestimates always the "spring constant" γ. For $n = 2$ one finds

$$W_{2,x}(x',t) = x^2 - \frac{\alpha}{\beta - \gamma}\left(1 - e^{2(\beta-\gamma)t}\right) + x'^2 e^{2(\beta-\gamma)t} - 2xx'e^{-\gamma t} \tag{18.27}$$

which yields the "apparent" diffusion coefficient

$$D_{\tau}^{(2)}(x) = \frac{1}{2\tau} W_{2,x}(x' = x, t = \tau) = \frac{1}{2\tau}\left[x^2\left(1 + e^{2(\beta-\gamma)\tau} - 2e^{-\gamma\tau}\right) - \frac{\alpha}{\beta - \gamma}\left(1 - e^{2(\beta-\gamma)\tau}\right)\right] \tag{18.28}$$

---

[2] To compute $W_{n,x}(x',t)$, with initial condition $W_{n,x}(x',0) = (x' - x)^n$, we note that in Eq. (18.22) acting the operator $\exp\{\tau\, \mathcal{L}_{FP}^{\dagger}(x')\}$ will keep or reduce the order of n in $(x' - x)^n$, then we can write a series for $W_{n,x}(x',t)$ as

$$W_{n,x}(x',t) = \sum_{i,j=0:i+j=n} c_{ij}(t)(x)^i(x')^j$$

where $c_{ij}(t)$ are time dependent coefficients. Inserting the series in (18.23) gives a set of first order differential equations $\dot{c_{ij}}(t)$ (by equating the coefficients of $(x)^i(x')^j$ at both sides). At $t = 0$ one should finds $W_{n,x}(x',0) = (x' - x)^n$, which fixes all unknown constants in the integrations.

For $\beta = 0$, we obtain the corresponding result for an Ornstein-Uhlenbeck (OU) process,

$$D_{\tau}^{(2)}(x) = \frac{1}{2\tau}\left[x^2\left(1-e^{-\gamma\tau}\right)^2 + \frac{\alpha}{\gamma}\left(1-e^{-2\gamma\tau}\right)\right] \tag{18.29}$$

also the apparent drift for the OU process is the same as Eq. (18.26).

The OU process has linear drift and constant diffusion coefficient. Equations (18.26) and (18.29), show that as sampling interval τ increases, the apparent drift coefficient remains linear but decreases in slope, while a quadratic component appears in the apparent diffusion. This causes the diffusion estimate to be larger at the tails than at the center, an effect first noticed by Ragwitz and Kantz [14], see also [15].

18.4.1.2 Estimation of Drift and Diffusion Coefficients in Vanishing Time Interval Limit for Time Series with Finite Time Interval or with Finite *Markov–Einstein* Time Scale

In practice one can choose some functional form of drift and diffusion coefficients, for instance in the form $D^{(1)}(x) = -\gamma x$ and $D^{(2)}(x) = \alpha + \beta x^2$, and compute the expressions for $W_{1,x}(x', t)$ and $W_{2,x}(x', t)$ via solving Eq. (18.23). By fitting (or using some optimization procedure) the estimated KM conditional moments $K_{\mathrm{d}}^{(1)}(x, \tau)$ and $K_{\mathrm{d}}^{(2)}(x, \tau)$ for finite τ (also for time series with finite *Markov–Einstein* scale where $\tau \equiv t_M$) with expressions $W_{1,x}(x' = x, t = \tau)$ and $W_{2,x}(x' = x, t = \tau)$, one can extract the coefficients, for instances parameters γ, α and β, which provide us the estimation for drift and diffusion coefficients in the vanishing limit $\tau \to 0$.

18.4.2 The Optimization Procedure to Extract Kramers–Moyal Coefficients in Vanishing Time Interval Limit

Let us describe how method proposed in Sect. 18.4.1 (to obtain the "true" KM coefficients from estimated finite time KM coefficient) can be reformulate to an optimization problem. The basic idea is as follows [8, 9, 16]. Assume that "true" KM coefficients, for instance drift and diffusion coefficients for diffusion processes have polynomial expansion (polynomial ansatz) as

$$D^{(1)}(x, \sigma) = a_0 + a_1 x + \cdots + a_n x^n,$$

$$D^{(2)}(x, \sigma) = b_0 + b_1 x + \cdots + b_m x^m$$

with a set of free parameters

$$\sigma = \{a_0, a_1, \cdots, a_n, b_0, b_1, \cdots, b_m\} \quad .$$

For a specific set of parameters σ, we can compute the apparent (τ-dependent) KM conditional moments $W_{1,x}(x' = x, t = \tau, \sigma)$ and $W_{2,x}(x' = x, t = \tau, \sigma)$. The aim of optimisation procedure is to find the set of parameters σ^* such that $W_{1,x}(x' = x, t = \tau, \sigma)$ and $W_{2,x}(x' = x, t = \tau, \sigma)$ are as close as possible to the estimated KM conditional moments from time series $K_{\mathrm{d}}^{(1)}(x, \tau)$ and $K_{\mathrm{d}}^{(2)}(x, \tau)$, by employing for instance, least square distance [8] or in general maximum likelihood framework [16]. The parameter set σ^* that minimizes the least square distance corresponds to the maximum likelihood estimation. In [16] reader will find Python codes for estimation of set σ^* in maximum likelihood framework.

All the obtained results up to now are related to diffusion processes and extension of current approaches to jump-diffusion processes need further investigations.

Problems

18.1 The Kramers–Moyal conditional moments

Fill in the details in the derivation of Eqs. (18.10) and (18.20).

18.2 Numerical verification of the Pawula theorem

Reproduce the results presented in Figs. 18.2 and 18.3. To integrate related Langevin equations use the Euler–Maruyama scheme.

18.3 "Apparent" drift and diffusion coefficients

Fill in the details in the derivation of Eqs. (18.26) and (18.28).

References

1. K. Lehnertz, L. Zabawa, M.R.R. Tabar, New J. Phys. **20**, 113043 (2018)
2. M. Ragwitz, H. Kantz, Phys. Rev. Lett. **87**, 254501 (2001)
3. R. Friedrich, C. Renner, M. Siefert, J. Peinke, Phys. Rev. Lett. **89**, 217 (2002)
4. P. Sura, J. Barsugli, Phys. Lett. A **305**, 304 (2002)
5. D. Kleinhans, R. Friedrich, A. Nawroth, J. Peinke, Phys. Lett. A **346**, 42 (2005)
6. J. Gottschall, J. Peinke, J. New, Physics **10**, 083034 (2008)
7. S.J. Lade, Phys. Lett. A **373**, 3705 (2009)
8. C. Honisch, R. Friedrich, Phys. Rev. E **83**, 066701 (2011)
9. C. Honisch, R. Friedrich, F. Horner, C. Denz, Phys. Rev. E **86**, 026702 (2012)
10. K. Tang, P. Ao, B. Yuan, Europhys. Lett. **102**, 40003 (2013)
11. J. Prusseit, K. Lehnertz, Phys. Rev. E **77**, 041914 (2008)
12. M. Anvari, K. Lehnertz, M.R. Rahimi Tabar, J. Peinke, Sci. Rep. **6**, 35435 (2016)

13. C. Honisch, Analysis of complex systems: from stochastic time series to pattern formation in microscopic fluidic films. University Dissertation, Münster (Westfalen), 2014
14. M. Ragwitz, H. Kantz, Phys. Rev. Lett. **89**, 149402 (2002)
15. R. Friedrich, Ch. Renner, M. Siefert, J. Peinke, Phys. Rev. Lett. **89**, 149401 (2002)
16. D. Kleinhans, Phys. Rev. E **85**, 026705 (2012)

Chapter 19
Distinguishing Diffusive and Jumpy Behaviors in Real-World Time Series

Jumps are discontinuous variations in time series and with large amplitude can be considered as an extreme event. We expect the higher the jump activity to cause higher uncertainty in the stochastic behaviour of measured time series. Therefore, building statistical evidence to detect real jump seems of primary importance. In addition jump events can participate in the observed non-Gaussian feature of the increments' (ramp on and ramp down) statistics of many time series [1]. This is the reason that most of the jump detection techniques are based on threshold values for differential of time series. There is not, however, a robust method for detection and characterisation of such discontinuous events that is able to estimate time-dependence of the "jump rate" and their amplitudes, etc.

In this chapter, using finite sampling τ expansion of the Kramers–Moyal conditional moments for the Langevin and jump-diffusion dynamics, we introduce a criterion (as a necessary condition) that allows one to check whether for a given, even noisy time series the underlying process has a continuous (diffusion) trajectory, or has jump discontinuity. This enables one to detect and characterize the abrupt changes (here jump events) in time series. The proposed criterion is validated numerically using synthetic continuous time series and those with jump discontinuity. We demonstrate the applicability of the criterion for distinguishing diffusive and jumpy behavior by a data-driven inference of higher-order KM conditional moments from empirical observations [2].

19.1 Distinguishing Diffusive from Jumpy Stochastic Behavior in Complex Time Series

Many empirical time series exhibit fluctuations that are interrupted by jumps in very short time between different states of a system [3]. Examples include dynamics of charge transport in various materials [4–15], stochastic resonance [16], moving fronts [17], light curves of variable astronomical objects [18], fluctuations of wind

M. R. Rahimi Tabar, *Analysis and Data-Based Reconstruction of Complex Nonlinear Dynamical Systems*, Understanding Complex Systems,
https://doi.org/10.1007/978-3-030-18472-8_19

and solar power systems [1], early warning signals of systems near to their tipping points [19, 20], transitions in financial [21, 22] and climate data [23], ion channel dynamics [24, 25] eye movements [26, 27], or movement and foraging paths of animals [28]. A number of recent empirical and theoretical studies proved the existence of jumps and their substantial impact on risk management, see [29, 30] and references therein.

Despite advances in this field of research, the studies have found that jumps are empirically difficult to identify, because in practice only discrete data are available from continuous-time models. In fact data sampled at discrete times appears as a succession of discontinuous jumps, even if the underlying trajectory is continuous.

The state of the art techniques for jump detections are, (i) parametric methods, includes classical t-test where one should knows the probability distribution of time series, (ii) non-parametric methods such as U-test of Mann-Whitney and Wilcoxon test of sum. In these methods knowing of the probability distribution of data is not necessary. Other methods are curve fitting method, Bayesian methods, linear or non-linear regression methods, wavelets methods, cumulative methods (CUMSUM), sequential methods (ARMA, ARIMA, Kalman filter, differential methods), etc. See [31] and references therein.

19.2 A Jump Detection Measure Q(x)

With KM conditional moments $K^{(4)}_{\bullet}(x,\tau)$ and $K^{(6)}_{\bullet}(x,\tau)$ (the subscript $\bullet$ is a placeholder for either "d" or "j"), we now derive a function $Q(x)$ that— for small time intervals τ—allows one to detect and to distinguish diffusive and jumpy behavior in time series. Using Eqs. (18.10) and (18.20), in small limit τ we find [2],[1]

$$Q(x,\tau)=\frac{K^{(6)}_{\bullet}(x,\tau)}{5K^{(4)}_{\bullet}(x,\tau)}\approx\begin{cases}b^2(x)\tau, & \text{diffusive } (\bullet=\text{“}d'') \\ \sigma_{\xi}^2(x), & \text{jumpy } (\bullet=\text{“}j'')\end{cases} . \tag{19.1}$$

Here $b^2(x)/2$ is the diffusion coefficients in Langevin dynamics and $\sigma_{\xi}^2(x)$ is the jump amplitude in jump-diffusion modeling. For non-stationary time series function Q, as well as b will be time dependent, see below. We note that use of the second- and fourth-order KM conditional moments to define a measure for distinguishing the jumps will be problematic in small time interval limit τ, as we have seen in the Sect. 18.3.

For the exemplary time series of linear and nonlinear continuous stochastic processes generated with Langevin equation, $Q(x)$ exhibits a vanishing behavior with decreasing time interval τ, where the slope depends on the diffusion coefficients b (see Fig. 19.1). The details of linear, nonlinear and jump-diffusion are given in the Sects. 18.2 and 18.3. For the exemplary time series of the jump-diffusion process

[1]We note in variance gamma model for jump amplitude introduced in Chap. 12, one finds $K^{(6)}_j(x,\tau)/5K^{(4)}_j(x,\tau)\approx\sigma_{\xi}^2(x,\hat{b})(1+2\hat{b})$, where $\hat{b}$ is given by Eq. (12.15).

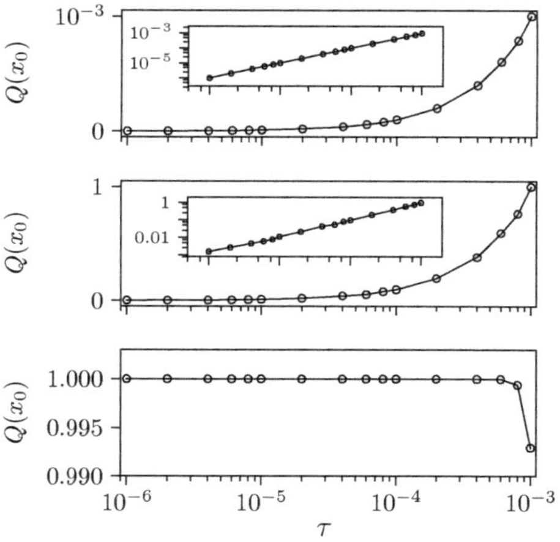

Fig. 19.1 Dependence of $Q(x_0)$ on time interval τ for exemplary time series of a linear (top) and nonlinear continuous stochastic process (middle) and of a jump-diffusion process with jump amplitude $\sigma_\xi^2 = 1$ (bottom). Insets demonstrate the $\mathcal{O}(\tau)$ behavior for the linear and nonlinear processes. We chose x_0 around the mean of the respective time series. Details of stochastic models are presented in Sects. 18.2 and 18.3. Source from [2]

generated with Eq. (18.12), $Q(x)$ converges to the jump amplitude σ_ξ^2. We note that $Q(x)$ will approach $\langle \xi^6 \rangle / 5 \langle \xi^4 \rangle$ for jump amplitudes with a non-Gaussian distribution.

As we have mentioned in Sect. 17.2, KM conditional moments of noisy empirical time series will have extra τ-independent terms contributing to the conditional moments and these constants are related to different statistical moments of the microstructure (measurement) noise. Using $1/\tau$ behaviour of KM conditional moments one can find the averaged higher-order statistical moments of noise from the knowledge of the lower-order ones. Having estimated the averaged statistical moments of the noise allows one to find conditional moments for the original time series, and with Eq. (19.1) one can check whether the underlying process has a continuous (diffusive) trajectory or possess jump discontinuity.

19.3 Application to Real-World Time Series

In this section, we demonstrate applicability of introduced criterion to distinguish diffusive and jumpy behavior (Eq. (19.1)) by a data-driven inference of higher-order conditional moments from empirical observations. Here we assume that measured data can be modeled by a Langevin or jump-diffusion precess.

We note that in the case of empirically derived time series that were sampled with a fixed sampling interval Δ, Eq. (19.1) can be verified by scaling the time interval as $\tau = \alpha\Delta$, where $\alpha = 1, 2, 3, \ldots$ (i.e., by considering data points $\{x(0), x(\alpha\Delta), \ldots\}$

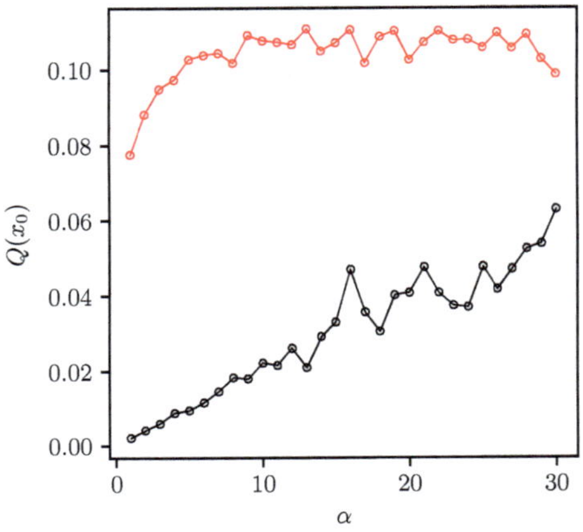

Fig. 19.2 Dependence of $Q(x_0)$ on the scaling α for time series of spatial positions of a bead trapped in optical tweezers (black) and of clear-sky index data (red). We chose x_0 around the peak of the respective probability distribution functions of the time series. It appears that for time-scales greater than 8 s ($\alpha = 8$) the time series of clear-sky index data contains jumps with amplitude $\sigma_\xi^2 \simeq 0.11$ for the chosen bin. Source from [2]

only).[2] This scaling changes the estimated drift coefficient a to αa, the diffusion coefficient b to $\sqrt{\alpha}b$, and the estimated jump amplitude σ_ξ^2 to $\alpha^2\sigma_\xi^2$, so that $Q(x)$ scales as $\alpha^2 Q(x)$. Checking the dependence of $Q(x)$ on $\tau = \alpha\Delta$ (with $\tau <$ minimum [correlation time scale of process: mean waiting time of two jumps, i.e. $1/\lambda$]), then indicates a possibly diffusive or jumpy behavior [2].

For coarse scales ($\alpha \gg 1$), we expect $Q(x)$ to take on non-vanishing values. For small scales ($\alpha = \mathcal{O}(1)$) and diffusive processes, $Q(x)$ approaches zero since the Brownian-type (Wiener-type) behavior of the process produces a continuous trajectory [32]. Thus, the small-scale behavior of $Q(x)$ for $\alpha = \mathcal{O}(1)$ is an indicator for rapid changes or jumps in a given time series.

The details of data we analyze here have been presented in [1, 33–35]. The first time series is a measurement of the spatial position (in one-direction) of a dielectric bead trapped in optical tweezers. The positional time series were recorded over a period of 3 s each with a sampling rate of 22 kHz (see [34] for further details), and a data-driven estimation of the first- and second-order conditional moments confirmed the diffusive nature of the bead dynamics [34]. In Fig. 19.2, we show the dependence of $Q(x)$ on the scaling coefficient α, which clearly indicates a continuous diffusion process, as expected.

The second time series is 12-months measurement (with a sampling rate of 1Hz, measured in Hawaii) of clear-sky index data $x(t) = I(t)/I_{cs}$, where $I(t)$ and I_{cs} are the measured solar irradiance and its theoretical prediction under clear-sky conditions

---

[2]We note that for $\alpha = 2$, one can start from $x(0)$ and $x(\Delta)$, then find two coarse samples of time series which are not statistically independent, of course. We can average the estimated KM conditional moments from two series. For $\alpha = n$, we have n possible coarse samples.

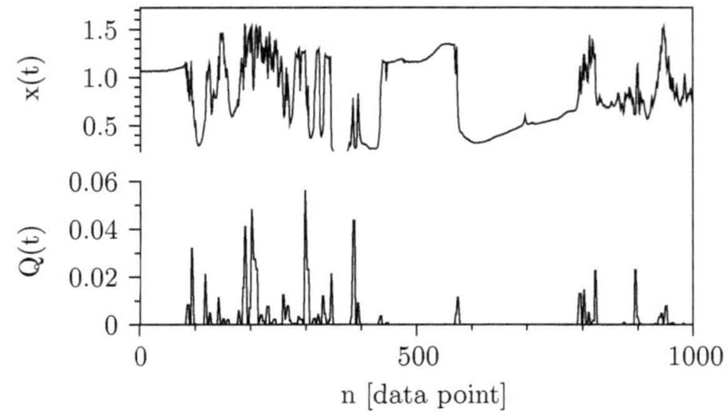

Fig. 19.3 Time-dependent $Q(t)$ with $\alpha = 1$ for time series of clear-sky index data

at a given latitude and longitude, respectively (see Refs. [1, 33] for further details). A jumpy stochastic behavior of solar irradiance represents a major obstacle in the power production as it influences not only the availability of energy, but also the stability of the entire power grid [1]. Moreover, it requires expensive technical solutions, such as fast reserves or storage systems in power supply to overcome and compensate such fluctuations. For this time series, our analyses indicate a discontinuous trajectory (see Fig. 19.3) with jumps that can be traced back to on-off fluctuations of the cloud structure generated by turbulence in the atmosphere [1]. It appears that for time-scales greater than 8 s ($\alpha = 8$) the time series of clear-sky index data contains jumps with amplitude $\sigma_\xi^2 \simeq 0.11$.

19.3.1 Jump Discontinuity in Non-stationary Time Series

Using kernel-based estimation of higher-order KM conditional moments, we can find time dependent $Q(t) = Q(x(t))$, which will be applicable also for nonstationary time series [32, 36, 37]. As we have seen in Chap. 17, this allows also for a time-resolved estimation of local characteristics of a time series, such as drift and diffusion coefficients as well as the jump rate and jump amplitude.

In Fig. 19.3, using a Gaussian kernel with bandwidth $h = 0.1$, we plot the time dependent $Q(t)$ for solar clear-sky data, with $\alpha = 1$ as well as the original time series. The jump amplitudes at this scale are presented in the figure. In Fig. 19.4, same plot for positional time series of trapped particle in optical tweezers is given, which amplitude of $Q(t)$ shows the absence of jumps (jumps with very small amplitudes).

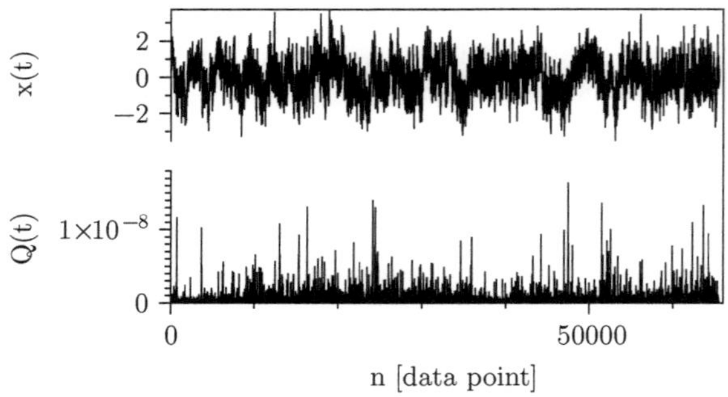

Fig. 19.4 Time-dependent $Q(t)$ with $\alpha = 1$ for normalised time series of tapped particle in optical tweezers is given. The amplitude of $Q(t)$ shows the jumps with very small amplitudes

Problems

19.1 Jump Detection Measure Q(x)

Reproduce the plots in Fig. 19.1. The details of linear, nonlinear and jump-diffusion are given in the caption of Figs. 18.1 and 18.2. Choose $\tau \in \{10^{-6}, 2 \times 10^{-6}, 4 \times 10^{-6}, 6 \times 10^{-6}, \ldots, 10^{-3}\}$, in integration of corresponding stochastic equations in Euler–Maruyama scheme.

References

1. M. Anvari, G. Lohmann, M. Wächter, P. Milan, E. Lorenz, D. Heinemann, M.R. Rahimi Tabar, J. Peinke, New J. Phys. **18**, 063027 (2016)
2. K. Lehnertz, L. Zabawa, M.R.R. Tabar, New J. Phys. **20**, 113043 (2018)
3. M.B. Weissman, Rev. Mod. Phys. **60**, 537 (1988)
4. R.T. Wakai, D.J.V. Harlingen, Appl. Phys. Lett. **49**, 593 (1986)
5. C.T. Rogers, R.A. Buhrman, H. Kroger, L.N. Smith, Appl. Phys. Lett. **49**, 1107 (1986)
6. M. Matsuda, S. Kuriki, Appl. Phys. Lett. **53**, 621 (1988)
7. K.S. Ralls, R.A. Buhrman, Phys. Rev. Lett. **60**, 2434 (1988)
8. D.H. Cobden, A. Savchenko, M. Pepper, N.K. Patel, Rev. Lett. **69**, 502 (1992)
9. I. Bloom, A.C. Marley, M.B. Weissman, Phys. Rev. Lett. **71**, 4385 (1993)
10. P.D. Dresselhaus, L. Ji, S. Han, J.E. Lukens, K.K. Likharev, Phys. Rev. Lett. **72**, 3226 (1994)
11. M.J. Ferrari, M. Johnson, F.C. Wellstood, J.J. Kingston, T.J. Shaw, J. Clarke, J. Low Temp. Phys. **94**, 15 (1994)
12. R.J.P. Keijsers, O.I. Shklyarevskii, H. van Kempen, Phys. Rev. Lett. **77**, 3411 (1996)
13. A.L. Efros, M. Rosen, Phys. Rev. Lett. **78**, 1110 (1997)
14. E. Shung, T.F. Rosenbaum, S.N. Coppersmith, G.W. Crabtree, W. Kwok, Phys. Rev. B **56**, R11431 (1997)
15. M. Einax, W. Dieterich, P. Maass, Rev. Mod. Phys. **85**, 921 (2013)
16. L. Gammaitoni, P. Hänggi, P. Jung, F. Marchesoni, Rev. Mod. Phys. **70**, 223 (1998)
17. D. Brockmann, L. Hufnagel, Phys. Rev. Lett. **98**, 178301 (2007)
18. S. Vaughan, Philos. Trans. R. Soc. Lond. A: Math. Phys. Eng. Sci. **371**, 20110549 (2013)
19. T.M. Lenton, Nat. Clim. Chang. **1**, 201 (2011)
20. M. Scheffer et al., Science **338**, 344 (2012)
21. Y. Aït-Sahalia, J. Financ. **57**, 2075 (2002)

22. S.S. Lee, P.A. Mykland, Rev. Financ. Stud. **21**, 2535 (2008)
23. B. Goswami, N. Boers, A. Rheinwalt, N. Marwan, J. Heitzig, S.F. Breitenbach, J. Kurths, Nat. Commun. **9**, 48 (2018)
24. D. Colquhoun, A. Hawkes, Proc. R. Soc. Lond. B **211**, 205 (1981)
25. L.S. Liebovitch, Ann. Biomed. Eng. **16**, 483 (1988)
26. S. Martinez-Conde, S.L. Macknik, D.H. Hubel, Nat. Rev. Neurosci. **5**, 229 (2004)
27. H.F. Credidio, E.N. Teixeira, S.D. Reis, A.A. Moreira, J.S. Andrade, Sci. Rep. **2**, 920 (2012)
28. M.J. Plank, E.A. Codling, Ecology **90**, 3546 (2009)
29. Y. Aït-Sahalia, J. Jacod, Testing for jumps in a discretely observed process. Ann. Stat. **37**, 184 (2009)
30. S. Lee, P.A. Mykland, Jumps in financial markets: A new nonparametric test and jump clustering. Rev. Financ. Stud. **21**, 2535 (2008)
31. M. Gupta, J. Gao, C. Aggawal, J. Han, *Outlier Detection for Temporal Data, Systhesis Lectures on Data Mining and Knowledge Discovery* (Morgan and Claypool Publishers, San Rafael, 2014)
32. M. Anvari, K. Lehnertz, M.R. Rahimi Tabar, J. Peinke, Sci. Rep. **6**, 35435 (2016)
33. A. Madanchi, M. Absalan, G. Lohmann, M. Anvari, Tabar M. Rahimi, Sol. Energy **144**, 1 (2017)
34. S.M. Mousavi et al., Sci. Rep. **7**, 4832 (2017)
35. M.R.R. Tabar, M. Anvari, G. Lohmann, D. Heinemann, M. Wächte, P. Milan, E. Lorenz, J. Peinke, Eur. Phys. J. Spec. Top. **223**, 1 (2014)
36. A.M. Van Mourik, A. Daffertshofer, P.J. Beek, Biol. Cybern. **94**, 233 (2006)
37. D. Lamouroux, K. Lehnertz, Phys. Lett. A **373**, 3507 (2009)

Chapter 20
Reconstruction Procedure for Writing Down the Langevin and Jump-Diffusion Dynamics from Empirical Uni- and Bivariate Time Series

In this chapter we present the steps of *reconstruction procedure* for writing down the Langevin and jump diffusion stochastic dynamical equations for uni- and bivariate time series, sampled with time intervals τ.

20.1 The Reconstruction Procedure, Diffusion Processes

20.1.1 One Dimensional Time Series

The reconstruction procedure for diffusion processes—the Langevin dynamics—in one dimension has the following steps:

- Identification of the order parameters, such as the state variable of a system or its increment, in a given lag (scale) or its log return.
- Verifying the Markovianity.
  Use the Wilcoxon test or check the Chapman–Kolmogorov equation, Eq. (16.12); see Sect. 16.4 to verify that the time series is Markov and estimate the *Markov–Einstein* time scale.
- Estimate the Kramers–Moyal conditional moments. For stationary time series sampled with time intervals τ, discretize the state space of the process x into n bins, which span its complete range, and then estimate the Kramers–Moyal (KM) conditional moments

$$K^{(m)}(x,\tau) = \langle (x(t+\tau) - x(t))^m |_{x(t)=x} \rangle \quad .$$

  The conditional average must be calculated for each bin separately, and the brackets indicate time averaging.

The original version of this chapter was revised: Belated correction has been incorporated. The correction to this chapter is available at https://doi.org/10.1007/978-3-030-18472-8_24

M. R. Rahimi Tabar, *Analysis and Data-Based Reconstruction of Complex Nonlinear Dynamical Systems*, Understanding Complex Systems,
https://doi.org/10.1007/978-3-030-18472-8_20

- Check the presence of microstructure (measurement) noise.
  Use the method described in Sect. 17.2 to check the presence of microstructure (measurement) noise and, if it is present, estimate the statistical moments of the microstructural noise and the KM conditional moments of the original time series, after removing of the noise contribution.
- Check the Pawula theorem.
  Use the criterion $K^{(4)}(x,\tau) \simeq 3(K^{(2)}(x,\tau))^2$ [Eq. (18.11)] to judge whether the fourth-order KM coefficients approaches zero. By plotting the x-dependence of $K^{(4)}(x,\tau)$ and $3(K^{(2)}(x,\tau))^2$ and considering their statistical errors, decide whether the time series is diffusive over the entire range of x.
- Estimate the "apparent" drift and diffusion coefficients.
  The "apparent" (finite τ) drift and diffusion coefficients are determined by the KM conditional moments $K^{(m)}(x,\tau)$ as,

$$D_\tau^{(m)}(x) \simeq \frac{1}{m!\tau} K^{(m)}(x,\tau),$$

  for $m = 1, 2, \cdots$, respectively. We recall that $D_\tau^{(m)}(x)$ is related to $M_\tau^{(m)}(x)$ by, $D_\tau^{(m)}(x) = M_\tau^{(m)}(x)/m!$.
- Estimate the "true" (vanishing limit of τ) drift and diffusion coefficients.

  Invoke, for example, polynomial ansatzes for the "true" drift and diffusion coefficients, say first- or third-order polynomials for the drift and quadratic or quartic function for the diffusion coefficients. Based on the assumed functional form of the drift and diffusion coefficients solve Eq. (18.23) and compute $W_{n,x}(x',t)$ for $n = 1$ and 2. For linear drift and quadratic diffusion coefficients, one uses Eqs. (18.26) and (18.27). By fitting (or by using an optimization procedure) the estimated KM conditional moments, $K^{(1)}(x,\tau)$ and $K^{(2)}(x,\tau)$ for finite τ with expressions $W_{1,x}(x'=x, t=\tau)$ and $W_{2,x}(x'=x, t=\tau)$, one extracts the unknown coefficients in the polynomial ansatzes. Therefore, the drift and diffusion coefficients are estimated in the vanishing limit τ.
- Derive an effective stochastic equation.
  Use the estimated drift and diffusion coefficients in the vanishing limit τ to write down the (Itô) Langevin dynamical equation,

$$\frac{dx}{dt} = D^{(1)}(x) + \sqrt{2D^{(2)}(x)}\Gamma(t)$$

  where $\Gamma(t)$ is zero-mean Gaussian white noise.
- Check for self-consistency.
  See Sect. 16.7 for various possible self-consistency checks.
- Non-stationary time series.
  Use Nadaraya–Watson (kernel) estimator to estimate time-dependent conditional moments $K^{(m)}(x,t,\tau)$ and check at each step of the time the criterion $K^{(4)}(x,t,\tau) \simeq 3(K^{(2)}(x,t,\tau))^2$ to ensure that the process is diffusive. Use the esti-

mated $K^{(1)}(x,t,\tau)$ and $K^{(2)}(x,t,\tau)$ to estimate the drift and diffusion coefficients and write down the Langevin dynamics with time-dependent drift and diffusion coefficients. In this way, the computed drift and diffusion coefficients will be the apparent ones.

20.1.2 Two Dimensional (Bivariate) Time Series

Two-dimensional Langevin dynamics is given by,

$$\begin{pmatrix} \mathrm{d}x_1(t) \\ \mathrm{d}x_2(t) \end{pmatrix} = \underbrace{\begin{pmatrix} D_1^{(1)}(\mathbf{x},t) \\ D_2^{(1)}(\mathbf{x},t) \end{pmatrix}}_{\text{drift}} \mathrm{d}t + \underbrace{\begin{pmatrix} g_{11}(\mathbf{x},t) & g_{12}(\mathbf{x},t) \\ g_{21}(\mathbf{x},t) & g_{22}(\mathbf{x},t) \end{pmatrix}}_{\text{diffusion matrix}} \begin{pmatrix} \mathrm{d}W_1 \\ \mathrm{d}W_2 \end{pmatrix} \tag{20.1}$$

where $W_1(t)$ and $W_2(t)$ are two independent Wiener processes. The vector $\mathbf{D}^{(1)}(\mathbf{x},t) = (D_1^{(1)}, D_2^{(1)})$ is a two-dimensional drift vector and the diffusion coefficient is a 2×2-matrix given by $\mathbf{D}^{(2)}(\mathbf{x},t) = \mathbf{g}\mathbf{g}^T$, i.e., $D_{ij}^{(2)} = \sum_k g_{ik} g_{jk}$. There are 6 unknown functions in Eq. (21.1), namely, $D_i^{(1)}(\mathbf{x},t)$ and $g_{ij}(\mathbf{x},t)$ with $i, j = 1$ and 2.

- Estimate the "apparent" drift vector and diffusion matrix.
  Partition the two-dimensional state space into a set of finite, but small two-dimensional volume elements. Each bin α is located at point $\mathbf{x}_\alpha$. The points $\mathbf{x}(t)$ are taken from the bin located at $\mathbf{x}_\alpha$.
  Next, use, as a first approximation, the following relations to estimate the drift vector and diffusion matrix; see Eqs. (9.7) and (9.8):

$$D_i^{(1)}(\mathbf{x},t,\mathcal{O}(\tau)) \approx \frac{1}{\tau}\langle x_i(t+\tau) - x_i(t)\rangle\Big|_{\mathbf{x}(t)=\mathbf{x}}$$
$$D_{ij}^{(2)}(\mathbf{x},t,\mathcal{O}(\tau)) \approx \frac{1}{\tau}\langle (x_i(t+\tau) - x_i(t))\,(x_j(t+\tau) - x_j(t))\rangle\Big|_{\mathbf{x}(t)=\mathbf{x}} \quad .$$

  The corrections will be of order of $\mathcal{O}(\tau)$ for drift vector and diffusion matrix.
  As the second approximation we use the following expressions for the drift vector and diffusion matrix; see Eqs. (16.14) and (16.18) [1]

$$D_i^{(1)}(\mathbf{x},t,\mathcal{O}(\tau)) \approx \frac{1}{\tau}\langle x_i(t+\tau) - x_i(t)\rangle\Big|_{\mathbf{x}(t)=\mathbf{x}}$$
$$D_{ij}^{(2)}(\mathbf{x},t,\mathcal{O}(\tau^2)) \approx \frac{1}{\tau}\langle (x_i(t+\tau) - x_i(t) - \tau D_i^{(1)}(\mathbf{x},t,\mathcal{O}(\tau)))$$
$$(x_j(t+\tau) - x_j(t) - \tau D_j^{(1)}(\mathbf{x},t,\mathcal{O}(\tau)))\rangle\Big|_{\mathbf{x}(t)=\mathbf{x}} \quad . \tag{20.2}$$

Here, the corrections will be of the order of $\mathcal{O}(\tau)$ and $\mathcal{O}(\tau^2)$ for the drift vector and diffusion matrix, respectively. To estimate the drift vector components in high precisions one may use the expressions presented in Eqs. (21.9), (21.10) and (21.11) (with $j = 1$).

The finite τ expansions of the KM conditional moments $K_i^{(1)}$ and $K_{ij}^{(2)}$ with known $D_i^{(1)}(\mathbf{x}, t)$ and $D_{ij}^{(2)}(\mathbf{x}, t)$ are obtained in [2].[1]
In what follows, we use the expressions given by Eq. (20.2) to estimate the drift vector and diffusion coefficient. We omit the x-, t- and τ-dependence of $D_i^{(1)}$ and $D_{ij}^{(2)}$ to enhance readability.

- The drift vector and components of diffusion matrix **g**.
  The components of drift vector are determined from the time series using the first relation in (20.2). However, from second relation in (20.2) for two-dimensional time series we estimate $D_{11}^{(2)}$, $D_{12}^{(2)} = D_{21}^{(2)}$ and $D_{22}^{(2)}$, where by construction $D_{ij}^{(2)}(\mathbf{x}, t)$ is a symmetric matrix. Therefore, we should determine the four unknown functions $g_{ij}(\mathbf{x})$ from three estimated functions $D_{ij}^{(2)}(\mathbf{x})$. This means that for bivariate time series the inverse transformation, i.e., one from the time series to the Langevin dynamics, is not so direct.

[1] The finite τ expansions of the KM conditional moments $K_i^{(1)}$ and $K_{ij}^{(2)}$ with known $D_i^{(1)}(\mathbf{x}, t)$ and $D_{ij}^{(2)}(\mathbf{x}, t)$ are

$$K_i^{(1)} = \tau D_i^{(1)} + \frac{\tau^2}{2}\left(\sum_k D_k^{(1)} \partial_{x_k} D_i^{(1)} + \sum_{kl} D_{kl}^{(2)} \partial_{x_l} \partial_{x_k} D_i^{(1)}\right) + \mathcal{O}(\tau^3)$$

$$\begin{aligned} K_{ij}^{(2)} = 2\tau D_{ij}^{(2)} + \tau^2 \Big[& D_i^{(1)} D_j^{(1)} \\ & + \sum_k \left(D_{jk}^{(2)} \partial_{x_k} D_i^{(1)} + D_{ik}^{(2)} \partial_{x_k} D_j^{(1)}\right) \\ & + \sum_k D_k^{(1)} \partial_{x_k} D_{ij}^{(2)} + \sum_{kl} D_{kl}^{(2)} \partial_{x_l} \partial_{x_k} D_{ij}^{(2)} \Big] \\ + \mathcal{O}(\tau^3). \end{aligned}$$

where

$$K_i^{(1)}(\mathbf{x}, t, \tau) = \langle [\mathbf{x}(t+\tau) - \mathbf{x}(t)]_i \rangle |_{\mathbf{x}(t)=\mathbf{x}}$$

$$K_{ij}^{(2)}(\mathbf{x}, t, \tau) = \big\langle [\mathbf{x}(t+\tau) - \mathbf{x}(t)]_i \, [\mathbf{x}(t+\tau) - \mathbf{x}(t)]_j \big\rangle |_{\mathbf{x}(t)=\mathbf{x}}$$

where, $\partial_{x_i} = \partial/\partial x_i$.

We present two approaches to deal with this issue. Let us assume that the symmetric diffusion matrix $D_{ij}^{(2)}(\mathbf{x}, t)$ is a positive definite matrix. For given x and t, the 2×2 diffusion matrix

$$D_{ij}^{(2)} = \begin{pmatrix} D_{11}^{(2)} & D_{12}^{(2)} \\ D_{12}^{(2)} & D_{22}^{(2)} \end{pmatrix}$$

is positive definite if its two eigenvalues λ_1 and λ_2 are positive, or $D_{11}^{(2)} > 0$ and $D_{11}^{(2)} D_{22}^{(2)} - (D_{12}^{(2)})^2 > 0$ (see Problem 9.3; in fact the diffusion matrix is semi-positive definite).

Method I for Estimation of Matrix g, Diagonalization

As the first method of deriving the matrix $\mathbf{g}$ from $\mathbf{D}^{(2)}$, one can compute a root of the diffusion matrix $\mathbf{D}^{(2)}$ [3–5]. A root $\mathbf{R}$ of a diagonalizable matrix $\mathbf{M}$ is a matrix satisfying, $\mathbf{R}^2 = \mathbf{M}$. With diagonalization $\mathbf{M} = \mathbf{PAP}^{-1}$, where $\mathbf{P}$ is an invertible matrix and $\mathbf{A}$ is a diagonal matrix, a root $\mathbf{R}$ is computed by:

$$\mathbf{R} = \mathbf{P}\left(\sqrt{\mathbf{A}}\right)\mathbf{P}^{-1} \tag{20.3}$$

where the root of a diagonal matrix is the matrix of the roots of all the entries. The root of a matrix is not unique, because we have the choice of $\mathbf{P}$, and for every orthonormal matrix $\mathbf{O}$ the matrix $\widetilde{\mathbf{P}} = \mathbf{OP}$ yields a diagonalization $\mathbf{M} = \widetilde{\mathbf{P}}\widetilde{\mathbf{A}}\widetilde{\mathbf{P}}^{-1}$, too. It follows that for every Fokker–Planck equation we have a corresponding family of Langevin equations for various $\mathbf{O}$ with $\mathbf{g} = \widetilde{\mathbf{g}}\mathbf{O}$. Fortunately, this does not lead to any difficulty, as the Langevin equations specify the same stochastic process that can be explained by the fact that the rotated two-dimensional Wiener processes $d\widetilde{\mathbf{W}} = \mathbf{O}\, d\mathbf{W}$ is the same as the process $d\mathbf{W}$.

In practice, the square-root $\sqrt{\mathbf{D}^{(2)}}$ is to be calculated by diagonalizing the matrix $\mathbf{D}^{(2)}$ (by an orthogonal transformation), taking the square root of each element of the diagonalized matrix and transforming the result back into the original system of coordinates. This should be done in each bin, when the state space is decomposed into, for example, $n \times n$ bins.

Method II for Estimation of Matrix g

Via Eq. (20.1), several possibilities exist for constructing the matrix $\mathbf{g}$ from the symmetric diffusion matrix $\mathbf{D}^{(2)}$, i.e., $\mathbf{D}^{(2)}(\mathbf{x}, t) = \mathbf{g}\mathbf{g}^T$. For 2×2 diffusion matrix with

$$\begin{pmatrix} a & c \\ c & b \end{pmatrix} = \mathbf{g}\mathbf{g}^T$$

the matrix $\mathbf{g}$ can be produced by the lower triangular matrix $\mathbf{g}_l$, the upper triangular matrix $\mathbf{g}_u$, and symmetric matrix $\mathbf{g}_s$ as,

$$
\mathbf{g}_l = \begin{pmatrix} \sqrt{a} & 0 \\ \frac{c}{\sqrt{a}} & \sqrt{b - \frac{c^2}{a}} \end{pmatrix},
$$

$$
\mathbf{g}_u = \begin{pmatrix} \sqrt{b - \frac{c^2}{a}} & \frac{c}{\sqrt{a}} \\ 0 & \sqrt{b} \end{pmatrix},
$$

$$
\mathbf{g}_s = \begin{pmatrix} h & k \\ k & l \end{pmatrix} \tag{20.4}
$$

respectively. In the symmetric case we need to have $h^2 + k^2 = a$, $k^2 + l^2 = b$ and $k(l + h) = c$, which have positive solutions for k, h and l.[2] The generalization of the results for higher-dimensional diffusion matrix $d \geq 3$ is readily shown for $\mathbf{g}_l$ and $\mathbf{g}_u$ [6].
For example, the lower triangular matrix may be determined by the Cholesky decomposition method, which is also applicable to higher dimensions; see, for example, [7] and references therein. Therefore, using the relation between the diffusion matrix $\mathbf{D}^{(2)}$ and the matrix $\mathbf{g}$, we may choose one of the possible forms for $\mathbf{g}$ via relations (20.4) to compute its entries.

As mentioned earlier, the solution is not unique. We note that in the Fokker–Planck equation we need to estimate the drift vector and diffusion matrix, and different choices for $\mathbf{g}$ will not influence its solution for the conditional probability distribution function; see Eq. (9.4). This means that all two-dimensional Langevin equations with given $\mathbf{g}$, which satisfy $\mathbf{D}^{(2)} = \mathbf{g}\mathbf{g}^T$, will produce same conditional probability distribution function, as will be determined from two-dimensional Fokker–Planck equation; see Eqs. (9.20) and (9.21). It is evident that deriving the matrix $\mathbf{g}$ from relations (20.4) (or via Cholesky decomposition in higher dimensions) needs less efforts than using the diagonalization method, which should be done in each bin in the two-dimensional state space. For nonstationary bivariate time series one can use two-dimensional kernels to estimate time-dependent KM coefficients.

2

$$
h = \frac{ac^2\sqrt{\frac{-2\sqrt{-c^4(c^2-ab)}+ac^2+bc^2}{a^2-2ab+b^2+4c^2}} + \sqrt{-c^4(c^2-ab)}\sqrt{\frac{-2\sqrt{-c^4(c^2-ab)}+ac^2+bc^2}{a^2-2ab+b^2+4c^2}}}{c^3}
$$

$$
l = \frac{bc^2\sqrt{\frac{-2\sqrt{-c^4(c^2-ab)}+ac^2+bc^2}{a^2-2ab+b^2+4c^2}} + \sqrt{-c^4(c^2-ab)}\sqrt{\frac{-2\sqrt{-c^4(c^2-ab)}+ac^2+bc^2}{a^2-2ab+b^2+4c^2}}}{c^3}
$$

$$
k = \sqrt{\frac{-2\sqrt{c^4(ab-c^2)} + ac^2 + bc^2}{a^2 - 2ab + b^2 + 4c^2}},
$$

with $a^2 - 2ab + b^2 + 4c^2 \neq 0$, $ab - c^2 \geq 0$ and $c \neq 0$. There are 9 other real solutions for h, l amd k, depending on the values of a, b and c.

20.1.3 An R Package for Reconstruction of One- and Two-Dimensional Stochastic Diffusion Processes: White Noise-Driven Langevin Dynamics

An *R* package has been developed by the research group, "Turbulence, wind energy and stochastic (TWiSt)" at Carl von Ossietzky University of Oldenburg, which extracts the (stochastic) evolution equation underlying a set of uni- and bi-variate time series [8]. The authors assumed that the processes under study are continuous, and estimated the drift (vector) and diffusion (matrix) coefficients.

20.2 The Reconstruction Procedure for the Lévy Noise-Driven Langevin Equation

Follow the steps described in Sect. 10.6 to construct a Lévy noise-driven Langevin equation for one-dimensional time series and estimate the Lévy stability index α, as well as deterministic drift coefficient $D^{(1)}(x,t)$ and multiplicative "diffusion" $g(x,t)$, see also Sect. 21.3.

20.3 The Reconstruction Procedure and Jump-Diffusion Stochastic Dynamics

20.3.1 One-Dimensional Time Series

After identification of the order parameter, verifying the Markov nature of a time series, and checking the presence of microstructure (measurement) noise, one needs to validate the Pawula theorem. For the processes that are not diffusive, the criterion $K^{(4)}(x,\tau) \simeq 3(K^{(2)}(x,\tau))^2$ will not be satisfied in the entire range or some portions of the x interval. By plotting the x-dependence of $K^{(4)}(x,\tau)$ and $3(K^{(2)}(x,\tau))^2$ and considering their statistical errors, we will find substantial deviation from the criterion $K^{(4)}(x,\tau) \simeq 3(K^{(2)}(x,\tau))^2$, which indicates that the fourth-order KM coefficient does not vanish. Therefore, this provides a hint that the time series may have jump discontinuity.

The reconstruction procedure for time series that has jump discontinuity consists of the following steps:

- Estimate the Kramers–Moyal conditional moments.
  Estimate the KM conditional moments

$$K^{(m)}(x,\tau) = \langle (x(t+\tau) - x(t))^m|_{x(t)=x} \rangle$$

for $m = 1, \cdots, 6$, where the brackets indicate time averaging for stationary time series.

- Checke the presence of jump discontinuity.
  Use the method describe in Sect. 19.2 to check the necessary condition for having jump discontinuity. For some x_0, such as around the mean of time series, plot $Q(x_0, \alpha)$

$$Q(x_0, \alpha) = \frac{K^{(6)}(x_0, \alpha)}{5K^{(4)}(x_0, \alpha)}$$

  verses the scaling factor α; see Fig. 19.2 for example.
- Estimate the "apparent" KM coefficients.
  Estimate the KM coefficients $M^{(m)}(x)$ as the first approximation by

$$M^{(m)}(x) \approx \frac{1}{\tau} K^{(m)}(x, \tau)$$

  where τ is the sampling time interval.
  Suppose that we would like to model the time series via the following jump-diffusion dynamics

$$dx(t) = N(x, t)dt + \sqrt{D(x, t)}dW(t) + \xi \, dJ(t),$$

  where $\{W(t), t \geq 0\}$ is a scalar Wiener process, $N(x, t)$ is the drift function, $D(x, t)$ is the diffusion function, and $J(t)$ is a time-homogeneous Poisson jump process. The jump has the rate $\lambda(x)$ and can be state-dependent and size ξ that we assume it to have zero mean Gaussian distribution with variance $\sigma_\xi^2(x)$. In this modeling one needs to estimate the KM coefficients $m \in \{1, 2, \cdots, 6\}$. If one would like to model the jumps with the variance gamma model, then one needs to estimate the KM coefficients of order $m \in \{1, 2, \cdots, 8\}$; see Sect. 12.3.
- The functions in jump-diffusion modeling.
  The unknown functions $N(x, t)$, $D(x, t)$, $\lambda(x)$, and $\sigma_\xi^2(x)$ are determined in term of the KM coefficients $M^{(m)}(x)$ through

$$M^{(1)}(x) = N(x, t)$$

$$M^{(2)}(x) = [D(x, t) + \langle \xi^2 \rangle \lambda(x)]$$

$$M^{(2m)}(x) = \langle \xi^{2m} \rangle \lambda(x), \quad \text{for } 2m > 2 \quad . \tag{20.5}$$

As we saw in Sect. 12.2, with Gaussian jump amplitude with variance σ_ξ^2, i.e., $\xi \sim \mathcal{N}(0, \sigma_\xi^2)$, we use the relation $\langle \xi^{2n} \rangle = \frac{(2n)!}{2^n n!} \langle \xi^2 \rangle^n$ and $\langle \xi^{2n-1} \rangle = 0$, for $n = 1, 2, \cdots$. Now, using the last relation in Eq. (20.5) with $m = 2$ and 3, we estimate the jump amplitude $\sigma_\xi^2(x)$ and the jump rate $\lambda(x)$ by

$$\sigma_\xi^2(x) = \frac{M^{(6)}(x)}{5M^{(4)}(x)}, \qquad \lambda(x) = \frac{M^{(4)}(x)}{3\sigma_\xi^4(x)} \quad . \tag{20.6}$$

Once the jump components are identified, the second moment $M^{(2)}(x)$ identifies the diffusion function $D(x)$ and the first moment gives us the estimate for the drift function $N(x)$. In practice, we can also define the averaged jump amplitude as $\bar{\sigma_\xi}^2 = \frac{1}{n}\sum_{n=1}^{N}\sigma_\xi^2(x_n)$, where we have assumed that the state space of the process is discretized. The conditional average must be calculated for every x_n (with binning into N bins) separately.

- Checking self-consistency.
  See Sect. 16.7 and choose an appropriate self-consistency check.
- Non-stationary time series.
  Use Nadaraya–Watson (kernel) estimator to estimate the time-dependent KM conditional moments $K^{(m)}(x, t, \tau)$ and, therefore, the time-dependent $N(x)$, $D(x)$ and $\lambda(t)$. Finally, one can estimate at each step of the time the function $Q(t, \alpha = 1)$ to determine the time-dependent jump amplitude σ_ξ^2.

20.3.2 Two-Dimensional (Bivariate) Time Series

Two-dimensional jump-diffusion dynamics is given by [9, 10]:

$$\begin{pmatrix} \mathrm{d}x_1(t) \\ \mathrm{d}x_2(t) \end{pmatrix} = \underbrace{\begin{pmatrix} D_1^{(1)}(\mathbf{x}, t) \\ D_2^{(1)}(\mathbf{x}, t) \end{pmatrix}}_{\text{drift}} \mathrm{d}t + \underbrace{\begin{pmatrix} g_{11}(\mathbf{x}, t) & g_{12}(\mathbf{x}, t) \\ g_{21}(\mathbf{x}, t) & g_{22}(\mathbf{x}, t) \end{pmatrix}}_{\text{diffusion matrix}} \begin{pmatrix} \mathrm{d}W_1 \\ \mathrm{d}W_2 \end{pmatrix} + \underbrace{\begin{pmatrix} \xi_{11} & \xi_{12} \\ \xi_{21} & \xi_{22} \end{pmatrix} \begin{pmatrix} \mathrm{d}J_1 \\ \mathrm{d}J_2 \end{pmatrix}}_{\text{jumps}} . \tag{20.7}$$

The functions $D_1^{(1)}(\mathbf{x}, t)$ and $D_2^{(1)}(\mathbf{x}, t)$ are the components of the drift vector, which may depend on both (x_1, x_2), while g_{ij} are the elements of the diffusion matrix. Generally speaking, g_{ij} are functions of (x_1, x_2). We assume that the jump amplitudes ξ_{ij} are sampled from a Gaussian distributions with zero mean $\langle \xi_{ij} \rangle = 0$ and variance $\langle (\xi_{ij})^2 \rangle = \sigma_{\xi_{ij}}^2(\mathbf{x}, t) \equiv s_{ij}$, and are independent. The two Poisson processes $J_{1,2}$ are characterised by the jump rates λ_1 and λ_2, which may depend on (x_1, x_2). The Wiener processes $\{W_{1,2}(t)$ with $t \geq 0\}$ are assumed to be independent, and the Poisson-distributed jumps $J_{1,2}$ with rates $\lambda_{1,2}$ are uncorrelated. The eight stochastic variables, namely, two Wiener processes, four jump amplitudes and two Poisson jump processes are assumed to be independent of each other. Stochastic jump-diffusion equation, Eq. (20.7), has 12 unknown functions, $D_i^{(1)}$, g_{ij}, s_{ij} and λ_i, with $i, j = 1$ and 2, which in general are state dependent.

The two-dimensional KM coefficients of order $[k_1, k_2]$, are defined by

$$\mathcal{M}^{[k_1,k_2]}(\mathbf{x}, t) = \lim_{\tau \to 0} \frac{1}{\tau} \langle dx_1^{k_1}(t) dx_2^{k_2}(t) \rangle_t |_{x_1(t)=x_1, x_2(t)=x_2} \tag{20.8}$$

$$dx_i(t) = x_i(t+\tau) - x_i(t), \tag{20.9}$$

and the drift vector, the elements of the diffusion matrix, and the jump components are given in terms of the KM coefficients as follows,

$$\mathcal{M}^{[1,0]}(\mathbf{x},t) = N_1 \tag{20.10}$$

$$\mathcal{M}^{[0,1]}(\mathbf{x},t) = N_2 \tag{20.11}$$

$$\mathcal{M}^{[1,1]}(\mathbf{x},t) = g_{11}g_{21} + g_{12}g_{22} \tag{20.12}$$

$$\mathcal{M}^{[2,0]}(\mathbf{x},t) = g_{11}^2 + s_{11}\lambda_1 + g_{12}^2 + s_{12}\lambda_2 \tag{20.13}$$

$$\mathcal{M}^{[0,2]}(\mathbf{x},t) = g_{21}^2 + s_{21}\lambda_1 + g_{22}^2 + s_{22}\lambda_2 \tag{20.14}$$

$$\mathcal{M}^{[2,2]}(\mathbf{x},t) = s_{11}s_{21}\lambda_1 + s_{12}s_{22}\lambda_2 \tag{20.15}$$

$$\mathcal{M}^{[k_1,k_2]}(\mathbf{x},t) = [s_{11}^n s_{21}^m\lambda_1 + s_{12}^n s_{22}^m\lambda_2]\,\frac{(2n)!}{2^n n!}\frac{(2m)!}{2^m m!} \tag{20.16}$$

where $k_1 = 2n,\ k_2 = 2m,\quad n, m \in \mathbb{N}^+$. In the functions on the right side of Eqs. (20.10)–(20.16), we omitted the $\mathbf{x}$- and t-dependence to enhance readability.

Using the indices $[k_1, k_2]$ in the relations (20.16) as $[2, 2]$, $[4, 0]$, $[0, 4]$, $[0, 6]$ $[6, 0]$, $[2, 4]$ or $[4, 2]$, we estimate six unknown functions s_{ij} and λ_i, with $i, j = 1$ and 2 [10]. One can use some optimisation techniques, such as that proposed in Sect. 18.4.2, and can estimate these unknown functions.

Once the jump components are identified, using the conditional moments with indices $[k_1, k_2]$ as $[0, 2]$, $[2, 0]$ and $[1, 1]$, one computes g_{ij} by

$$M(\mathbf{x},t) = \begin{pmatrix} \mathcal{M}^{[2,0]}(\mathbf{x},t) - (s_{11}\lambda_1 + s_{12}\lambda_2) & \mathcal{M}^{[1,1]}(\mathbf{x},t) \\ \mathcal{M}^{[1,1]}(\mathbf{x},t) & \mathcal{M}^{[0,2]}(\mathbf{x},t) - (s_{21}\lambda_1 + s_{22}\lambda_2) \end{pmatrix}$$
$$= \begin{pmatrix} g_{11} & g_{12} \\ g_{21} & g_{22} \end{pmatrix} \cdot \begin{pmatrix} g_{11} & g_{21} \\ g_{12} & g_{22} \end{pmatrix} .$$

For positive definite $M(\mathbf{x},t)$ we use the methods presented in Sect. 20.1.2 to determine four unknown functions g_{ij} from the entries of the matrix $M(\mathbf{x},t)$. Finally, the two unknown drift functions $N_1(x_1, x_2)$ and $N_2(x_1, x_2)$ are determined from the conditional moments with indices $[1, 0]$ as $[0, 1]$, respectively, using Eqs. (20.10) and (20.11); see [10] for more details.

20.4 Other Methods for Reconstruction of Time Series

What we have learned so far is that using Markov properties of time series, one can study deterministic dynamics, as well as their stochastic contributions, including diffusion and jump parts. The constructed Langevin dynamics or jump-diffusion modeling enables us to reconstruct a time series that is similar to the original one in the statistical sense. There are, however, other methods for reconstruction of time series. Here, we give brief introduction to more recents ones.

20.4.1 Multiscale Reconstruction of Time Series

Nowroth et al. introduced a method that allows a reconstruction of time series based on higher-order multiscale statistics given by a hierarchical process [11–15]. The method is able to model time series, not only on a specific scale, but also for a range of scales. It is possible to generate complete new time series, or to model the next steps for a given sequence of data. The method itself is based on the joint n-point probability density that can be extracted directly from given data and, thus, no estimation of the parameters is necessary. The unconditional and conditional probability densities of the original and reconstructed time series are compared, and the ability to reproduce both is demonstrated, see [11] for more details.

20.4.2 Mapping Stochastic Processes onto Complex Networks

A general method was proposed in [16] by which a given stochastic process is mapped onto a complex network with distinct geometrical properties. The relation between the statistical properties of the stochastic time series, such as the intermittency and correlation length, and their stochastic characteristics, as well as the properties of their equivalent networks, were then studied. The KM coefficients are found in terms of the weights in the links of corresponding network. All possible dynamics, diffusive and jumpy behaviours are transformed into the network properties, such as the mean length, diameter, clustering, and average number of connections per node, and degree distribution of the vertices.

It is shown that mapping onto complex network is possible for time series $x(t)$ with given *Markov–Einstein* time scale t_M.[3] The important aspect in this method is that the corresponding time series can be reconstructed with high precision by means of a random walk on the network equivalent, which means that mapping

[3]Now let us consider a discrete stationary process $x(t)$, with unit *Markov–Einstein* time scale $t_M = 1$ (in units of the data lag). If k is the number of bins needed to precisely represent, or evaluate, the PDF of $x(t)$, the data are partitioned into k bins, with each bin having an equal number of data points. Each bin is represented by a node of the equivalent complex network of the series, with k being the number of the nodes in the network. Nodes i and j are linked if as the time increases the value of $x(t)$ in bin i changes to that in bin j in one time step. A weight w_{ij} is attributed to a link ij, which is the number of times that a given value of $x(t)$ changes from its value in bin i to bin j in one time step, and is normalized at each node. The transition matrix $\mathbf{W} = [w_{ij}]$ is, in general, not symmetric. Thus, the network is both *directed* and *weighted*. For some time series with a finite *Markov–Einstein* time scale $t_M > 1$, one can construct the transition matrix with entries given by $p[x(t)|x(t - t_M)]$, and attribute to each node *a set of arrays of data with length* t_M, see [16] for more details.

a time series onto complex networks is statistically bi-direction. This is the unique advantage of the method proposed in [16]; see also [17–27]. We believe that mapping stochastic processes and stochastic dynamical equations, such as the Langevin and jump diffusion, onto complex networks and vice versa provides a general avenue to deeper understanding of the dynamics of complex systems.

References

1. L. Rydin Gorjão, F. Meirinhos, Python KM, a Python package to calculate Kramers–Moyal coefficients, https://github.com/LRydin/KramersMoyal
2. J. Prusseit, K. Lehnertz, Phys. Rev. E **77**, 041914 (2008)
3. H. Risken, *The Fokker-Planck Equation* (Springer, Berlin, 1989)
4. F. Lenz, Statistical analysis and stochastic modelling of foraging Bumblebees, Doctoral thesis, Queen Mary University of London
5. A. Bahraminasab, D. Kenwright, A. Stefanovska, F. Ghasemi, P.V.E. McClintock, IET Syst. Biol. **2**, 48 (2008)
6. N. Schaudinnus, Stochastic modeling of biomolecular systems using the data-driven Langevin equation, Doctoral thesis, Albert-Ludwigs-Universität
7. F. Nikakhtar, M. Ayromlou, S. Baghram, S. Rahvar, M.R. Rahimi Tabar, R.K. Sheth, Mon. Not. R. Astron. Soc. **478**, 5296 (2018)
8. https://cran.r-project.org/web/packages/Langevin/index.html
9. M. Anvari, K. Lehnertz, M.R. Rahimi Tabar, J. Peinke, Sci. Rep. **6**, 35435 (2016)
10. L.R. Gorjão, J. Heysel, K. Lehnertz, M.R. Rahimi Tabar, arXiv:1907.05371
11. J. Peinke, M.R. Rahimi Tabar, M. Wächter, Annu. Rev. Condens. Matter Phys. **10** (2019)
12. A. Nawroth, J. Peinke, Phys. Lett. A **360**, 234 (2006)
13. R. Stresing, J. Peinke, New J. Phys. **12**, 103046 (2010)
14. A. Nawroth, R. Friedrich, J. Peinke, New J. Phys. **12**, 083021 (2010)
15. A. Hadjihosseini, M. Wächter, N.P. Hoffmann, J. Peinke, New J. Phys. **18**, 013017 (2016)
16. A.H. Shirazi, G.R. Jafari, J. Davoudi, J. Peinke, M.R. Rahimi Tabar, M. Sahimi, J. Stat. Mech. **P07046** (2009)
17. J. Zhang, M. Small, Phys. Rev. Lett. **96**, 238701 (2006)
18. A. Shreim, P. Grassberger, W. Nadler, B. Samuelsson, J.E.S. Socolar, M. Paczuski, Phys. Rev. Lett. **98**, 198701 (2007)
19. L. Lacasa, B. Luque, F. Ballesteros, J. Luque, J.C. Nuño, Proc. Nat. Acad. Sci. **105**, 4972 (2008)
20. P. Manshour, M.R. Rahimi Tabar, J. Peinke, J. Stat. Mech. **8**, P08031 (2015)
21. R.-Q. Su, W.-X. Wang, X. Wang, Y.-C. Lai, R. Soc. Open Sci. **3**, 150577 (2016)
22. Z. Gao, M. Small, J. Kurths, Europhys. Lett. **116**, 50001 (2016)
23. M. Jiang, X. Gao, H. An, H. Li, B. Sun, Sci. Rep. **7**, 10486 (2017)
24. A. Campanharo, M. Sirer, R. Malmgren, F. Ramos, L. Nunes Amaral, PLoS ONE **6**, e23378 (2011)
25. S. Karimi, A.H. Darooneh, Phys. A **392**, 287 (2013)
26. A.M. Nuñez, L. Lacasa, J.P. Gomez, B. Luque, Visibility algorithms: a short review, in *New Frontiers in Graph Theory* ed. by Y. Zhang (InTech, Rijeka, Croatia, 2012), pp. 119–152
27. C. Ma, H.S. Chen, Y.C. Lai, H.F. Zhang, Phys. Rev. E **97**, 022301 (2018)

Chapter 21
Reconstruction of Stochastic Dynamical Equations: Exemplary Diffusion, Jump-Diffusion Processes and Lévy Noise-Driven Langevin Dynamics

In this chapter we reconstruct stochastic dynamical equations with known drift and diffusion coefficients, as well as known properties of jumps, jump amplitude and jump rate from synthetic time series, sampled with time interval τ. The examples have Langevin (white noise- and Lévy noise-driven) and jump-diffusion dynamical equations. We also study the estimation of the Kramers–Moyal coefficients for "phase" dynamics that enable us to investigate the phenomenon of synchronisation in systems with interaction.

21.1 One and Two-Dimensional Diffusion Processes

21.1.1 Bistable Potential

To demonstrate the outlined mathematical procedure described in the Chap. 20 in practice, we consider a dynamical system with a cubic nonlinearity. In its deterministic form, the dynamics reads [1]:

$$\frac{dx}{dt} = \epsilon x - x^3 = -\frac{dV(x)}{dx}, \qquad V(x) = -\frac{\epsilon}{2}x^2 + \frac{1}{4}x^4. \tag{21.1}$$

Here $V(x)$ represents the potential function. For $\epsilon > 0$, the system has three fixed points, i.e., $x_0 = 0$ as unstable fixed point and $x_{1,2} = \pm\sqrt{\epsilon}$ as stable ones (see Fig. 21.1). By adding white noise, system (21.1) becomes a stochastic dynamics of the form:

$$\frac{dx}{dt} = \epsilon x - x^3 + \sqrt{2Q}\Gamma(t) \tag{21.2}$$

M. R. Rahimi Tabar, *Analysis and Data-Based Reconstruction of Complex Nonlinear Dynamical Systems*, Understanding Complex Systems,
https://doi.org/10.1007/978-3-030-18472-8_21

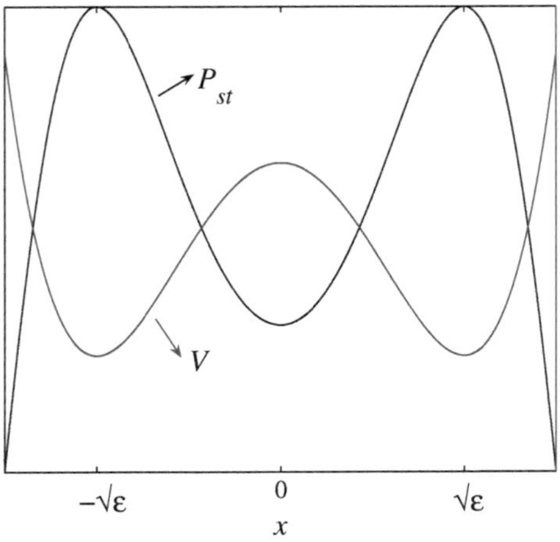

Fig. 21.1 Sketch of the probability distribution p_{st} and the potential function $V(x)$. Source from [1]

where Γ is zero mean Gaussian white noise. The constant parameter Q is diffusion coefficient. The corresponding Fokker–Planck equation can be written as:

$$\frac{\partial p}{\partial t} = -\frac{\partial}{\partial x}[\epsilon x - x^3 - Q\frac{\partial}{\partial x}]p(x,t). \tag{21.3}$$

One can explicitly compute the stationary solution of (21.3) by substituting $\frac{\partial p_{st}}{\partial t} = 0$ yielding $p_{st}(x) \sim \exp(-V(x)/Q)$.

To illustrate the corresponding reconstruction of drift and diffusion coefficients, one can simulated a time series via the numerical form of Eq. (21.2) using the Euler–Maruyama scheme, yielding 10^6 data samples with ($\tau = 0.01$ and $Q = 1$, $\epsilon = 1$) (see Fig. 21.2, left panel). Figure 21.2 (middle and right panel) shows the numerical values for both the reconstructed deterministic component and its potential function. As can be appreciated from these figures, the extracted dynamics match the original analytical function very well.

21.1.2 Reconstruction of Bivariate Data Sets

Consider two stochastic variables $x_1(t)$ and $x_2(t)$, satisfy the following coupled Langevin equations,

$$\frac{d}{dt}\begin{bmatrix} x_1 \\ x_2 \end{bmatrix} = \begin{bmatrix} D_1^{(1)}(x_1, x_2) \\ D_2^{(1)}(x_1, x_2) \end{bmatrix} + \begin{bmatrix} g_{11}(x_1, x_2) & g_{12}(x_1, x_2) \\ g_{21}(x_1, x_2) & g_{22}(x_1, x_2) \end{bmatrix} \begin{bmatrix} \Gamma_1(t) \\ \Gamma_2(t) \end{bmatrix} \tag{21.4}$$

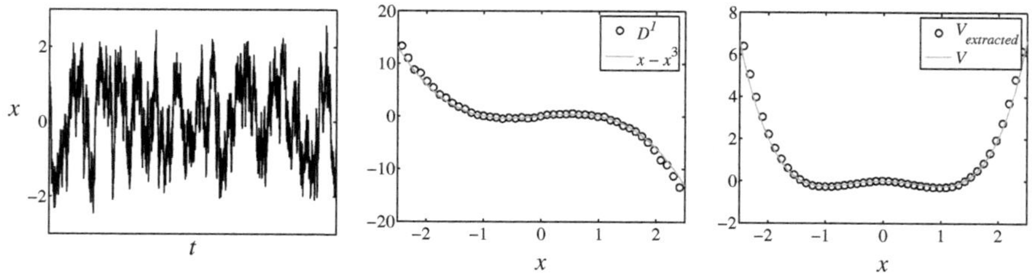

Fig. 21.2 Left panel: Part of the time series simulated via Eq. (21.2) with $Q = 1$, $\epsilon = 1$ and time step of $\tau = 0.01$. Middle panel: Circles deterministic component of Eq. (21.2) obtained from the simulated time series via the reconstruction procedure. Lines are original analytical function. Right panel: Circles potential function (21.2) obtained from the simulated time series via the reconstruction procedure and subsequent integration. Source from [1]

the drift vector $\mathbf{D}^{(1)} = (D_1^{(1)}, D_2^{(1)})$ is a two-dimensional vector and the diffusion coefficient is a 2×2-matrix given by $\mathbf{D}^{(2)} = \mathbf{g}\mathbf{g}^T$, which means that $D_{ij}^{(2)} = \sum_k g_{ik} g_{jk}$. Integration of Eq. (21.4) can be done in Euler–Maruyama scheme as follows:

$$\begin{aligned}\begin{bmatrix} x_1(t+\tau) \\ x_2(t+\tau) \end{bmatrix} = \begin{bmatrix} x_1(t) \\ x_2(t) \end{bmatrix} + \tau \begin{bmatrix} D_1^{(1)}(x_1, x_2) \\ D_2^{(1)}(x_1, x_2) \end{bmatrix} \\ + \sqrt{\tau} \begin{bmatrix} g_{11}(x_1, x_2) & g_{12}(x_1, x_2) \\ g_{21}(x_1, x_2) & g_{22}(x_1, x_2) \end{bmatrix} \begin{bmatrix} \eta_1(t) \\ \eta_2(t) \end{bmatrix}\end{aligned} \tag{21.5}$$

where $\eta_1(t)$ and $\eta_2(t)$ are two independent zero mean, unit variance and normally distributed random variables.

As an example, we integrate the following coupled Langevin equations,

$$\frac{dx_1}{dt} = x_2 + a\Gamma_1(t) \tag{21.6a}$$

$$\frac{dx_2}{dt} = 0.02x_1 + 0.03x_2 - x_1^3 - x_1^2 x_2 + a\Gamma_2(t), \tag{21.6b}$$

where a is a real constant. Figure 21.3a shows the integrated trajectory (scatter–plot) (x_1, x_2) for $a = 0$, a case where no stochastic contribution is present, whereas in Fig. 21.3b the same trajectory is plotted with stochastic forces having a constant amplitude of $a = 0.05$.

The results for reconstruction of the components of drift vector and diffusion tensor are shown in Fig. 21.4. The numerical results can be properly fitted through the functions used for the integration in Eq. (21.6), namely: $D_1^{(1)} = x_2$, $D_2^{(1)} = 0.02x_1 + 0.03x_2 - x_1^3 - x_1^2 x_2$, $D_{11}^{(2)} = D_{22}^{(2)} = 0.05^2$ and $D_{12}^{(2)} = D_{21}^{(2)} = 0$. Notice that the large deviations in the boundaries are due to the finite length of the time series and thus the lower population in the boundary bins resulting in a poor estimation of the drift and the diffusion coefficients.

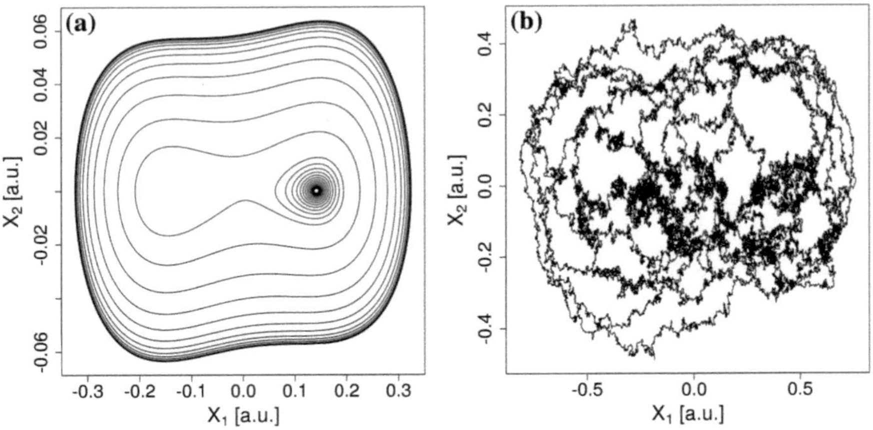

Fig. 21.3 (**a**) Scatter–plot $(x_1(t), x_2(t))$ from Eq. (21.6) with $a = 0$ and (**b**) the same plot integrating the same equations with non-zero stochastic terms ($a = 0.05$). Source from [2]

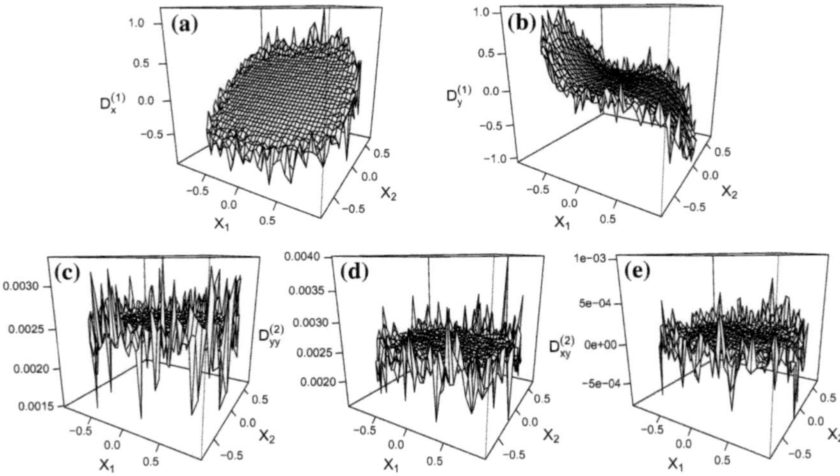

Fig. 21.4 Drift coefficient of (**a**) the x_1 component, $D_1^{(1)}$, and (**b**) the x_2 component, $D_2^{(1)}$, together with all diffusion coefficients, namely (**c**) $D_{11}^{(2)}$, (**d**) $D_{22}^{(2)}$, (**e**) $D_{12}^{(2)} = D_{21}^{(2)}$, see Eq. (21.4). The KM coefficients are estimated with additive noise, i.e., $a = 0.05$ in Eq. (21.6). Source from [2]

21.2 Jump-Diffusion Processes

21.2.1 Reconstruction of an Ornstein–Uhlenbeck Process with Jumps

Consider jump-diffusion stochastic dynamics

$$dx(t) = N(x,t)dt + \sqrt{D(x,t)}dW(t) + \xi \, dJ(t), \tag{21.7}$$

where $\{W(t), t \geq 0\}$ is a scalar Wiener process, $N(x,t)$ the drift function, $D(x,t)$ is related to the second order Kramers–Moyal coefficients $M^{(2)}(x)$ (see Eq. (12.2)), and $J(t)$ is a time-homogeneous Poisson jump process. The jump has rate $\lambda(x)$, can be state dependent and size ξ which we assume it has zero mean gaussian distribution with variance $\sigma_\xi^2(x)$.

As an example for jump-diffusion process, we consider an Ornstein–Uhlenbeck process in the presence of jumps, so that $N(x) = -x$, $D(x) = 1$, unit jump amplitude (with gaussian distribution) $\sigma_\xi^2(x) = 1$ and constant jump rate $\lambda = 0.6$ (jump rate per data point). We generate a synthetic time series by a numerical simulation of the corresponding dynamical Eq. (21.7) using a sampling interval $\tau = 0.01$ in Euler–Maruyama scheme. We note that finite-N synthetic data from Eq. (21.7) may not have jump number $n_j \simeq \lambda N$, see Problem 14.12. Counting the jumps in each run gives the corresponding correct jump rate for each simulation. The KM coefficients $M^{(j)}$, for $j = 1, 2, 4$, and 6 are then estimated. We used the Eqs. (12.2) and (12.8) to estimate unknown functions and parameters in Eq. (21.7).

The estimated functions and coefficients for *normalized* (which means that $x \equiv (x - \langle x \rangle)/\sigma$, where σ^2 is the variance of x) synthetic time series are plotted in Fig. 21.5 and show a very good agreement with theory. We note that the diffusion coefficient and jump amplitude estimated from normalized time series (with original variance σ) should multiply to σ^2 to get the original diffusion coefficient and jump amplitude.

21.2.2 Reconstruction of a Black-Scholes Process with Jumps

As another example, we investigate a Black-Scholes process [3] in the presence of jumps [4]. We consider a linear drift ($N(x) = -10x$), a quadratic diffusion ($D(x) = x^2$), a unit jump amplitude with gaussian distribution ($\sigma_\xi^2(x) = 1$), and two constant jump rates $\lambda = 0.1$ and $\lambda = 0.4$. We generate synthetic time series by a numerical simulation of the corresponding dynamical equation using a sampling interval $\tau = 0.001$ in Euler–Maruyama scheme.

For all estimated functions and coefficients, we obtain a very good agreement with theory (see Fig. 21.6). The fourth-order KM coefficient $M^{(4)}$ does not vanish,

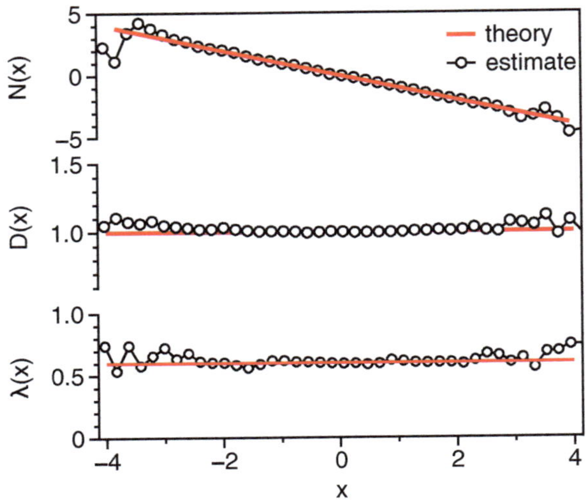

Fig. 21.5 Estimated drift term (top), coefficient $D(x)$ (middle), and jump rate (bottom) for a normalized synthetic Ornstein–Uhlenbeck process with jumps. The analyzed time series consisted of 10^6 data points with $\tau = 0.01$, and we here find $\sigma_\xi^2(x) \simeq 1.02$. For shorter time series (5×10^4 – 10^5 data points), $\sigma_\xi^2(x)$ deviates from the expected value by a few percent

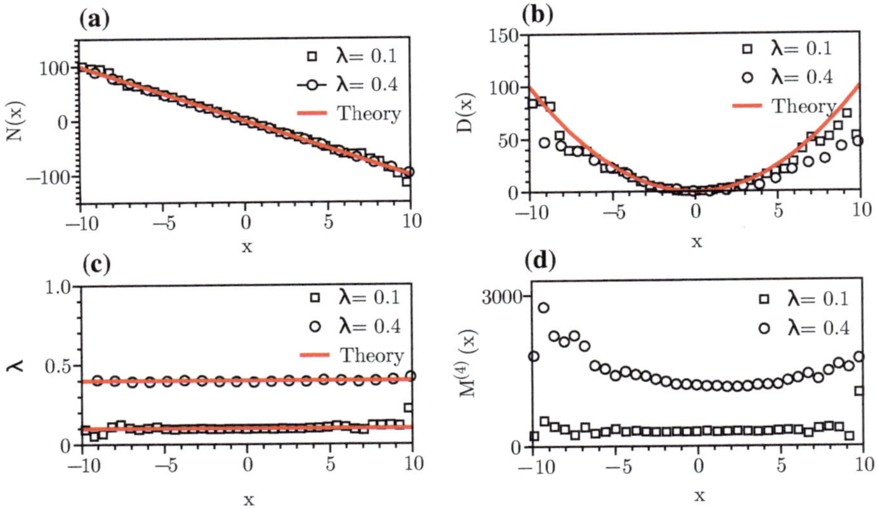

Fig. 21.6 Reconstruction of a Black-Scholes process with jumps. Estimated drift term (**a**), coefficient $D(x)$ (**b**), jump rate (**c**), and fourth-order Kramers–Moyal coefficient (**d**) for different jump rates. The time series consisted of 3×10^6 data points, and we here find $\sigma_\xi^2(x) \simeq 1.01$. Source from [5]

and increases with an increasing jump rate λ, see Fig. 21.6. In addition, one can verify that it approaches zero for $\lambda \to 0$. We note that the coefficient $D(x)$ and the jump characteristics $\lambda(x)$ and $\sigma_\xi^2(x) = \langle \xi^2 \rangle$ contribute to the second-order KM coefficient (see Eq. 12.2). Thus, with a Langevin-type modeling, it is not possible to separate these diffusive from jump contributions. This nonlinear example indicates the importance and physical meanings of higher-order KM coefficients.

21.3 Lévy Noise-Driven Langevin Dynamics

A procedure to reconstruct Lévy noise-driven Langevin dynamics Eq. (10.4) can be done to determine Lévy stability parameter α and drift and multiplicative functions, along the same lines which has been explained in Sect. 10.6 and [6].

Consider a one dimensional Lévy noise-driven Langevin dynamics

$$x(t_{i+1}) = x(t_i) + D^{(1)}(x(t_i), t_i)\tau + g(x(t_i), t_i)\tau^{1/\alpha}\eta_\alpha^{\gamma,\beta,\mu}(t_i) \tag{21.8}$$

where $D^{(1)}(x(t_i), t_i)$ and $\eta_\alpha^{\gamma,\beta,\mu}(t_i)$ are the drift function and a random quantity, distributed according to the Lévy stable law, characterized by the Lévy parameters α, γ, β and μ, see Sect. 10.1.1. Here $g(x(t_i), t_i)$ acts as a multiplicative noise. As an example, consider the Lévy noise-driven Langevin dynamics for process $x(t)$ with the following functions,

$$D^{(1)}(x, t) = 0.1x - x^3, \qquad g(x, t) = 1$$

and set the mean of Lévy noise $\mu = 0$, the scale parameter $\gamma = 1$ and Lévy stability index $\alpha = 1.2$.

Now generate Lévy noise $\eta_{\alpha=1.2}^{\gamma=1,\beta=0,\mu=0}$ (using the algorithm presented in Sect. 10.6) and synthetic time series by a numerical simulation of the corresponding dynamical Eq. (21.8) using sampling interval $\tau_0 = 0.001$. In Fig. 21.7, the generated Lévy noise as well as synthetic time series are plotted.

To reconstruct the drift function $D^{(1)}(x, t)$, coefficient $g(x, t)$ and Lévy stability index α, we employ the Eqs. (10.22) and (10.23). Log-log plot of $T^{(2)}(x, \tau)$ over time lag $\tau = k\tau_0$ where $k = 10, 20, 40, 80, 160$ (i.e., by considering data points $\{x(0), x(k\tau_0), \ldots\}$ only) and for the bins in the interval (–1,1), provides an estimation for Lévy stability index $\alpha = 1.2 \pm 0.1$, as shown in Fig. 21.8 for instance for central bin, i.e. $x \simeq 0$ around the mean. The estimated deterministic function $D^{(1)}(x, t)$ and coefficient $g(x, t)$ for synthetic time series are plotted in Fig. 21.9. The figures show a very good agreement with theory. See [6] for more examples.

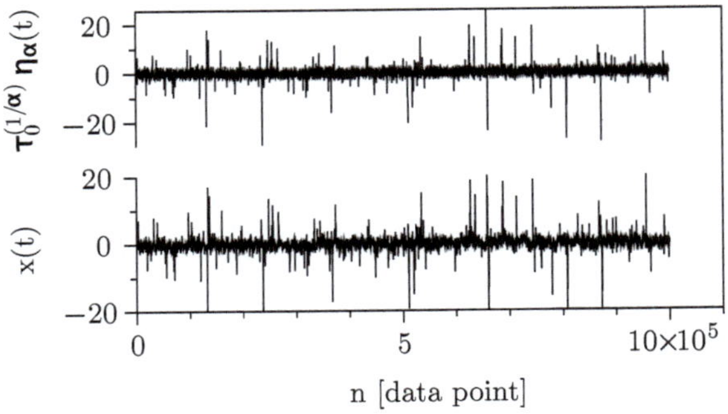

Fig. 21.7 Generated scaled Lévy noise $\tau_0^{(1/\alpha)}\ \eta_{\alpha=1.2}^{\gamma=1,\beta=0,\mu=0}$ and time series $x(t)$. The time series consist of 10^6 data points

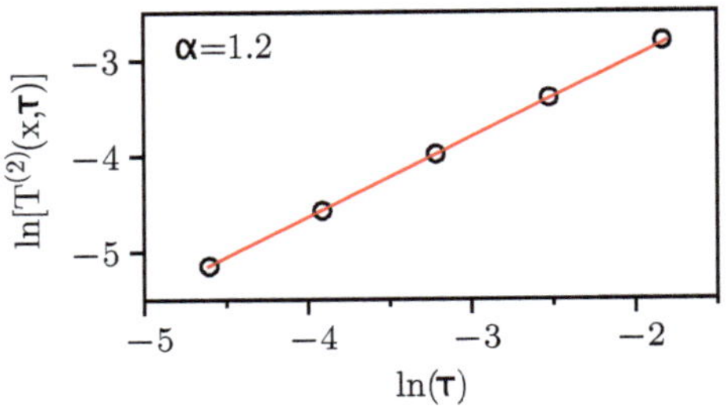

Fig. 21.8 Logarithm of the numerically estimated average $T^{(2)}(x, \tau) = \langle |x(t+\tau) - x(t) - D^{(1)}(x)\tau| \rangle|_{x(t)=x}$ of the stochastic variable $x(t)$ over the logarithm of the time lag $\tau = k\tau_0$, where $k = 10, 20, 40, 80, 160$ (i.e., by considering data points $\{x(0), x(k\tau_0), \ldots\}$ only). Here $\tau_0 = 0.001$ and $x \simeq 0$ is around the mean. The straight line was fitted to the numerically estimated values, where the slope is $1/\alpha$

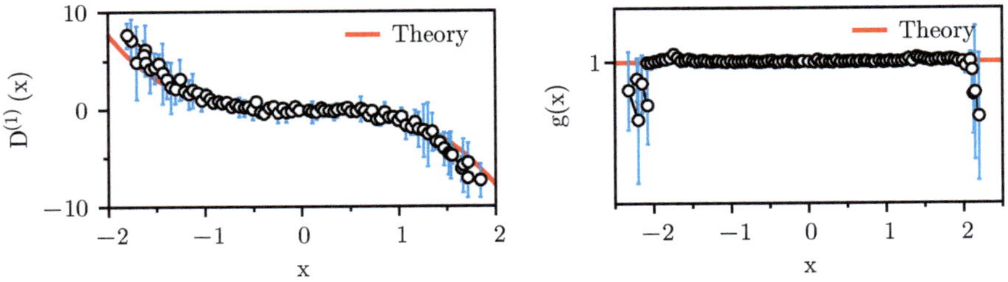

Fig. 21.9 Numerically estimated values for the deterministic function $D^{(1)}(x)$ and coefficient $g(x)$ in state variable x. The values were estimated directly by data analysis of the time series shown in Fig. 21.7, were compared with the theoretical curve

21.4 Phase Dynamics and Synchronization

In what follows we describe how to identify the dynamical equation for *phase* difference time series, which enables us to study the phase synchronization phenomenon. It is well-known that when two interacting oscillators are locked together in a relationship between their phases or amplitudes, exhibit synchronization phenomenon [7, 8]. Synchronization is a direct indicator of interactions in complex systems. Here is a brief summary of different types of synchronization [7, 8].

- Complete synchronization
  In complete synchronization, coupled systems are identical but not necessarily have the same initial conditions. In this case, the systems become synchronized when their states are the same, at this point the coupling terms vanish as the systems no longer need to be influenced by each other to remain synchronized.
- Generalised synchronization
  The generalised synchronization occurs when the state of one system can be determined directly via a functional relationship to the state of the other. This typically occurs when the influence of a unidirectional coupling from one system to the other causes the maximal Lyapunov exponent of the system to become negative [8].
- Phase synchronization
  The other main type of synchronization is phase synchronization, which is defined as the case where the phases of the two systems are locked in an $n : m$ relationship so that the phase difference $|n\phi_1 - m\phi_2|$ is constant, while the amplitudes may still vary independently.

Many complex systems can be modelled as ensembles of interacting oscillators with negligible amplitude dynamics, where only phase dynamics is relevant [9]. Amplitude dynamics will be a result of resonance when the oscillators become phase synchronized. In other cases, amplitude dynamics can sometimes be separated into additional oscillatory components, which can then be investigated individually for phase synchronization [7].

Let us consider two processes $x_1(t)$ and $x_2(t)$, and determine their local phases using for instance Hilbert transform or marked events.[1]

---

[1] *Hilbert transform:* To determine local "phase" of time series apply the Hilbert transform to process x as;

$$y(t) = \frac{1}{\pi} P.V. \int_{-\infty}^{\infty} \frac{x(t')}{t - t'} dt'$$

where $P.V.$ is the Cauchy principal value of integral. Define complex (analytical) signal $z(t) = x(t) + iy(t) = a(t)e^{i\phi(t)}$, where $a(t) = [x(t)^2 + y(t)^2]^{1/2}$ and therefore local phase is given by $\phi(t) = tan^{-1}(y(t)/x(t))$. For example if $x(t) = \cos(\omega t)$ with $\omega > 0$, its Hilbert transform will be $y(t) = \sin(\omega t)$ and one finds local phase as $\phi(t) = \omega t$.

Marked event: Suppose at t_n time series cross (up or down crossing) the level zero (zero-crossing for a mean-subtracted oscillation) and the next crossing occurs at time t_{n+1}, therefore we can define instant phase as

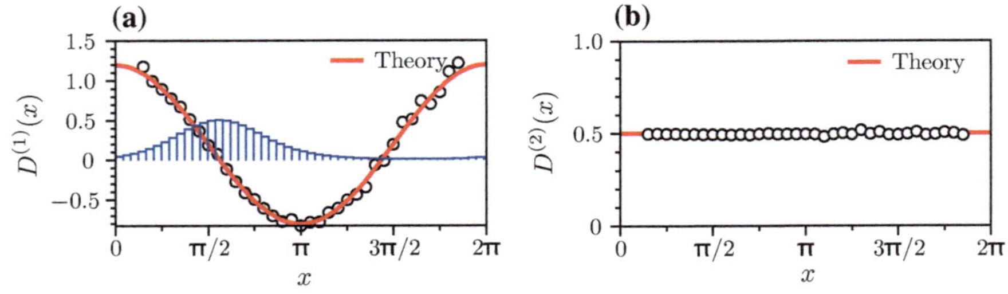

Fig. 21.10 The drift and diffusion coefficients are estimated from analyzing phase time series consisted of 10^7 data points with $\tau = 0.01$. The blue graph is the probability distribution function (PDF) of the phase. The peak of PDF shows the presence of an attractive fixed point at $x \approx 1.77$. We note that deviations from theory in the boundaries are due to the spurious jumps because of *mod* function (data not shown). Also population of data points around $x = \pi$ are not symmetric, and data has lower populations in the bins with $x > \pi$, resulting in a poor estimation of the drift and the diffusion coefficients. At the repelling fixed point $x \approx 4.5$, the highest error is present

Therefore, one finds the local phases $\phi_1(t)$ and $\phi_2(t)$. To study the phase synchronization, one can also study different order parameters such as phase coherence, synchronization index, etc. [7].

Here to analyze the phase synchronization, tendency of a pair of coupled noisy phase oscillators ϕ_1 and ϕ_2, we estimate the drift and diffusion coefficients for the phase difference $x \equiv \phi_1 - \phi_2$. We have chosen the case $n = 1, m = 1$. However, one can apply the following approach to estimate Kramers–Moyal coefficients for given n and m. If one can identify a stable fixed point in the drift term, or the potential of the dynamics, both oscillators tend to phase synchronize. The dynamics of process should be invariant under change $x \to x + 2\pi$, therefore the KM coefficients must be 2π periodic. This knowledge can be exploited by defining the KM coefficients as [10],

$$D^{(n)}(x, t) = \lim_{\tau \to 0} \frac{1}{n!\tau} \langle [x(t+\tau) - x(t)]^n \rangle \Big|_{x(t)\ (mod\ 2\pi) = x}.$$

As an example, let us consider a Langevin dynamics for process x with drift and diffusion coefficients as,

$$D^{(1)}(x, t) = 0.2 + \cos(x) \qquad D^{(2)}(x, t) = 0.5.$$

The Langevin equation is integrated in Euler–Maruyama scheme for 10^7 data points and a sampling interval $\tau = 0.01$. The estimated drift and diffusion coefficients are represented in Fig. 21.10. Using the drift coefficient one can see that there is an attractive fixed point at $x \approx 1.77$ and a repelling fixed point at $x \approx 4.5$. For state dependent diffusion coefficients one can use the potential introduced by Eq. (4.17),

---

$$\phi(t) = \pi \frac{t - t_n}{t_{n+1} - t_n} + \pi(n-1), \qquad t_n \le t < t_{n+1}$$

where the phase $\phi(t)$ should be evaluated with mod 2π.

to study the fixed points of phase dynamics. We should remark that when we use the *mod* function for time series x, then this produces spurious jumps which contradict the fact that the original process is given by Langevin dynamics.

We note that using the method described in Sect. 21.1.2, one can estimate the coupled dynamical equations of $\dot{\phi}_1$ and $\dot{\phi}_2$ as bivariate time series and find drift vector and diffusion matrix of processes.

21.5 Estimation of Kramers–Moyal Coefficients for Time Series with Finite *Markov–Einstein* Time Scale

There are various available approximate methods to estimate Kramers–Moyal coefficients in vanishing limit of sampling interval τ for time series with finite *Markov–Einstein* time scale t_M (estimated for instance using Eq. (16.12) or Wilcoxon test [11, 12]) [13]. Among of them, one may estimate for given x (with some binning), the Kramers–Moyal conditional moments (divided by τ) $K^{(j)}(x, \tau)/\tau$ for different ratio τ/t_M. Then extrapolate to vanishing limit $\tau \to 0$ to obtain $K^{(j)}(x = 0, \tau)/\tau$ for $\tau \to 0$, as shown in Fig. 21.11 for instance for Kramers–Moyal coefficient order $j = 2$. This is a naive approximation that estimates the KM coefficients in the time interval that the process is not Markov process. As another approximate method one can use the following expression for estimation of "apparent" KM coefficients

$$D^{(m)}(x) \approx \frac{1}{m!\tau} K^{(m)}(x, \tau)|_{\tau=t_M}.$$

In this approximation, the KM coefficients are estimated in time lag $\tau = t_M$.

To make this approximation precise, we can study finite τ expansion of Kramers–Moyal conditional moments as has been discussed in details in Sect. 18.4.1. In this way, we can deal with the conditional KM moments in presence of finite t_M.

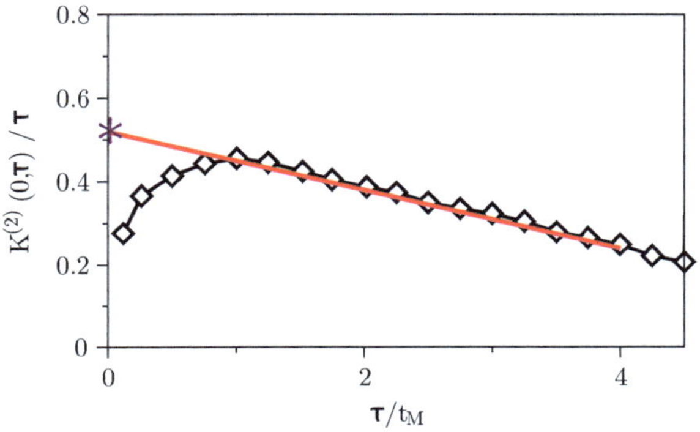

Fig. 21.11 Estimation of Kramers–Moyal coefficient in vanishing limit of sampling interval. The zero τ limit of the second order Kramers–Moyal conditional moments for time series with *Markov–Einstein* time scale t_M. The star marks the values of second order Kramers–Moyal coefficient obtained from the limit $K^{(j)}(x = 0, \tau)/\tau$ for $\tau \to 0$

We note that if one consider Kramers–Moyal conditional moments as measure for quantifying stochastic dynamics of given time series, then can use τ as the sampling interval of measured time series [13]. In any chosen approximate method after estimation of KM coefficients one should use appropriate self-consistency check to validate the reconstruction procedure.

21.6 Estimation of Kramers–Moyal Conditional Moments for Diffusion Processes with Different Precision

Now suppose we have a time series with t_M equals to sampling interval τ (Markov process) and higher order KM coefficients with $n > 2$ vanishes, see Sect. 18.2. Therefore we can model the time series with Langevin dynamics. To estimate drift and diffusion coefficients in Langevin dynamics one can use the following relations to estimate KM conditional moments with different precision.

Using Itô-Taylor expansion for Langevin dynamics (see Sect. 6.4), it is shown by Stanton [14] that the first-order approximation (for $j = 1, 2$) for Kramers–Moyal conditional moments is given by

$$K^{(j)}(x,\tau) = \langle (x(t+\tau) - x(t))^j|_{x(t)=x}\rangle + \mathcal{O}(\tau)^2 \tag{21.9}$$

and in the second-order approximation one finds

$$\begin{aligned} K^{(j)}(x,\tau) = \frac{1}{2}\Big[&4\langle (x(t+\tau) - x(t))^j|_{x(t)=x}\rangle \\ &- \langle (x(t+2\tau) - x(t))^j|_{x(t)=x}\rangle\Big] + \mathcal{O}(\tau)^3 \end{aligned} \tag{21.10}$$

and in the third-order approximation we have

$$\begin{aligned} K^{(j)}(x,\tau) = \frac{1}{6}\Big[&18\langle (x(t+\tau) - x(t))^j|_{x(t)=x}\rangle \\ &- 9\langle (x(t+2\tau) - x(t))^j|_{x(t)=x}\rangle \\ &+ 2\langle (x(t+3\tau) - x(t))^j|_{x(t)=x}\rangle\Big] + \mathcal{O}(\tau)^4 \end{aligned} \tag{21.11}$$

etc. Equations (21.9)–(21.11) allow us to estimate order $j = 1, 2$ of the KM conditional moments and therefore KM coefficients (drift and diffusion) with different precision.

Problems

21.1 Reconstruction of Ornstein–Uhlenbeck process

Consider one dimensional Langevin equation

$$dx = D^{(1)}(x)dt + \sqrt{2D^{(2)}(x)}dW(t)$$

where $D^{(1)}(x)$ and $D^{(2)}(x)$ are drift and diffusion coefficients, respectively and $W(t)$ is the Wiener process.

(a) Integrate the Langevin equation with $D^{(1)}(x) = -x$, $D^{(2)}(x) = 1$ (Ornstein–Uhlenbeck process) with $\tau = 0.01$ in Euler–Maruyama scheme and initial condition $x(t = 0) = 0$ and generate a synthetic time series with 10^6 data points.

(b) Show that the increments of the Ornstein–Uhlenbeck process in the stationary state are negatively correlated on disjoint time intervals,

$$\langle [x(t_4) - x(t_3)][x(t_2) - x(t_1)]\rangle < 0$$

for any $t_4 > t_3 > t_2 > t_1$.

(c) Estimate Kramers–Moyal coefficients $D^{(1)}(x)$ and $D^{(2)}(x)$ from synthetic data and verify the Pawula theorem by estimation of $K^{(4)}(x, \tau)$, (use the criterion $K^{(4)}(x, \tau) \simeq 3(K^{(2)}(x, \tau))^2$ to judge whether the fourth-order KM coefficient is ignorable).

(d) For a self-consistency check, derive η from Eq. (16.27) with estimated drift and diffusion coefficients and verify that it is an uncorrelated noise with normal distribution.

21.2 Reconstruction of Ornstein–Uhlenbeck process with jump

Consider one dimensional (jump-diffusion) Langevin equation

$$dx = D^{(1)}(x)dt + \sqrt{D(x)}dW(t) + \xi dJ(t)$$

where $D^{(1)}(x)$ is drift coefficient and $D(x)$ is related to the second order Kramers–Moyal coefficients $M^{(2)}(x)$ (see Eq. 12.2) and $W(t)$ is the Wiener process. Here $J(t)$ is jump Poisson process and jump amplitude ξ has zero mean Gaussian distribution with variance $\sigma_\xi^2 = 1$ which can symmetrically jumps up and down, i.e. $\xi \sim \mathcal{N}(0, \sigma_\xi^2 = 1)$.

(a) Integrate the Langevin equation with $D^{(1)}(x) = -x$, $D(x) = 1$, unit jump amplitude $\sigma_\xi^2(x) = 1$ and constant jump rates $\lambda = 0.1, 0.6$ and with sampling interval $\tau = 0.001$. Use the Euler–Maruyama scheme, with initial condition $x(t = 0) = 0$ and generate a synthetic time series with 10^7 data points.

(b) Estimate Kramers–Moyal coefficients $M^{(1)}(x)$, $M^{(2)}(x)$, $M^{(4)}(x)$ and $M^{(6)}(x)$ from synthetic data and use the relations (12.2) and (12.8) to find drift, diffusion and jump properties of the synthetic time series.

21.3 Reconstruction of two-dimensional coupled equations

Consider two coupled Langevin equations (21.6) and construct the results presented in Figs. 21.3a, b and 21.4a, b.

21.4 Reconstruction of Lévy noise-driven Langevin equation

Consider the following Lévy noise-driven Langevin equation

$$dx = D^{(1)}(x)dt + dL_{\alpha}^{\gamma=1,\beta=0,\mu=0}(t)$$

where $dL(t)$ is Lévy noise increment and drift is given by $D^{(1)}(x) = -10x$. Use the discrete time evolution law (10.10) to approximate this stochastic differential equation and generate 10^7 data points of a $x(t_n)$ with $x(t_0) = x_0 = 0$ and $\tau = 0.001$, for $\alpha = 1.8$.

Use the method presented in Sect. 10.6, to reconstruct drift term $D^{(1)}(x)$ and Lévy stability index α.

21.5 Phase dynamics

Reconstruct the results presented in Fig. 21.10.

21.6 Estimation of drift and diffusion coefficients with different precision

Prove the relations (21.9)–(21.11).

References

1. A.M. Van Mourik, A. Daffertshofer, P.J. Beek, Biol. Cybern. **94**, 233 (2006)
2. P. Rinn, P.G. Lind, M. Wächter, J. Peinke, The Langevin approach: an r package for modeling markov processes. J. Open Res. Softw. **4**, e34 (2016). https://doi.org/10.5334/jors.123, https://cran.r-project.org/web/packages/Langevin/index.html
3. F. Black, M. Scholes, The pricing of options and corporate liabilities. J. Polit. Econ. **81**, 637 (1973)
4. R.C. Merton, Option pricing when underlying stock returns are discontinuous. J. Financ. Econ. **3**, 125 (1976)
5. M. Anvari, K. Lehnertz, M.R. Rahimi Tabar, J. Peinke, Sci. Rep. **6**, 35435 (2016)
6. S. Siegert, R. Friedrich, Phys. Rev. E **64**, 041107 (2001)
7. P.T. Clemson, A. Stefanovska, Phy. Rep. **542**, 297 (2014)
8. A. Pikovsky, M. Rosenblum, J. Kurths, *Synchronization,: A Universal Concept in Nonlinear Sciences* (Cambridge University Press, Cambridge, 2001)
9. A.Y. Kuramoto, *Chemical Oscillations, Waves, and Turbulence* (Springer, Berlin, 1984)

10. C. Honisch, *Analysis of Complex Systems: From Stochastic Time Series to Pattern Formation in Microscopic Fluidic Films* (Diss, Universität, Münster, (Westfalen), 2014)
11. F. Wilcoxon, Biometrics **1**, 80 (1945)
12. C. Renner, J. Peinke, R. Friedrich, J. Fluid Mech. **433**, 383 (2001)
13. R. Friedrich, J. Peinke, M. Sahimi, M.R. Rahimi Tabar, Phys. Rep. **506**, 87 (2011)
14. R. Stanton, A nonparametric model of term structure dynamics and the market price of interest rate risk. J. Financ. **52**, 1973 (1997)

Chapter 22
Applications and Outlook

The method outlined in the Chaps. 15–21 has been used for revealing nonlinear deterministic and stochastic behaviors in a variety of problems, ranging from physics, to neuroscience, biology and medicine. In most cases, alternative procedures with strong emphasis on deterministic features have been only partly successful, due to their inappropriate treatment of the dynamical fluctuations [1]. In this chapter, we provide a list of the investigated phenomena using the introduced reconstruction method. In the "outlook" possible research directions for future are discussed.

22.1 Applications

The following list (with some exemplary citations) provides an overview of the investigated phenomena using the methods described in this book.

- Chatter in cutting processes [2, 3].
- Identification of bifurcations towards drifting solitary structures in a gas-discharge system [4–6].
- Electric circuits [7–9].
- Rough surfaces [10–14].
- Wind energy convertors [15–19].
- Traffic flow [20].
- Cosmic microwave background radiation [21].
- Heart rhythms [22–27].
- Tremor data [9].
- Meteorological data, such as data for El NINO [28–30].
- Seismic time series [31–33].
- Epileptic brain dynamics [34, 35].
- Inverse grading of granular flows [36].

M. R. Rahimi Tabar, *Analysis and Data-Based Reconstruction of Complex Nonlinear Dynamical Systems*, Understanding Complex Systems,
https://doi.org/10.1007/978-3-030-18472-8_22

- Experimental chaotic time series and synchronization [37–39].
- Rhythmic human movement [40–42].
- Solar wind turbulence [43].
- Decoding the time evolution of the control parameters in a system of neurons [44].
- Porous media [45].
- Gel and glass transitions [46].
- Amorphous organic films [47].
- Electrical discharge current fluctuations in plasma [48].
- Light scattering intensity fluctuations [49].
- Brain dynamics and synchronization [50].
- Cellular automaton model of tumor-immune interactions [51].
- Random ameboid motion [52].
- Fractal-generated turbulence [53].
- Nanocrystalline thin films [54].
- Solar granulation [55].
- Biological macromolecules [56].
- Optical tweezers [57, 58].
- Single particle trajectories in cell biology [59].
- Stochastic model of Bumblebee flights [60].
- Stochastic dynamics and the predictability of big hits in online videos [61].
- Probing small-scale intermittency with a fluctuation theorem [62].
- Rogue (monster or freak) waves [63, 64].
- Boundary layer turbulence [65, 66].
- Test of stationarity for time series [67].
- Characterization of complex surface structures [68].
- Stochastic equation of earthquakes series [69].
- The dynamic brain [70, 71].
- A solution algorithm for the fluid dynamic equations based on a stochastic model for molecular motion [72].
- Reconstruction of Langevin equation from persistent time series [73, 74].
- Intrinsic fluctuations of cell migration [75].
- Robust identification of harmonic oscillator parameters [76].
- Data analysis for nonlinear stochastic systems with time delays [77].
- Stochastic modelling of non-stationary financial assets [78–80].
- Estimation of stochastic equations with sparse Gaussian processes [81].
- Master equation for multifractal scaling [82].
- The magnetic field in the quiet solar photosphere [83].
- Ballistic deposition and restricted solid-on-solid models [84].
- Solar energy fluctuations [85].
- Fluctuations in the "waiting times" [86].
- Wake modeling in wind farm [87].
- Stochastic process driven by Ornstein–Uhlenbeck noise [88].
- Granger causality [89, 90].
- Parameter-free resolution of the superposition of stochastic signals [91].
- Spontaneous transitions between active and inactive behavioral states [92, 93].

- Langevin equation for systems with negative temperatures. [94].
- Learning stochastic differential equations with gaussian processes [95].
- Granular media [96].

See also [97–115] for more applications of the reconstruction method. As we have seen, the main advantage of the method of stochastic data analysis is its independence from the modeling assumptions, as it is purely data driven.

Before presenting the outlook, we would like to make the following comment. The method of analysis that we describe removes from the data dynamical and measurement noise, and provides the drift vector field, i.e., it yields the underlying deterministic dynamical system. In turn, the system can be analyzed by the methods that are used to analyze nonlinear time series. One can determine the proper embedding, the Lyapunov exponents, the dimensions, the fixed points, stable and unstable limit cycles, etc. [116, 117].

22.2 Outlook

Advanced sensor techniques enable scientists to collect huge data sets measured with high precision. Based on the stochastic approach we have described so far, the issue is no longer how to invest much efforts into reduction of dynamical noise but, to the contrary, how to use the noise in order to derive a better characterization and, thus, gain a better understanding of the system under study. Indeed, there are many applications in the inanimate and the animate world, ranging from technical applications over socio-economic systems, to biomedical applications. An interesting feature is the extraction of higher correlations, such as the question of the cause and effect chain, which may be unfolded by asymmetric determinism and the reconstructed noise from data.

The description of complex systems on the basis of stochastic processes, which include nonlinear dynamics, seems to be a promising approach. The challenge will be to extend the understanding to more complex processes, such as mixture of diffusion and the Lévy processes, processes with no white noise, extension of Langevin dynamics with Wiener process to non-Markovin fractional Brownian motion (fBm), or higher-dimensional processes, to name but a few. As has been shown in this book, for these problems it should be possible to derive from precise mathematical results general methods of data series analysis.

Besides further improvement of the method, we are convinced that there is still a wide range of further applications, especially for the systems that higher order Kramers–Moyal coefficients are non-vanishing.

As the outlook we present in what follows some interesting research directions that need further investigations.

22.2.1 Representation of Jump-Diffusion Dynamics in Terms of Fractional Brownian Motion of Order k

As we have seen throughout this book, the analytical calculations for the Poisson jump process are more severe than for the Wiener process. This was due to the fact that the singularities of the Poisson process arise at the zeroth order with the value of $J(t)$ jumping, while for Wiener process $W(t)$ arises at the first-order derivative. Therefore, it would be useful to replace the jump contributions with some stochastic dynamics that, for instance with fractional Brownian motion (fBm)[1] which enables one to interplay between Wiener and Poisson jump processes.

In what follows, we show that by combining the Kramers–Moyal (KM) expansion with the fBm of order k, one obtains an "equivalent" stochastic differential equation involving both Gaussian white noise and jumps with Poisson distribution, in terms of rotating Gaussian white noise on the circle defined by the complex roots of the unity [118]. Consider the jump-diffusion dynamics

$$dx(t) = N(x,t)dt + \sqrt{D(x,t)}dW(t) + \xi\, dJ(t), \tag{22.1}$$

where $\{W(t), t \geq 0\}$ is a scalar Wiener process, $N(x,t)$ the drift function, $D(x,t)$ the diffusion function, and $J(t)$ is a time-homogeneous Poisson jump process. The jump has the rate $\lambda(x)$, and can be state dependent and the size ξ that we assume it to have zero mean (symmetric) with finite statistical moments $\langle \xi^{2m} \rangle$. The KM coefficients, in terms of the functions $N(x,t)$, $D(x,t)$, $\lambda(x)$, and $\langle \xi^{2m} \rangle$ are given by

---

[1]The fBm is a Gaussian stochastic process $B_H(t)$ with Hurst exponent $0 < H < 1$ and the following properties

(i) $\langle B_H(t) \rangle = 0$,

(ii) $\langle B_H(t) B_H(t') \rangle = \frac{\sigma^2}{2}(t^{2H} + t'^{2H} - |t - t'|^{2H})$,
where σ is the variance parameter.

(iii) $B_H(\lambda t) \overset{\text{law}}{=} \lambda^H B_H(t) \quad \lambda > 0$.

(iv) The fBm can be constructed from the Wiener process (classical Brownian motion), $W(t) = B_{H=1/2}(t)$, by a linear transformation of the form,

$$B_H(t) = B_H(0) + \frac{1}{\Gamma(H+1/2)} \left\{ \int_{-\infty}^{0} \left[(t-s)^{H-1/2} - (-s)^{H-1/2} \right] dW(s) + \int_0^t (t-s)^{H-1/2} dW(s) \right\}.$$

(v) Successive increments of the fBm are dependent. The following relationship holds for $t_1 < t_2 < t_3 < t_4$

$$\langle [B_H(t_4) - B_H(t_3)][B_H(t_2) - B_H(t_1)] \rangle = H(2H-1) \int_{t_1}^{t_2} \int_{t_3}^{t_4} (u-v)^{2H-2} du dv,$$

for $H > 1/2$ ($H < 1/2$) the increments of the process are positively (negatively) correlated. The process $B_H(t)$ has independent increments if and only if $H = \frac{1}{2}$.

$$M^{(1)}(x) = N(x,t)$$

$$M^{(2)}(x) = [D(x,t) + \langle \xi^2 \rangle \lambda(x)]$$

$$M^{(2m)}(x) = \langle \xi^{2m} \rangle \lambda(x), \quad \text{for } 2m > 2\,. \tag{22.2}$$

The conditional probability density $p(x,t|x_0,t_0)$ satisfies the following KM differential equation,

$$\frac{\partial}{\partial t} p(x,t|x_0,t_0) = \sum_{j=1}^{\infty} \frac{(-1)^j}{j!} \frac{\partial^j}{\partial x^j} \left[M^{(j)}(x) p(x,t|x_0,t_0) \right] \tag{22.3}$$

where $M^{(j)}(x)$ are given by Eq. (22.2). The expressions for KM coefficients suggest that it is possible replace Eq. (22.1) as follows. Define a fBm of order k, $b(t,k)$ as a stochastic process that satisfies the following conditions [118]:

$$db(t,k) = \frac{1}{(c_k)^{1/k}} \Gamma(t,k)(dt)^{1/k},$$

$$\langle \Gamma^j(t,k) \rangle = 0, \quad \text{for } 1 \le j \le k-1,$$

$$\langle \Gamma^k(t,k) \rangle = c_k \tag{22.4}$$

where $\Gamma(t,k)$ is a white noise of order k, and c_k are normalization constants. The stochastic process $\Gamma(t,k)$ can be thought of as an extension of the Gaussian white noise $\Gamma(t,k=2) \equiv \eta(t)$, (where $\eta(t)$ is uncorrelated gaussian zero mean noise with unit variance) and will be called as Gaussian white noise of order k [118]. The Wiener process will be $W(t) \equiv b(t,2)$. The process $b(t,k)$ for $t \ge 0$ has continuous trajectories, with stationary *independent* increments whilst the classical fBm has *dependent* increments.

Using the relations in (22.2) and the properties of the noise $\Gamma(t,k)$, the jump diffusion dynamics Eq. (22.1) is rewritten as

$$dx(t) = N(x,t)dt + (D(x,t) + \langle \xi^2 \rangle \lambda(x))^{1/2} db(t,2)$$

$$+ \sum_{m=2}^{\infty} [\langle \xi^{2m} \rangle \lambda(x)]^{1/2m} \, db(t,2m). \tag{22.5}$$

In this representation we call the term $\langle \xi^2 \rangle \lambda(x)$ the *jump-induced diffusion*. Itô stochastic calculus can be formally generalized to the order k in order to deal with the fBm so introduced by using mean-squared value convergence of order k [118]. In this framework, we shall have the equality

$$\int_0^t [db(t,k)]^k = t, \tag{22.6}$$

which allows to write

$$[db(t,k)]^k \overset{\text{dt}}{=} dt.$$

To construct the noise $\Gamma(t,k)$ one considers the dynamical equation (22.5) in the complex plane. An algorithm to construct $\Gamma(t,k)$ for $k = 2m$ is as follows.

- Refer to the complex root of order $2m$ of the unity, $\omega_l(2m) = 2\pi i l/2m$, $l = 0, \cdots, 2m-1$, and define the white noise $R(t, 2m)$, known as the *Rademacher white noise*, which takes on the values $\omega_l(2m)$ at time instance t, with the uniform probability $1/2m$. For example, for the case $2m = 4$, we have $R(t, 2m) = \exp(2\pi i l/4)$, where l is a random variable that takes on the values $l \in \{0, 1, 2, 3\}$ with the same probability $1/4$. Therefore, we choose $\Gamma(t, 2m) = R(t, 2m)\,\eta(t)$.
- Then, a model for $b(t, 2m)$ is selected in the form

$$db(t, 2m) = \frac{1}{(d_{2m})^{1/2m}} R(t, 2m)\eta(t)(dt)^{1/2m}.$$

here $d_{2m} = (2m)!/2^m m!$ is a normalization constant. We note that two noises $R(t, 2m)$ and $\eta(t)$ are independent and by virtue of the properties of the complex roots of unity, we have

$$\langle R^j(t, 2m)\rangle = \delta_{j,2m}.$$

Truncation of the expansion in Eq. (22.5) allows one to determine approximate trajectories of $x(t)$ that do not have jump discontinuity, but simulate trajectories with higher or slower speeds that are unlikely to be generated by statistical continuous, diffusion-type processes. In addition, the representation (22.5) paves the way towards fractional KM expansion for jump-diffusion processes [119]. Generalisation of the approach for higher dimensional jump-diffusion processes would be interesting where, for instance, the KM coefficients are given by Eqs. (20.10)–(20.16) in two dimensions.

22.2.2 Langevin Dynamics Driven by Fractional Brownian Motion

Consider the fBm-driven Langevin dynamics,

$$dx(t) = N(x,t)dt + g(x,t)dB_H(t) \tag{22.7}$$

where $N(x, t)$ and $g(x, t)$ are drift and multiplicative functions, and $B_H(t)$ is the fBm with the Hurst exponent H. The fBm $B_H(t)$ is a non-Markov process and for $1/2 < H < 1$ has infinite-range memory.

It is pointed out in [120] that integration of the function $g(x, t)$ with $dB_H(t)$ can be defined as

$$\int_a^b g(x,t)dB_H(t) = \lim_{|\Delta|\to 0} \sum_{k=0}^{n-1} g(x,t_k)(B_H(t_{k+1}) - B_H(t_k)), \tag{22.8}$$

where $a = t_0 < t_1 < \cdots < t_n = b$ and Δ is a partition of the interval $[a, b]$ and $|\Delta| = max_{\,0\le k\le n-1}(t_{k+1} - t_k)$. The integrals do not have zero average value, are called *fractional path-wise* integrals.

For the integral in Eq. (22.8) to have vanishing averaged value, it can be defined as [121],

$$\int_a^b g(x,t)dB_H(t) = \lim_{|\Delta|\to 0} \sum_{k=0}^{n-1} g(x,t_k) \diamond (B_H(t_{k+1}) - B_H(t_k)), \tag{22.9}$$

where $\diamond$ denotes the Wick product [121, 122].[2] These integrals have zero averages. For instance, using Wick product one can show that

$$\int_0^t B_H(s)dB_H(s) = \frac{1}{2}B_H^2(t) - \frac{1}{2}t^{2H}. \tag{22.10}$$

Another example is the Black–Scholes process driven by fBm, given by

$$dx(t) = \mu x(t)dt + \sigma x(t)dB_H(t), \quad x(0) = x_0 > 0, \tag{22.11}$$

where x_0, μ and σ are constants. Using the Wick product, it can be shown that the stochastic solution is given by

[2]The Wick product is related to fractional path-wise for $H > 1/2$ as

$$\sum_{i=1}^{n} g(B_H(t_{i-1})) \diamond (B_H(t_i) - B_H(t_{i-1})) = \sum_{i=1}^{n} g(B_H(t_{i-1}))(B_H(t_i) - B_H(t_{i-1}))$$
$$-\sum_{i=1}^{n} \frac{1}{2} g'(B_H(t_{i-1}))[(t_i)^{2H} - (t_{i-1})^{2H} - (t_i - t_{i-1})^{2H}].$$

In the limit $n \to \infty$ we find

$$\int_0^t g(B_H(t)) \diamond dB_H(t) = \int_0^t g(B_H(t))dB_H(t) - H\int_0^t g'(B_H(t))t^{2H-1}\,dt$$

where $\int_0^t g(B_H(t))dB_H(t)$ is the path-wise *Riemann–Stieltjes* integral.

$$x(t) = x_0 \exp\left\{\sigma B_H(t) + \mu t - \frac{1}{2}\sigma^2 t^{2H}\right\}. \tag{22.12}$$

Non-parametric estimation of the Hurst exponent H, as well as the drift and "diffusion" functions $N(x, t)$ and $g(x, t)$ from time series has been a subject of recent investigations, which would be helpful to deal with the time series with large *Markov–Einstein* time scale. For instance, recent works indicate that for the fBm-driven Ornstein–Uhlenbeck process, it is possible to estimate the parameters in the dynamical equation from time series; see, for example, [123] and references therein.

22.2.3 The Integral Fluctuation Theorem for Diffusion Processes (Cascade Processes)

The thermodynamic interpretation of non-equilibrium diffusion processes is a subject of stochastic thermodynamics [124, 125]. Formally, an information theoretic entropy production ΔS for a single stochastic trajectory can be defined. As a mathematical fact, the averaged value of the exponential of $-\Delta S$ must amount to unity,

$$\left\langle e^{-\Delta S}\right\rangle = 1 , \tag{22.13}$$

which is known as the integral Fluctuation Theorem (iFT) for the total entropy production [126–128]. In order to achieve an exponential average of one, many positive ΔS must outbalance a few negative ΔS [129, 130].

Now, the natural question that arises is, are the stochastic nature of complex non-linear dynamical systems or cascade processes suitable for thermodynamic interpretation?

Recently in [62], the entropy production ΔS was determined from the reconstructed Langevin equations in scale for the velocity increment data $u_r = u(x + r) - u(x)$ for a turbulent free jet experiment [131, 132]. It turns out that the exponential average in the iFT converges faster to unity than most examples known from stochastic thermodynamics, endorsing the stochastic approach to the underlying dynamical system given by the Navier-Stokes equation. It was further found that entropy is produced for typical realizations where the flow velocity increments decrease along the cascade process, whereas realisations with increasing increments towards smaller scales give rise to entropy consumption. In this picture, the entropy-consuming realisations will be origin for the phenomenon of small-scale intermittency. Since the iFT demands a delicate balance between the two kinds of realizations, it constitutes a convenient sum rule for validating stochastic equations with a particular emphasis on correct modelling of small-scale intermittency.

The stochastic reproduction used in [62, 131] involves low-order polynomials for the drift and diffusion coefficients in scale. In order to better interpret the model equations, a minimal model that captures the essential features of the turbulent cascade is desired. In [133, 134] the iFT was used to validate various stochastic models for a variety of flows, including cylinder, grid, jet and fractal-generated turbulence. The studies show that the stochastic model equations are not universal across the various types of flow and, thus, a minimal model was pinned down to comprise a linear drift with uncorrelated multiplicative and additive noise. The additive noise acts on large scales and is responsible for energy injection, whereas the multiplicative noise extenuates fluctuations paired with large excursions, giving rise to both energy dissipation and small-scale intermittency. The drift function fixes the deterministic tendency of the cascade responsible for energy transport from large to small scales.

The fact that the multiplicative and additive noise terms in [134] appear to be uncorrelated allows for further interpretation of two noise sources to be taken as two distinct reservoirs. Turbulent energy is transported from one reservoir to the other in a non-equilibrium stationary process. If this process could be properly identified, the entropy production can be split up into two contribution, a house-keeping entropy produced by maintaining the stationary state, and the remaining excess entropy production. Each contribution would separately satisfy an iFT as known from stochastic thermodynamics [126, 135, 136].

The split-up of the iFT into two iFTs is part of a general methodology [126]. Mathematically, a fluctuation theorem is proven through reference to a conjugate dynamics. The conjugate dynamics for the iFT used for the turbulent cascade arises from scale reversal, the iFT for the house-keeping entropy is linked to current reversal. The key for the iFT to be meaningful is that the conjugate dynamics must be meaningful.

Despite the aforementioned advances using the iFT, the thermodynamic interpretation of entropy production of stochastic realizations lacks a theoretical foundation. One reason is that the stochastic reproduction used in [133, 134] is based on realistic flows facing the complete complexity of the Navier-Stokes equation. Although clearly favourable, deriving stochastic equations for the cascade process directly from the Navier-Stokes equation is an unsolved and fierce challenge. Some progress could, however, be made by reducing the analysis to ideal flows described by various models of turbulence: the iFT of the log-normal model [137] was already analysed in [62], generalisations to multiplicative cascades [138–140] and the field-theoretic approach [141–143] were investigated in [147, 148]. However, the iFT resulting from the mentioned turbulence models did not prove to hold for real flows. It is an open problem whether other models, e.g. multifractal models [149–152]), lead to an iFT that holds for realistic flows and offers a more profound thermodynamic interpretation.

An iFT that holds universally for an ideal flow (that is under some limiting assumptions, such as infinite Reynolds numbers, or homogeneity and isotropy), or even realistic flows clearly is of theoretical and practical interest. In this direction, a number of interesting questions need to be addressed.

- A problem arises when formulating the iFT for jump process [153]. While the entropy contribution for the jump part is known [126], the application to data is subtle, since a split-up into continuous and discontinuous components would be necessary.

- Another question is whether it is necessary to include a jump part in the cascade processes [153]. Considering that the field-theoretic approach [142] and the log-Poisson model [154] describe measured data particularly well, the answer might be yes, as it was shown in [143–146, 148] that such a model requires a jump component.
- Finally, it should be interesting to investigate the applicability of the iFT and the role of negative entropy production for other complex systems, including fluctuations in heart rhythm, neuroscience, seismology, financial markets, solar and wind power supply systems, etc.

22.2.4 Estimation of Memory Kernel from Time Series

Consider the generalized Langevin equation with memory kernel in the following form

$$\begin{cases} \dot{x} = v, \\ \dot{v} = F(x) - \displaystyle\int_0^t \theta(t-\tau)v(\tau)d\tau + R(t), \end{cases} \tag{22.14}$$

where x and v are the state variable and velocity, respectively, and $F(x)$ is an external force that we may assume to be a conservative force with potential $U(x)$, i.e., $F(x) = -\partial U(x)/\partial x$. Here, $\theta(t)$ denotes a memory kernel function, and the noise $R(t)$ is a stationary Gaussian process with zero mean, satisfying the second fluctuation-dissipation theorem [155]:

$$\langle R(t)R(t')\rangle = \beta^{-1}\theta(t-t'), \tag{22.15}$$

where β^{-1} is constant and can be interpreted as temperature. One can, for example, take the values of a particle velocity v from some measurements or from some many-body molecular dynamics simulations and try to estimate an effective dynamical equation that, in principle, will be non-Markovian process, such as Eq. (22.14). Such equations have been derived in previous works; see [156] and references therein, using the Mori-Zwanzig formalism [157, 158]. For simplicity here, we consider the Langevin equation in one dimension, while the generalization to higher dimensions is described in [156].

Estimation of the kernel $\theta(t)$ from time series is the main focus of this part. In what follows we describe an approximate method to derive a set of linear equations for estimating the memory kernel [156] from measured time series. Let us assume that we have a time series of v and force $F(x)$, such that the time series corresponds to a stationary random process. By multiply the second equation in (22.14) by $v(0)$ we obtain

$$g(t) = \int_0^t \theta(t-\tau)h(\tau)d\tau. \tag{22.16}$$

Here, the correlation functions are given by,

$$g(t) = \langle [\dot{v}(t) - F(x(t))]v(0)\rangle, \quad h(t) = \langle v(t)v(0)\rangle, \tag{22.17}$$

and we also assumed that $\langle R(t)v(0)\rangle = 0$. We note that from the data we can estimate the correlation functions $g(t)$ and $h(t)$.

Given the correlation functions, Eq. (22.16) can be regarded as an integral equation from which the memory function can be computed. However, this is an integral equation of the first kind and not well-posed, leading to unreliable solutions. Instead of determining the kernel function directly in the time-domain, one can instead parameterize its Laplace transform. Define the Laplace transform as

$$G(\lambda) = \int_0^{+\infty} g(t)e^{-t/\lambda}dt \tag{22.18}$$

and, similarly, denote the Laplace transforms of $h(t)$ and $\theta(t)$ by $H(\lambda)$ and $\Theta(\lambda)$, respectively. Taking the Laplace transform of Eq. (22.16), we find

$$G(\lambda) = -\Theta(\lambda)H(\lambda). \tag{22.19}$$

By taking $\lambda \to +\infty$, we obtain

$$\Theta(+\infty) = -G(+\infty)H(+\infty)^{-1}. \tag{22.20}$$

It is clear that

$$G(+\infty) = \int_0^{+\infty} g(t)dt, \quad H(+\infty) = \int_0^{+\infty} h(t)dt. \tag{22.21}$$

For short or intermediate time scales, one can use the point $\lambda = 0$. Using (22.19), one finds the limiting values of the kernel and its derivatives as $\lambda \to 0$. Direct calculations yield,

$$\Theta(0) = -G'(0)H'(0)^{-1}. \tag{22.22}$$

In addition, using the definitions of G and H, we find that $\Theta(0) = 0$, and

$$\begin{aligned} \Theta'(0) &= -\beta g'(0), \\ \Theta''(0) &= -2\beta\big[g''(0) + \beta g(0)h''(0)\big], \\ \Theta'''(0) &= -6\beta\big[g'''(0) + \beta g'(0)h''(0)\big]. \end{aligned} \tag{22.23}$$

Given limiting values extracted from the data, to construct $\Theta(\lambda)$ one can seek a rational function approximation for $\Theta(\lambda)$, which interpolates $\Theta(\lambda)$ with the interpolation conditions Eqs. (22.19) and (22.23) in the form of [156]

$$\Theta(\lambda) \approx \left[I - \lambda B_0 - \lambda^2 B_1 - \cdots - \lambda^n B_{n-1}\right]^{-1} \times \left[A_0 + \lambda A_1 + \cdots + \lambda^{n-1} A_{n-1}\right]\lambda. \tag{22.24}$$

The coefficients $\{A_i, B_i\}$ are determined by matching the limits of $\Theta(\lambda)$. The matching conditions lead to a linear system of equations, which can be solved analytically for small n or numerically for large n. The coefficients $\{A_i, B_i\}$ satisfy the following linear set of equations;

$$A_k + \sum_{i+j=k-1} B_i\, \Theta_j = \Theta_k \tag{22.25}$$

for $k = 0, \cdots, n-1$, and $\Theta_j = \frac{\Theta^{(j)}(0)}{j!}$, where $\Theta^{(j)}(0)$ denotes the jth-order derivative calculated at $\lambda = 0$ and are given by (22.23). Therefore, one can determine $\Theta(\lambda)$ by various approximations and in the time domain will be able to estimate the kernel $\theta(t)$.

For example, the zeroth-order approximation treats $\Theta(\lambda) \equiv \theta_0$ as a constant set to $\Theta(+\infty)$. Therefore, one gets a Markovian approximation by a Langevin dynamics with damping coefficient given by $\gamma = \theta_0$. The kernel $\theta(t)$ takes the Markovian approximation $\theta(t) = \theta_0 \delta(t)$. For the first-order approximation ($n = 1$), we have

$$\Theta(\lambda) = \left[I - \lambda B_0\right]^{-1} A_0 \lambda. \tag{22.26}$$

By matching Eqs. (22.20) and (22.23), we find that

$$A_0 = \theta(0), \quad B_0 = -\theta(0)\Theta(+\infty)^{-1}.$$

In this case, the memory function in the time domain is given by,

$$\theta(t) \approx e^{tB_0} A_0. \tag{22.27}$$

Depending on the value of B_0, the memory function exhibits exponential decay. By continuing this procedure using the correlations function $g(t)$ and $h(t)$ that can be determined directly from the time series, one will be able to estimate the memory kernel in various approximations. In this direction the dynamics of Langevin equation driven by Ornstein–Uhlenbeck noise has also been explored recently in [88], and a method for estimating the process parameters by a regression strategy has been presented.

22.2.5 *Anomalous Diffusion*

Anomalous diffusion is widely observed in complex systems, and in particular in biophysical systems where it has been established as an ubiquitous characteristic of cellular processes on a molecular level [159]. Theoretical models for anomalous diffusion are in general non-Markovian, such that the operational methods for the statistical analysis of such time series developed in [1] and presented in this book are not directly applicable. In particular, the two-point conditional PDF is no longer sufficient to characterize the underlying stochastic process, and multiple time correlations must be considered.

A commonly-used model for anomalous diffusion is the continuous-time random walk (CTRW), first introduced in [160] in which jump lengths and waiting times are drawn from given distributions that are possibly interdependent. In this case, the structure of the multi-point joint PDFs and associated multi-point correlation functions is well understood [161–167]. Using a representation of a CTRW in the diffusive limit this problem was first considered in [161], where the n-point joint PDF f_Y of a CTRW with power-law distributed waiting times was shown to exhibit the form

$$p_Y(\{x_n, t_n; ...; x_1, t_1\}) =$$

$$\int_0^\infty \mathrm{d}s_1 \cdots \int_0^\infty \mathrm{d}s_n \, h(\{s_n, t_n; ...; s_1, t_1\}) p_X(\{x_n, s_n; ...; x_1, s_1\}). \quad (22.28)$$

Here, p_X is the n-point joint PDF of a Markovian normal diffusive process in the operational time s, and h is the n-point joint PDF of an inverse Lévy-stable process. The Fokker–Planck-type evolution equations for p_Y can be derived, which contain fractional time derivatives [164], and it is shown that Eq. (22.28) is also valid for arbitrary waiting-time distributions. The calculation of h in this case was presented in [167]. These results confirmed with an alternative approach based on the Montroll-Weiss equation [163], in which the case of coupled jump and waiting-time distributions can also be treated systematically [166].

Apart from the CTRW, a wealth of theoretical models have also been proposed for anomalous diffusive systems, which incorporate temporal memory effects and non-Gaussian features by various mathematical techniques [168]. While a wide range of Markovian systems are successfully described by nonlinear Langevin equations and their associated Fokker–Planck equations with a unique mathematical structure, this does not seem to be the case in the non-Markovian regime. An important question is, therefore, whether the suitability of a model can already be assessed before any comparison with data, e.g., based on physical consistency. Recent work has shown that the concept of "weak Galilean invariance," imposing constraints on the form of a stochastic model in different inertial frames, can be used as an a-priori selection principle [169].

In the next chapter, we will review the application of the non-parametric method to the analysis of epileptic brain dynamics as an important example that measured intracranial electroencephalographic (iEEG) signals from electrodes in epileptic focus and distant have non-vanishing and vanishing higher order, $n > 2$ KM coefficients, respectively. More examples, such as analysing of heart rhythm, rough surfaces, financial markets, turbulence, passive scalar dynamics, etc., are reviewed in [1, 127].

References

1. R. Friedrich, J. Peinke, M. Sahimi, M.R. Rahimi Tabar, Phys. Rep. **506**, 87 (2011)
2. J. Gradisek, I. Grabec, S. Siegert, R. Friedrich, Mech. Syst. Signal Process. **16**(5), 831 (2002)
3. J. Gradisek, S. Siegert, R. Friedrich, I. Grabec, J. Sound Vib. **252**(3), 545 (2002)
4. H. Purwins, S. Amiranashvili, Phys. J. **6**, 21 (2007)
5. H.U. Bödeker, M. Röttger, A.W. Liehr, T.D. Frank, R. Friedrich, H.G. Purwins, Phys. Rev. E **67**, 056220 (2003)
6. A.W. Liehr, H.U. Bödeker, M. Röttger, T.D. Frank, R. Friedrich, H. Purwins, New J. Phys. **5**, 89 (2003)
7. M. Siefert, A. Kittel, R. Friedrich, J. Peinke, Europhys. Lett. **61**, 466 (2003)
8. M. Siefert, J. Peinke, Phys. Rev. E **70**, 015302 (2004)
9. R. Friedrich, S. Siegert, J. Peinke, S.T. Lück, M. Siefert, M. Lindemann, J. Raethjen, G. Deuschl, G. Pfister, Phys. Lett. A **271**, 217 (2000)
10. G.R. Jafari, S.M. Fazeli, F. Ghasemi, S.M. Vaez Allaei, M.R. Rahimi Tabar, A. Iraji Zad, G. Kavei, Phys. Rev. Lett. **91**, 226101 (2003)
11. M. Wächter, F. Riess, T. Schimmel, U. Wendt, J. Peinke, Eur. Phys. J. B **41**, 259 (2004)
12. G.R. Jafari, M.R. Rahimi Tabar, A. zad, G. Kavei, Phys. A **375**, 239 (2007)
13. S.M. Fazeli, A.H. Shirazi, G.R. Jafari, New J. Phys. **10**, 083020 (2008)
14. P. Sangpour, G.R. Jafari, O. Akhavan, A.Z. Moshfegh, M.R. Rahimi Tabar, Phys. Rev. B **71**, 155423 (2005)
15. E. Anahua, M. Lange, F. Böttcher, St. Barth, J. Peinke, *Stochastic Analysis of the Power Output for a Wind Turbine*, DEWEK (2004)
16. E. Anahua, St. Barth, J. Peinke, in *Wind Energy - Proceedings of the Euromech Colloquium*, ed. by J. Peinke, P. Schaumann, St. Barth (Springer, Berlin, 2007), p. 173
17. E. Anahua, St. Barth, J. Peinke, Markovian power curves for wind turbines. Wind Energy **11**, 219 (2008)
18. J. Gottschall, J. Peinke, Environ. Res. Lett. **3**, 015005 (2008)
19. P. Milan, M. Wächter, J. Peinke, Phys. Rev. Lett. **110**, 138701 (2013)
20. S. Kriso, R. Friedrich, J. Peinke, P. Wagner, Phys. Lett. A **299**, 287 (2002)
21. F. Ghasemi, A. Bahraminasab, M.S. Movahed, K.R. Sreenivasan, S. Rahvar, M.R. Rahimi Tabar, J. Stat. Mech. **P11008** (2006)
22. T. Kuusela, Phys. Rev. E **69**, 031916 (2004)
23. D.G. Luchinsky, M.M. Millonas, V.N. Smelyanskiy, A. Pershakova, A. Stefanovska, P.V. McClintock, Phys. Rev. E **72**, 021905 (2005)
24. F. Ghasemi, J. Peinke, M.R. Rahimi Tabar, M. Sahimi, Int. J. of Mod. Phys. C **17**, 571 (2006)
25. F. Ghasemi, M. Sahimi, J. Peinke, M.R. Rahimi Tabar, J. Biol. Phys. **32**, 117 (2006)
26. M.R. Rahimi Tabar, F. Ghasemi, J. Peinke, R. Friedrich, K. Kaviani, F. Taghavi, S. Sadeghi, G. Bijani, M. Sahimi, Comput. Sci. Eng. **8**, 54 (2006)
27. J. Kirchner, W. Meyer, M. Elsholz, B. Hensel, Phys. Rev. E **76**, 021110 (2007)
28. P. Sura, S.T. Gille, J. Mar. Res. **61**, 313 (2003)
29. P. Sura, J. Atmos. Sci. **60**, 654 (2003)
30. J. Egger, T. Jonsson, Tellus A **51**, 1 (2002)
31. M.R. Rahimi Tabar, et al., in *Modelling Critical and Catastrophic Phenomena in Geoscience: A Statistical Physics Approach*. Lecture Notes in Physics, vol. 705 (Springer, Berlin, 2007), p. 281
32. P. Manshour, S. Saberi, M. Sahimi, J. Peinke, A.F. Pacheco, M.R. Rahimi Tabar, Phys. Rev. Lett. **102**, 014101 (2009)
33. P. Manshour, F. Ghasemi, T. Matsumoto, G. Gómez, M. Sahimi, J. Peinke, A.F. Pacheco, M.R. Rahimi Tabar, Phys. Rev. E **82**, 036105 (2010)
34. J. Prusseit, K. Lehnertz, Phys. Rev. Lett. **98**, 138103 (2007)
35. K. Lehnertz, J. Biol. Phys. **34**, 253 (2008)
36. M. Kern, O. Buser, J. Peinke, M. Siefert, L. Vulliet, Phys. Lett. A **336**, 428 (2005)

37. T. Stemler, J.P. Werner, H. Benner, W. Just, Phys. Rev. Lett. **98**, 044102 (2007)
38. A. Bahraminasab, F. Ghasemi, A. Stefanovska, P.V.E. McClintock, H. Kantz, Phys. Rev. Lett. **100**, 084101 (2008)
39. N. Stepp, T.D. Frank, Eur. Phys. J. B **67**, 251 (2009)
40. A.M. Van Mourik, A. Daffertshofer, P.J. Beek, Biol. Cybern. **94**, 233 (2006)
41. J. Gottschall, J. Peinke, V. Lippens, V. Nagel, Phys. Lett. A **373**, 811 (2008)
42. A.M. Van Mourik, A. Daffertshofer, P.J. Beek, Phys. Lett. A **351**, 13 (2006)
43. M. Strumik, W. Macek, Phy. Rev. E **78**, 026414 (2008)
44. D.G. Luchinsky, V.N. Smelyanskiy, A. Duggento, P.V.E. McClintock, Phys. Rev. E **77**, 061105 (2008)
45. G.R. Jafari, M. Sahimi, M.R. Rasaei, M.R. Rahimi Tabar, Phys. Rev. E **83**, 026309 (2011)
46. F. Shayeganfar, S. Jabbari-Farouji, M.S. Movahed, G.R. Jafari, M.R. Rahimi Tabar, Phys. Rev. E **81**, 061404 (2010)
47. A. Farahzadi et al., Europhys. Lett. **90**, 10008 (2010)
48. S. Kimiagar, M.S. Movahed, S. Khorram, M.R. Rahimi Tabar, J. Stat. Phys. **143**, 148 (2011)
49. F. Shayeganfar, S. Jabbari-Farouji, M.S. Movahed, G.R. Jafari, M.R. Rahimi Tabar, Phys. Rev. E **80**, 061126 (2009)
50. A. Bahraminasab, F. Ghasemi, A. Stefanovska, P.V.E. McClintock, R. Friedrich, New J. Phys. **11**, 103051 (2009)
51. K. Patanarapeelert, T.D. Frank, I.M. Tang, Math. Comput. Model. **53**, 122 (2011)
52. H.U. Bödeker, C. Beta, T.D. Frank, E. Bodenschatz, Europhys. Lett. **90**, 28005 (2010)
53. R. Stresing, J. Peinke, R.E. Seoud, J.C. Vassilicos, Phys. Rev. Lett. **104**, 194501 (2010)
54. G. Nabiyouni, B.J. Farahani, Appl. Surface Sci. **256**, 674 (2009)
55. A.A. Ramos, Astron. Astrophys. **494**, 287 (2009)
56. M.H. Peters, J. Chem. Phys. **134**, 025105 (2011)
57. S.M. Mousavi et al., Sci. Rep. **7**, 4832 (2017)
58. C. Honisch, R. Friedrich, F. Hörner, C. Denz, Phys. Rev. E **86**, 026702 (2012)
59. N. Hozé, D. Holcman, Annu. Rev. Stat. Appl. **4**, 189 (2017)
60. F. Lenz, A.V. Chechkin, R. Klages, Constructing a stochastic model of bumblebee flights from experimental data. PLoS One **8**, e59036 (2013)
61. J.M. Miotto, H. Kantz, E.G. Altmann, Phys. Rev. **95**, 032311 (2017)
62. D. Nickelsen, A. Engel, Phys. Rev. Lett. **110**, 214501 (2013)
63. A. Hadjihosseini, J. Peinke, N.P. Hoffmann, New J. Phys. **16**(5), 053037 (2014)
64. A. Hadjihoseini, P.G. Lind, N. Mori, N.P. Hoffmann, J. Peinke, Europhys. Lett. **120**, 30008 (2018)
65. M. Tutkun, Phys. D **351**, 53 (2017)
66. M.S. Melius, M. Tutkun, R.B. Cal, J. Renew. Sustain. Energy **6**, 023121 (2014)
67. C. Erkal, A.A. Cecen, Phys. Rev. E **89**, 062907 (2014)
68. A. Mora, M. Haase, Nonlinear Dyn. **44**, 307 (2006)
69. M.C. Mariani, O.K. Tweneboah, H. Gonzalez-Huizar, L. Serpa, Pure Appl. Geophys. **173**, 2357 (2016)
70. G. Deco, V.K. Jirsa, P.A. Robinson, M. Breakspear, K. Friston, PLoS Comput. Biol. **4**, e1000092 (2008)
71. A. Melanson, Effective stochastic models of neuroscientific data with application to weakly electric fish, Doctoral thesis, University of Ottawa (2019)
72. P. Jenny, M. Torrilhon, S. Heinz, J. Comput. Phys. **229**, 1077 (2010)
73. Z. Czechowski, Chaos **26**, 053109 (2016)
74. Z. Czechowski, in *Complexity of Seismic Time Series*, ed. by T. Chelidze, F. Vallianatos, L. Telesca (Elsevier, Amsterdam, 2018), pp. 141–160
75. S. Chen, N. Li, S. Hsu, J. Zhang, P. Lai, C. Chan, W. Chen, Soft Matter **10**, 3421 (2014)
76. E. Boujo, N. Noiray, Proc. R. Soc. A **473**, 0894 (2016)
77. T.D. Frank, Phys. Rev. E **71**, 031106 (2005)
78. J. Estevens, P. Rocha, J. Boto, P.G. Lind, Chaos **27**, 113106 (2017)
79. Z. Farahpour et al., Phys. A **385**, 601 (2007)

80. F. Shayeganfar, M. Hölling, J. Peinke, M.R. Rahimi Tabar, Phys. A **391**, 209 (2012)
81. C. García, A. Otero, P. Félix, J. Presedo, D. Márquez, Phys. Rev. E **96**, 022104 (2017)
82. D. Nickelsen, J. Stat. Mech. **073209** (2017)
83. A.Y. Gorobets, J.M. Borrero, S. Berdyugina, Astrophys. J. Lett. **825**, L18 (2016)
84. S. Kimiagar, G.R. Jafari, M.R. Rahimi Tabar, J. Stat. Mech. **P0**, 2008 (2010)
85. M. Anvari, B. Werther, G. Lohmann, M. Wächter, J. Peinke, H.-P. Beck, Solar Energy **157**, 735 (2017)
86. M. Anvari et al., Phys. Rev. E **87**, 062139 (2013)
87. D. Bastine, L. Vollmer, M. Wächter, J. Peinke, Energies **11**, 612 (2018)
88. B. Lehle, J. Peinke, Phys. Rev. E **97**, 012113 (2018)
89. B. Wahl, U. Feudel, J. Hlinka, M. Wächter, J. Peinke, J.A. Freund, Eur. Phys. J. B **90**, 197 (2017)
90. L. Barnett, A. Seth, Detectability of Granger causality for subsampled continuous-time neurophysiological processes. J. Neurosci. Methods **275**, 93 (2017)
91. T. Scholz et al., Phys. Lett. A **381**, 194 (2017)
92. A. Melanson, J.F. Mejias, J. James Jun, L. Maler, A. Longtin, eNeuro **4**, ENEURO.0355-16 (2017)
93. R. Naud, A. Payeur, A. Longtin, Phys. Rev. X **7**, 031045 (2017)
94. M. Baldovin, A. Puglisi, A. Vulpiani, J. Stat. Mech. **043207** (2018)
95. C. Yildiz, M. Heinonen, J. Intosalmi, H. Mannerström, H. Lähdesmäki, *Machine Learning in Signal Processing, MLSP* (2018)
96. M. Baldovin, A. Puglisi, A. Vulpiani, PLoS ONE **14**(2), e0212135 (2019)
97. M. Greiner, J. Giesemann, P. Lipa, Phys. Rev. E **56**, 4263 (1997)
98. A. Riegert, N. Baba, K. Gelfert, W. Just, H. Kantz, Phys. Rev. Lett. **94**, 054103 (2005)
99. V.N. Smelyanskiy, D.G. Luchinsky, A. Stefanovska, P.V.E. McClintock, Phys. Rev. Lett. **94**, 098101 (2005)
100. V.N. Smelyanskiy, D.G. Luchinsky, D.A. Timuçin, A. Bandrivskyy, Phys. Rev. E **72**, 026202 (2005)
101. P.G. Lind, A. Mora, J.A. Gallas, M. Haase, Phys. Rev. E **72**, 056706 (2005)
102. S. Sato, T. Kitamura, Phys. Rev. E **73**, 026119 (2006)
103. I. Horenko, C. Hartmann, C. Schütte, F. Noe, Phys. Rev. E **76**, 016706 (2007)
104. E. Racca, F. Laio, D. Poggi, L. Ridolfi, Phys. Rev. E **75**, 011126 (2007)
105. T.L. Borgne, M. Dentz, J. Carrera, Phys. Rev. E **78**, 026308 (2008)
106. J. Masoliver, J. Perelló, Phys. Rev. E **78**, 056104 (2008)
107. A. Fulinski, Europhys. Lett. **118**, 60002 (2017)
108. S. Liao, High-dimensional problems in stochastic modelling of biological processes, Doctoral thesis, University of Oxford (2017)
109. N. Schaudinnus, Stochastic modeling of biomolecular systems using the data-driven Langevin equation, Doctoral thesis, Albert-Ludwigs-Universität
110. N. Schaudinnus, A.J. Rzepiela, R. Hegger, G. Stock, J. Chem. Phys. **138**, 204106 (2013)
111. N. Schaudinnus, B. Bastian, R. Hegger, G. Stock, Phys. Rev. Lett. **115**, 050602 (2015)
112. J.N. Pedersen, L. Li, C. Gradinaru, R.H. Austin, E.C. Cox, H. Flyvbjerg, Phys. Rev. E **94**, 062401 (2016)
113. D. Schnoerr, R. Grima, G. Sanguinetti, Cox process representation and inference for stochastic reaction-diffusion processes. Nat. Commun. **7**, 11729 (2016)
114. D. Schnoerr, G. Sanguinetti, R. Grima, Approximation and inference methods for stochastic biochemical kineticsa tutorial review. J. Phys. A **50**, 093001 (2017)
115. A.P. Browning, S.W. McCue, R.N. Binny, M.J. Plank, E.T. Shah, M.J. Simpson, Inferring parameters for a lattice-free model of cell migration and proliferation using experimental data. J. Theoret. Biol. **437**, 251 (2017)
116. H. Kantz, T. Schreiber, *Nonlinear Time Series Analysis* (Cambridge University Press, Cambridge, 2003)
117. J. Argyris, G. Faust, M. Haase, R. Friedrich, *An Exploration of Dynamical Systems and Chaos* (Springer, New York, 2015)

118. G. Jumarie, *Maximum Entropy, Information Without Probability and Complex Fractals, Classical and Quantum Approach*. Fundamental Theories of Physics, vol. 112 (Kluwer Academic Publishers, Dordrecht, 2000)
119. G. Jumarie, Formal calculus for real-valued fractional Brownian motions prospects in systems scienc. Kybernetes **35**, 1391 (2006)
120. T.E. Duncan, Y. Hu, B. Pasik-Duncan, Stochastic calculus for fractional Brownian motion, I. Theory, SIAM J. Control Optim. **38** 582 (2000)
121. Y. Hu, B. Oksendal, Fractional white noise calculus and applications to finance. Infin. Dimens. Anal. Quantum. Probab. Relat. Top. **6**, 1–32 (2003)
122. C. Bender, P. Parczewski, Approximating a geometric fractional Brownian motion and related processes via discrete Wick calculus. Bernoulli **16**, 389 (2010)
123. W. Xiao, J. Yu, Asymptotic theory for estimating drift parameters in the fractional Vasicek model. Econ. Theory **1–34**, (2018)
124. K. Sekimoto, Langevin equation and thermodynamics. Prog. Theor. Phys. Suppl. **130**, 17 (1998)
125. U. Seifert, From stochastic thermodynamics to thermodynamic inference. Annu. Rev. Condens. Matter Phys. **10** (2019)
126. U. Seifert, Stochastic thermodynamics, fluctuation theorems and molecular machines. Rep. Prog. Phys. **75**, 126001 (2012)
127. J. Peinke, M.R. Rahimi Tabar, M. Wächter, Annu. Rev. Condens. Matter Phys. **10** (2019)
128. U. Seifert, Entropy production along a stochastic trajectory and an integral fluctuation theorem. Phys. Rev. Lett. **95**, 040602 (2005)
129. C. Jarzynski, Rare events and the convergence of exponentially averaged work values. Phys. Rev. E **73**, 46105 (2006)
130. N. Abedpour, R. Asgari, M.R. Rahimi Tabar, Phys. Rev. Lett. **104**, 196804 (2010)
131. C. Renner, J. Peinke, R. Friedrich, On the interaction between velocity increment and energy dissipation in the turbulent cascade, arXiv:physics/0211121
132. P. Manshour, M. Anvari, N. Reinke, M. Sahimi, M.R. Rahimi Tabar, Sci. Rep. **6**, 27452 (2016)
133. N. Reinke, D. Nickelsen, A. Engel, J. Peinke, Application of an integral fluctuation theorem to turbulent flows. Springer Proc. Phys. **165**, 9–25 (2016)
134. N. Reinke, A. Fuchs, D. Nickelsen, J. Peinke, On universal features of the turbulent cascade in terms of non-equilibrium thermodynamics. J. Fluid Mech. **848**, 117 (2018)
135. T. Speck, U. Seifert, Integral fluctuation theorem for the housekeeping heat. J. Phys. **38**, L581 (2005)
136. M. Esposito, Ch. Van den Broeck, Three detailed fluctuation theorems. Phys. Rev. Lett. **104**, 090601 (2010)
137. A.N. Kolmogorov, A refinement of previous hypotheses concerning the local structure of turbulence in a viscous incompressible fluid at high Reynolds number. J. Fluid Mech. **13**, 82 (1962)
138. B. Castaing, The temperature of turbulent flows. Journal de Physique II **6**, 105 (1996)
139. F. Chillà, J. Peinke, B. Castaing, Multiplicative process in turbulent velocity statistics: a simplified analysis. Journal de Physique II **6**, 455 (1996)
140. P.-O. Amblard, J.-M. Brossier, On the cascade in fully developed turbulence. The propagator approach versus the Markovian description. Eur. Phys. J. B **12**, 579 (1999)
141. A. Polyakov, Turbulence without pressure. Phys. Rev. E **52**, 6183 (1995)
142. V. Yakhot, Probability density and scaling exponents of the moments of longitudinal velocity difference in strong turbulence. Phys. Rev. E **57**, 1737 (1998)
143. J. Davoudi, M.R. Rahimi Tabar, Phys. Rev. Lett. **82**, 1680 (1999)
144. J. Friedrich, R. Grauer, Generalized description of intermittency in turbulence via stochastic methods (2016), arXiv:1610.04432
145. J. Friedrich, Closure of the Lundgren-Monin-Novikov hierarchy in turbulence via a Markov property of velocity increments in scale, Doctoral thesis, Ruhr-Universität Bochum, Universitätsbibliothek (2017)
146. J. Friedrich, G. Margazoglou, L. Biferale, R. Grauer, Phys. Rev. E **98**, 023104 (2018)

147. D. Nickelsen, Markov processes linking thermodynamics and turbulence, Ph.D. thesis, University of Oldenburg (2014)
148. D. Nickelsen, Master equation for She-Leveque scaling and its classification in terms of other Markov models of developed turbulence. J. Stat. Mech.: Theory Exp. **073209** (2017)
149. E.A. Novikov, Infinitely divisible distributions in turbulence. Phys. Rev. E **50**, R3303 (1994)
150. B. Castaing, B. Dubrulle, Fully developed turbulence: a unifying point of view. Journal de Physique II **5**, 895 (1995)
151. Z. She, E.C. Waymire, Quantized energy cascade and log-poisson statistics in fully developed turbulence. Phys. Rev. Lett. **74**, 262 (1995)
152. A. Arneodo, J.F. Muzy, S.G. Roux, Experimental analysis of self-similarity and random cascade processes: application to fully developed turbulence data. Journal de Physique II **7**, 363 (1997)
153. M. Anvari, K. Lehnertz, M.R. Rahimi Tabar, J. Peinke, Sci. Rep. **6**, 35435 (2016)
154. Z. She, E. Leveque, Universal scaling laws in fully developed turbulence. Phys. Rev. Lett. **72**, 336 (1994)
155. R. Kubo, The fluctuation-dissipation theorem. Rep. Prog. Phys. **29**(1), 255 (1966)
156. H. Leia, N.A. Bakera, X. Li, Data-driven parameterization of the generalized Langevin equation. Proc. Natl. Acad. Sci. **113**, 14183 (2016)
157. H. Mori, Transport, collective motion, and Brownian motion. Prog. Theor. Phys. **33**, 423 (1965)
158. R. Zwanzig, Statistical mechanics of irreversibility. Lect. Theor. Phys. **3**, 106 (1961)
159. P.C. Bresslo, J.M. Newby, Rev. Mod. Phys. **85**, 135 (2013)
160. E.W. Montroll, G.H. Weiss, J. Math. Phys. **6**, 167 (1965)
161. A. Baule, A. Friedrich, Phys. Rev. E **71**, 026101 (2005)
162. F. Sanda, S. Mukamel, Phys. Rev. E **72**, 031108 (2005)
163. E. Barkai, I.M. Sokolov, J. Stat. Mech. **P08001** (2007)
164. A. Baule, R. Friedrich, Europhys. Lett. **77**, 10002 (2007)
165. A. Baule, R. Friedrich, Europhys. Lett. **79**, 60004 (2007)
166. M. Niemann, H. Kantz, Phys. Rev. E **78**, 051104 (2008)
167. A. Cairoli, A. Baule, Phys. Rev. Lett. **115**, 110601 (2015)
168. R. Metzler, J.H. Jeon, A.G. Cherstvy, E. Barkai, Phys. Chem. Chem. Phys. **16**, 24128 (2014)
169. A. Cairoli, R. Klages, A. Baule, Proc. Natl. Acad. Sci. USA **115**(22), 5714 (2018)

Chapter 23
Epileptic Brain Dynamics

As an example of the analysis and reconstruction of nonlinear stochastic time series, we consider an important medical problem, namely, epileptic brain dynamics.

23.1 Stochastic Qualifiers

There has been increasing interest over the last decades in the application of nonlinear dynamics theory (chaos theory [1]) to modeling of brain activity, from the level of single neurons to large-scale whole-brain dynamics [2–17]. Many studies have been concerned with electroencephalographic (EEG) signals recorded in animals or human subjects during normal physiological conditions (e.g., awake state, sleep, or cognition) and during pathophysiological conditions (e.g., epilepsy, Alzheimer's disease, or Parkinson's disease). A standard approach is using the repertoire of nonlinear time series analysis techniques [18] that allows a characterization of the dynamics of a complex systems from the analysis of a time series generated by the system. Such a time series typically consists of a sequential measurement in time of a pertinent and easily accessible variable.

EEG time series exhibit, in general, great irregularity that may have different origins, e.g., due to noise or, otherwise, a reflection of the presence of nonlinearities and stochastic behaviors. Chaos theory allows one to distinguish irregularities due to nonlinearities from those due to noise. In the presence of nonlinearities, the complexity of a dynamics can be quantified in terms of properties of the attractor in phase space, e.g., its fractal dimension. The fractal dimension may thus provide a classification of brain dynamics in terms of its complexity. However, a careful discussion is necessary in order to distinguish chaos from noise, because determining a fractal dimension is a necessary but not sufficient condition for chaos. In addition, a large

M. R. Rahimi Tabar, *Analysis and Data-Based Reconstruction of Complex Nonlinear Dynamical Systems*, Understanding Complex Systems,
https://doi.org/10.1007/978-3-030-18472-8_23

number of influencing factors have been identified and discounting them can lead to severe misinterpretations.

With the aforementioned pros and cons in mind, a large number of studies corroborated the high suitability of nonlinear EEG analysis, particularly in epilepsy [14]—a very common neurological disorder affecting approximately 65 million people worldwide [19]. The cardinal symptom of this disease—epileptic seizures—can not be controlled sufficiently with anti-epileptic drugs in approximately 30 % of people with epilepsy [20]. Among others, nonlinear EEG analysis has contributed to improving identification of the seizure-generating area—a prerequisite for the surgical treatment of otherwise non-controllable seizures—and has boosted research into the predictability of seizures [21], which opened up new vistas for the development of alternative treatment options [22]. Despite the many promising findings, there are a number of problems for which there are currently no satisfactory solutions. Although the epileptic process appears as a nonlinear deterministic dynamics in an otherwise stochastic environment [23, 24], in many cases crucial aspects of pathological brain dynamics must be regarded as a high-dimensional stochastic process. Such a dynamics may not be captured if time series analysis techniques are used that preferentially focus on the low-dimensional deterministic part of the dynamics.

Prusseit and Lehnertz [25] used the construction method in this book to analyze EEG time series recorded intracranially from eight patients with drug-resistant focal epilepsy, who had undergone evaluation for resective therapy. The multiday (5–12 days), multichannel (20–60 recording sites) EEG time series covered physiological and pathophysiological activities from the seizure-free interval. EEG data were recorded from the cortex and from within the relevant structures of the brain, hence with a high signal-to-noise ratio. The time series were sampled at 200 Hz using a 16 bit analog-to-digital converter, and filtered within a frequency band of 0.53–85 Hz.

The authors showed that the dissipative dynamical system *epileptic brain* under the influence of noise can often be successfully modeled by a Fokker–Planck or, equivalently, the associated Langevin equation. They also showed that using this approach, an improved characterization of pathological brain dynamics can be achieved by explicitly taking into account the stochastic part of the dynamics. Despite limitations that are attributed to the fact that the EEG time series may not entirely meet the prerequisites of the underlying theoretical framework, a one-dimensional Fokker–Planck model appeared to be appropriate for a description of physiological activities. In contrast, capturing all aspects of pathophysiological activities in such a model may not be possible. Nevertheless, Prusseit and Lehnertz derived stochastic qualifiers that allowed a more comprehensive characterization of the epileptic process. It was shown that the drift and diffusion coefficients appear to be quite useful characterizing quantities [26]. One expects that this approach, along with further improvements, can yield valuable information for diagnostic purposes, and may even advance our understanding of the complex dynamical system *epileptic brain*.

Figure 23.1 shows typical examples of the drift and diffusion coefficients estimated from EEG time series. Both coefficients are well approximated by low-order polynomials. As expected for this one-dimensional model, $D^{(1)}$ indicates an overall linear damping behavior. For EEG time series recorded from within the epileptic

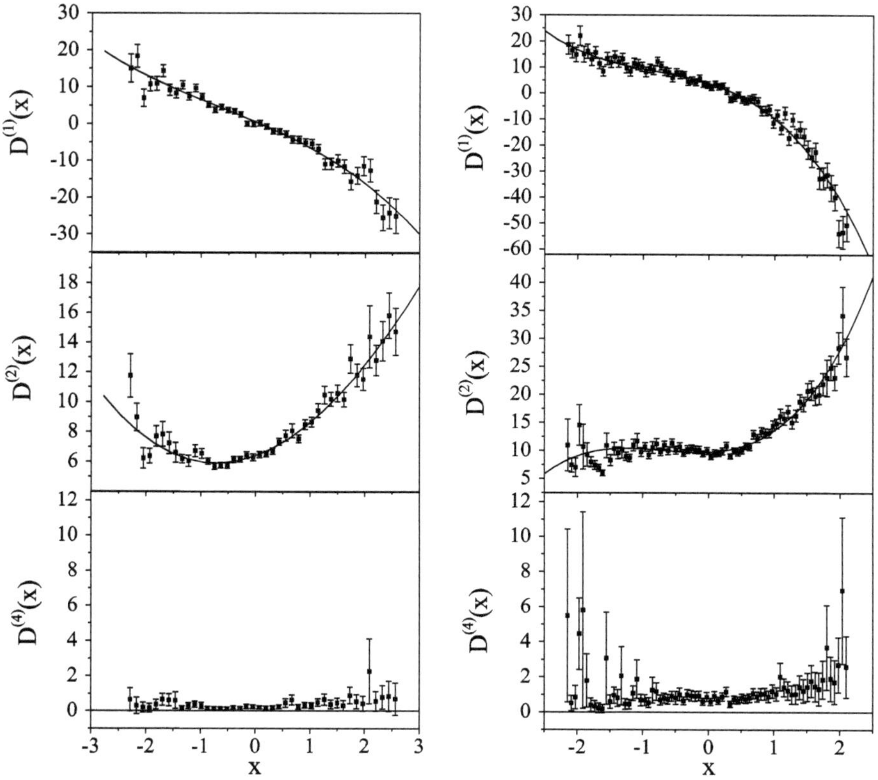

Fig. 23.1 Estimated coefficients $D^{(1)}$, $D^{(2)}$, and $D^{(4)}$ for exemplary EEG time series. Left: from a distant brain region; right: from within the epileptic focus. Shown are the estimates for time series consisting of 10^5 data points (squares) as well as fits with low-order polynomials (black lines). For epoch lengths ranging 2–8 min, the estimated Markov–Einstein time scale (using the Chapman–Kolmogorov equation) is unity (in units of the sampling interval). Source from [25]

focus, the authors of Ref. [25] observed small nonlinearities toward higher amplitude values, which is in line with the findings from studies that applied nonlinear time series analysis techniques [27]. The dependence of $D^{(2)}$ on the state variable x indicates a multiplicative influence of the noise. The fourth-order coefficient $D^{(4)}$ is also shown in Fig. 23.1, which allows one to determine whether the driving noise process $\Gamma(t)$ exhibits deviations from a Gaussian distribution [25]. For the EEG data recorded from a distant brain region, $D^{(4)}$ took on values slightly above zero, but its magnitude was less than $1/20$ of $D^{(2)}$. In contrast, for the EEG data recorded from within the epileptic focus, $D^{(4)}$ took on values clearly above zero. These findings indicate that a description of the pathological brain dynamics by a one-dimensional Fokker–Planck model may be inadequate, and one needs to also take into account an influence of higher-order Kramers–Moyal coefficients (see below).

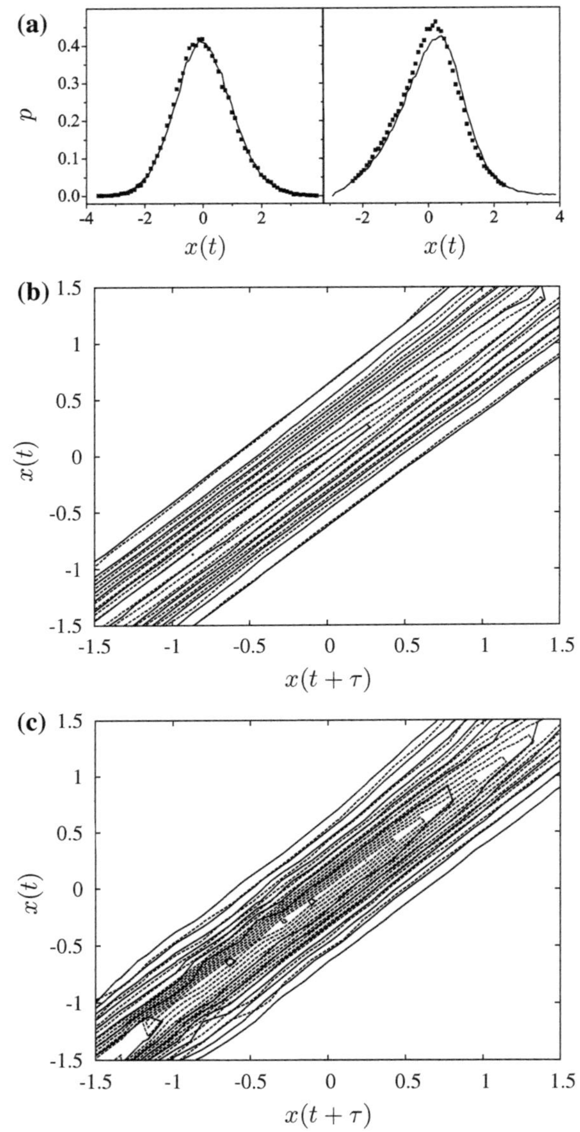

Fig. 23.2 Comparison of the stationary PDFs (**a**), and contour plots of conditional PDFs for lag $\tau = 1$ (**b**) and (**c**) for the EEG time series [squares in **a**, dashed lines in **b** and **c**] and time series generated by integrating the associated Langevin equations (straight lines). **a** left plot and **b**: from a distant brain region; **a** right plot and **c**: from within the epileptic focus. Contour plots were generated using an increment between contour lines of 0.02 in **b** and 0.012 in **c**. Source from [25]

This is further corroborated by the results obtained from integrating the Langevin equation using the estimated functions $D^{(1)}$ and $D^{(2)}$ for both EEG time series; see Fig. 23.2. While both the stationary and the conditional PDFs of the integrated model and of the EEG time series coincided quite well for the recording from a distant brain region, Prusseit and Lehnertz observed more pronounced deviations for the recordings from within the epileptic focus. These findings clearly indicate that

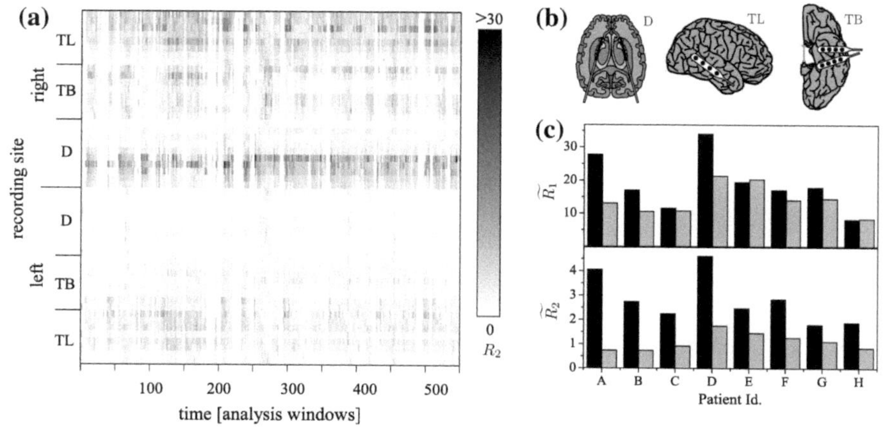

Fig. 23.3 (**a**) Time resolved estimates of R_2 calculated from a multichannel (52 contacts) EEG recording (approximately 19 h) of a patient suffering from a right-sided focal epilepsy (patient F). (**b**) Implantation scheme of intracranial electrodes: hippocampal depth electrodes (10 contacts each, D), lateral (4 × 16 contacts, TL), and basal (4 contacts each, TB) strip electrodes. (**c**) Spatiotemporal means of R_1 and R_2 for all investigated patients. Black bars denote values from the focal and gray bars from the non-focal hemisphere. Source from [25]

specific characteristics of the estimated drift and diffusion coefficients allow one to differentiate between physiological and pathophysiological activities.

One may define various quantities based on the estimated coefficients that serve as stochastic qualifiers of epileptic brain dynamics. As an example, consider the range covered by the values of the estimated coefficients $R_{1,2} = \left|\max D^{(1,2)} - \min D^{(1,2)}\right|$. The authors of Ref. [25] only took into account values of $D^{(1,2)}$ for which at least 100 data points were available for the estimation procedure. For the multichannel, multiday EEG recordings from all patients, they performed a time resolved estimation of $R_{1,2}$ using a moving-window technique. The windows were of size $N = 50000$ points, and windows overlapped by 50 %. This choice represents a compromise between sufficient statistics for a reliable estimation of $D^{(1,2)}$ and temporal resolution [28], which might be of interest for further EEG analysis.

Figure 23.3(a) shows a typical spatiotemporal distribution of R_2, calculated from a multichannel EEG recording from the seizure-free interval of a patient with an epileptic focus located in the right hemisphere of the brain. When comparing findings from the left and right brain hemisphere, one observes highest values of R_2 confined to brain regions close to or within the epileptic focus.

Since these stochastic qualifiers exhibited only little variance over time, for each contact, their temporal average, $\left(\langle R_{1,2}\rangle_t\right)$, was calculated and eventually averaged over all contacts c from each hemisphere $\left(\widetilde{R_{1,2}} = \langle\langle R_{1,2}\rangle_t\rangle_c\right)$. This allowed further condensing of the information contained in the spatiotemporal distribution of the stochastic qualifiers, and to investigate retrospectively whether they can provide diagnostically relevant information. One can refer to the brain hemisphere that contains the epileptic focus (determined by the presurgical workup and by the postoperative complete seizure control) as the focal side, whereas the opposite hemisphere is referred to as the non-focal side; see Fig. 23.3b. In six of the eight patients $\widetilde{R_1}$ was higher on the focal side. Interestingly, it was observed that $\widetilde{R_2}$ allowed correct identification of the focal side in all patients; see Fig. 23.3c.

23.2 Detailing the Stochastic Behavior of Epileptic Brain Dynamics

As we have seen in the previous section, stochastic qualifiers of epileptic brain dynamics that are based on specific characteristics of the first- and second-order Kramers–Moyal coefficients, that is drift and diffusion coefficients, estimated using the Langevin-type modeling of intracranially recorded EEG time series can yield valuable information for diagnostic purposes. Diffusion-coefficient-based qualifiers allowed a more detailed characterisation of spatial and temporal aspects of the epileptic process in the affected and the non-affected brain hemisphere. Nevertheless, the dynamics of the epileptic focus is characterised by a non-vanishing fourth-order Kramers–Moyal coefficient, in contrast to the dynamics of other, non-affected brain regions [25]. Thus, pathological brain dynamics appear to not belong to the class of continuous diffusion processes and consequently, the Langevin-type modeling may not capture all its aspects [25]. Under the consideration of the highly nonlinear character of this dynamics [6, 23, 25, 27, 29], recently Anvari et al. [30] explicitly estimated diffusion and jump characteristics and succeeded in disentangling the stochastic part of epileptic brain dynamics.

The authors analyzed EEG time series recorded intracranially from ten patients with drug-resistant focal epilepsy, who had undergone evaluation for resective therapy. The multiday (2–14 days), multichannel (16–88 recording sites) EEG time series covered physiological and pathophysiological activities from the seizure-free interval. EEG data were recorded from the cortex and from within the relevant structures of the brain, hence with a high signal-to-noise ratio. The time series were sampled at 200 Hz (sampling interval $\Delta t = 5$ ms) using a 16 bit analog-to-digital converter, and filtered within a frequency band of 0.1–70 Hz.

Exemplary findings obtained by investigating brain dynamics from within the epileptic focus and from a distant brain region recorded during the seizure-free interval of an epilepsy patient are shown in Fig. 23.4. For the KM conditional moments of orders 1, 2, 4, and 6 (calculated using the Nadaraya-Watson estimator with a Gaussian kernel), Anvari et al. obtained finite values in the limit of vanishing time increments, which allows one to conclude that the influence of the measurement noise can be neglected. They then considered the interval $x \in (-200, 200\,\mathrm{mV})$ and reported on averaged amplitudes of drift coefficient $D^{(1)}$, diffusion coefficient $D^{(2)}$, jump rate λ, and jump amplitude σ_ξ^2 for each brain region. The drift coefficients from both brain regions (data not shown here) indicated an overall linear damping behav-

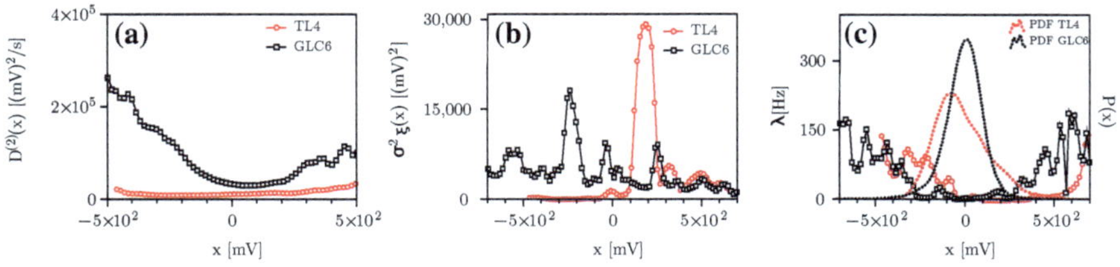

Fig. 23.4 Disentangling stochastic characteristics of epileptic brain dynamics I. Exemplary findings from an epilepsy patient with an epileptic focus in the left mesial temporal lobe. **a–c** Diffusion coefficients $D^{(2)}(x)$, jump amplitudes $\sigma_\xi^2(x)$, and jump rates $\lambda(x)$ together with the respective probability distribution functions $p(x)$ estimated from normalised EEG time series recorded during the seizure-free interval from within the epileptic focus and from a distant brain region. EEG time series had a duration of 2000 s, corresponding to $N = 4 \cdot 10^5$ data points. Source from [30]

ior, however, with small nonlinearities toward larger values of x for the dynamics within the epileptic focus [30]. The slopes of drift coefficients differed by about an order of magnitude (within epileptic focus: $\simeq -0.51 \pm 0.05$; distant brain region: $\simeq -3.60 \pm 0.12$). From the inverse of the slope of the linear part of the drift coefficients, the authors observed correlation time scales in the order of 1.96 ± 0.20s for the dynamics within the epileptic focus and of 0.28 ± 0.01s for the dynamics of the distant brain region. A comparable ratio was obtained for the averaged amplitudes of the diffusion coefficients, with the one for the former dynamics amounting to about a third of the one seen for the latter dynamics. The diffusion coefficient for the dynamics within the epileptic focus was largely independent of the state variable x and for the dynamics of the distant brain region, it depended parabolically on x (see Fig. 23.4a). Overall, these dependences on x indicate a multiplicative influence of the noise.

From Eq. (12.2), it is known that the second KM conditional moment contains contributions from the jumpy dynamics. One can thus further disentangle the stochastic part of epileptic brain dynamics and explicitly estimate jump characteristics. For the aforementioned exemplary data, it is found that the dynamics within the epileptic focus is characterised by a smaller averaged jump amplitude (within epileptic focus: $3400 \pm 270\,(\mathrm{mV})^2$, distant brain region: $4400 \pm 350\,(\mathrm{mV})^2$; see Fig. 23.4b) and by a higher averaged jump rate (within epileptic focus: 18 ± 3 Hz, distant brain region: 8 ± 1 Hz; see Fig. 23.4c).

Next, Anvari et al. demonstrated how the aforementioned exemplary findings translate to the long-term brain dynamics from all sampled brain regions. To this end, they performed a time-resolved analysis of the patient's EEG time series that were recorded over a period of more than eight days. They subdivided the time series into non-overlapping windows of size $N = 10^5$ data points, calculated $D^{(2)}$, σ_ξ^2, and λ for each window as outlined above, and by averaging over all windows they obtained their means and standard deviations for each recording site (see parts a–c of Fig. 23.5). Eventually, they derived—separately for the sites from within the epileptic focus and for the distant sites—spatial means and standard deviations of the tempo-

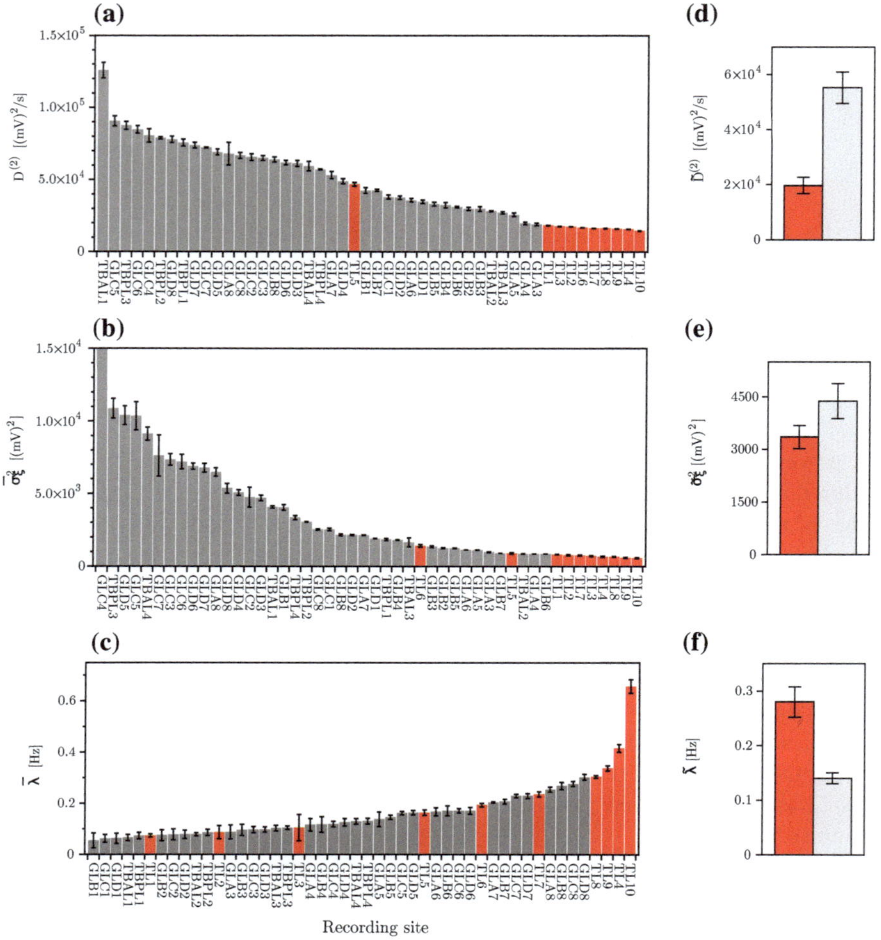

Fig. 23.5 Disentangling stochastic characteristics of epileptic brain dynamics II. Exemplary findings from an epilepsy patient with an epileptic focus in the left mesial temporal lobe. **a**–**c** Means and standard deviations of diffusion coefficients ($\bar{D}^{(2)}$), jump amplitudes ($\bar{\sigma}_{\xi}^{2}$), and jump rates ($\bar{\lambda}$) for EEG time series from all recording sites. Data from sites within the epileptic focus are colored red. **d**–**f** Spatial means and standard deviations of temporally averaged diffusion coefficients $\widetilde{D}^{(2)}$, jump amplitudes $\widetilde{\sigma}_{\xi}^{2}$, and jump rates $\widetilde{\lambda}$ (epileptic focus: red; distant sites: gray). Source from [30]

rally averaged estimators (marked with a tilde; see parts d–f of Fig. 23.5). In general, the dynamics of the epileptic focus in this patient can be characterised by a mean diffusion coefficient $\widetilde{D}^{(2)}$ whose amplitude was about half the one seen for the dynamics of the other brain areas. The same holds for the mean jump amplitude $\widetilde{\sigma}_{\xi}^{2}$, however, the differences between affected and non-affected brain regions were more pronounced, with a mean jump amplitude in the epileptic focus amounting to about a sixth of the one of the other brain areas. The mean jump rate $\bar{\lambda}$ attained high values

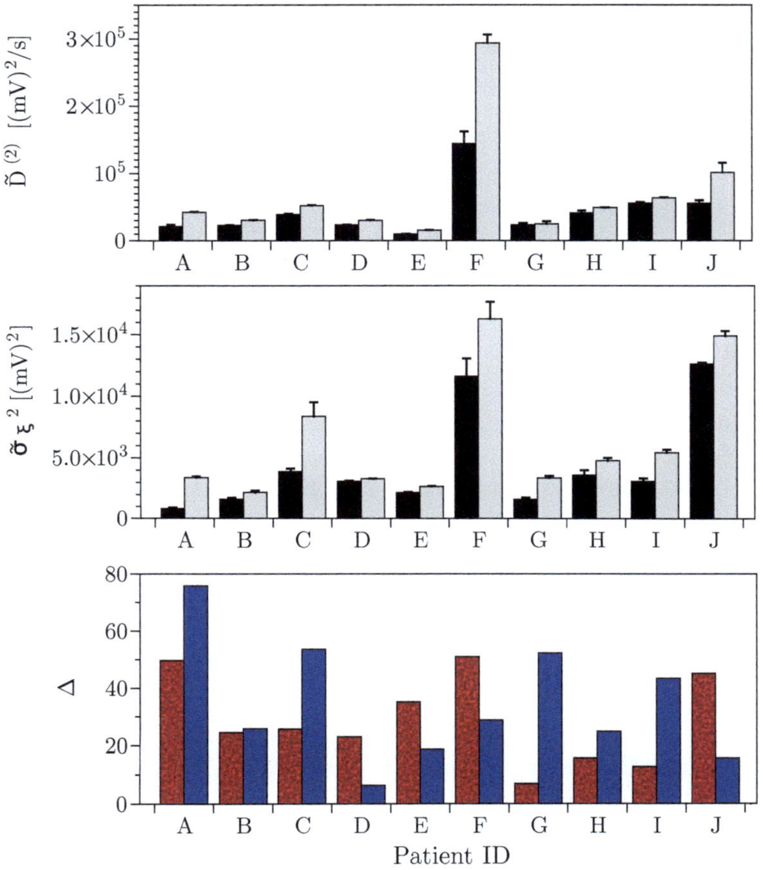

Fig. 23.6 Stochastic qualifiers of brain dynamics for each epilepsy patient. Spatial means and standard deviations of temporally averaged diffusion coefficients $\widetilde{D}^{(2)}$(top) and jump amplitudes $\widetilde{\sigma}_\xi^2$(middle), calculated separately for recordings from within the epileptic focus (black bars) and from distant sites (gray bars). Relative differentiability Δ (bottom) between non-affected (suffix d) and affected (suffix f) brain dynamics using diffusion coefficients $((\widetilde{D}_d^{(2)} - \widetilde{D}_f^{(2)})/\widetilde{D}_d^{(2)}$, red bars) and jump amplitudes $((\widetilde{\sigma}_{\xi\,d}^2 - \widetilde{\sigma}_{\xi\,f}^2)/\widetilde{\sigma}_{\xi\,d}^2$, blue bars). Source from [30]

at some—though not all—recording sites capturing the dynamics of the epileptic focus, and comparably high mean jump rates could also be observed at distant sites. Consequently, differentiability between affected and non-affected brain regions with the mean jump rate $\widetilde{\lambda}$ was insignificant.

Finally, the authors demonstrated extendability of their observations beyond exemplary data by applying the aforementioned steps of a time-resolved analysis to the long-term dynamics of all sampled brain regions from all patients. Figure 23.6 summarizes the main findings (differentiability between affected and non-affected brain regions with the mean jump rate $\widetilde{\lambda}$ was again insignificant and therefore not shown here). In general, both the mean diffusion coefficient $\widetilde{D}^{(2)}$ and the mean jump amplitude $\widetilde{\sigma}_{\xi}^{2}$ demonstrated a high variability between patients, both in terms of their sizes and with respect to differentiability between affected and non-affected brain regions. In 5 (out of 10) patients (see Fig. 23.6), differentiability with the mean jump amplitude clearly exceeded the one obtained with the mean diffusion coefficient ($\approx$28%), and in 4 patients the opposite holds true ($\approx$21%). In only one case (patient B), both estimators allowed for a comparable differentiability ($\approx$22%).

Summary

A stochastic dynamical jump-diffusion modelling allows one to separate the deterministic drift term as well as different stochastic behaviors, namely diffusive and jumpy behavior. It is argued that when the conditional moments of Kramers–Moyal coefficients of order larger than two are non-vanishing, jump events may play a significant role in a stochastic process. Indeed, these higher-order Kramers–Moyal coefficients carry information about the probability of arrival of jumps and about the features of the distribution of the jump size.

Through extensive analyses of multi-day, multi-channel intracranial electroencephalographic recordings from epilepsy patients it was demonstrated that the dynamics of the epileptic focus can be characterised as a stochastic process with a smaller mean diffusion coefficient and a smaller mean jump amplitude as compared to the dynamics of distant brain regions. Higher-order Kramers–Moyal coefficients thus provide extra information that can be regarded valuable for diagnostic purposes. Their exact relationship to actual physiological/pathophysiological activities, however, would need to be investigated in future studies. Taken together, generalising the Langevin-type modelling to a jump-diffusion modelling improves the characterisation of pathological brain dynamics beyond a continuous process.

References

1. H.G. Schuster, *Deterministic Chaos: An Introduction* (VCH, Weinheim, 1989)
2. D. Duke, W. Pritchard, *Measuring Chaos in the Human Brain* (World Scientific, Singapore, 1991)
3. T. Elbert, W.J. Ray, Z.J. Kowalik, J.E. Skinner, K.E. Graf, N. Birbaumer, Physiol. Rev. **74**, 1 (1994)
4. J. Wright, D. Liley, Behav. Brain Sci. **19**, 285 (1996)
5. K. Lehnertz, J. Arnhold, P. Grassberger, C.E. Elger, *Chaos in Brain?* (World Scientific, Singapore, 2000)
6. R.G. Andrzejak, K. Lehnertz, F. Mormann, C. Rieke, P. David, C.E. Elger, Phys. Rev. E **64**, 061907 (2001a)
7. J. Jeong, T.H. Chae, S.Y. Kim, S.H. Han, J. Clin. Neurophysiol. **18**, 58 (2001)
8. H. Korn, P. Faure, C. R. Biol. **326**, 787 (2003)

9. E. Pereda, R. Quian Quiroga, J. Bhattacharya, Prog. Neurobiol. **77**, 1 (2005)
10. C.J. Stam, Clin. Neurophysiol. **116**, 2266 (2005)
11. M.I. Rabinovich, P. Varona, A.I. Selverston, H.D.I. Abarbanel, Rev. Mod. Phys. **78**, 1213 (2006)
12. E.M. Izhikevich, *Dynamical Systems in Neuroscience: The Geometry of Excitability and Bursting* (MIT Press, Cambridge, 2007)
13. H. Osterhage, K. Lehnertz, Int. J. Bifurc. Chaos Appl. Sci. Eng. **17**, 3305 (2007)
14. K. Lehnertz et al., J. Clin. Neurophysiol. **18**, 209 (2001)
15. D.R. Chialvo, Nat. Phys. **6**, 744 (2010)
16. L. Glass, Chaos **25**, 097603 (2015)
17. M. Breakspear, Nat. Neurosci. **20**, 340 (2017)
18. H. Kantz, T. Schreiber, *Nonlinear Time Series Analysis* (Cambridge University Press, Cambridge, 2003)
19. S.L. Moshé, E. Perucca, P. Ryvlin, T. Tomson, Epilepsy: new advances. Lancet **385**, 884, 326 (2015)
20. P. Kwan, S.C. Schachter, M.J. Brodie, Drug-resistant epilepsy. N. Engl. J. Med. **365**(328), 919–926 (2011)
21. L. Kuhlmann, K. Lehnertz, M.P. Richardson, B. Schelter, H.P. Zaveri, Nat. Rev. Neurol. **14**, 618 (2018)
22. I. Kiral-Kornek et al., EBioMedicine **27**, 103 (2018)
23. R.G. Andrzejak, G. Widman, K. Lehnertz, P. David, C.E. Elger, Epilepsy Res. **44**, 129 (2001)
24. R.G. Andrzejak, K. Schindler, C. Rummel, Phys. Rev. E **86**, 046206 (2012)
25. J. Prusseit, K. Lehnertz, Phys. Rev. Lett. **98**, 138103 (2007)
26. T. Stemler, J.P. Werner, H. Benner, W. Just, Phys. Rev. Lett. **98**, 044102 (2007)
27. R.G. Andrzejak, F. Mormann, G. Widman, T. Kreuz, C.E. Elger, K. Lehnertz, Epilepsy Res. **69**, 30 (2006)
28. D. Lamouroux, K. Lehnertz, Phys. Lett. A **373**, 3507 (2009)
29. J.P. Pijn, J. van Neerven, A. Noest, F.H. Lopes da Silva, Electroencephalogr. Clin. Neurophysiol. **79**, 371 (1991)
30. M. Anvari, K. Lehnertz, M.R. Rahimi Tabar, J. Peinke, Sci. Rep. **6**, 35435 (2016)

Correction to: Analysis and Data-Based Reconstruction of Complex Nonlinear Dynamical Systems

Correction to:
M. R. Rahimi Tabar, *Analysis and Data-Based Reconstruction of Complex Nonlinear Dynamical Systems*, Understanding Complex Systems, https://doi.org/10.1007/978-3-030-18472-8

In the original version of the book, the following post-publication corrections have been incorporated in Chapters 8, 13, 16, 17 and 20:

In chapter 8, "(Cox–Ingersoll–Ross)" has been incorporated for problem 8.3 in the section (c).

In chapter 13, reference 2 has been modified from "In prepration" to "arXiv:1907.05371".

In chapter 16, reference 18 has been modified from "P0,2006 (2003)" to "P02003 (2006)".

In chapter 17, reference 3 has been modified from "Bierens, H.J." to "H.J. Bierens,".

In chapter 20, the reference "L. Rydin Gorjão …" has been added as 1st and cited in section 20.1.2. Hence, all the reference has been renumbered accordingly and the reference 10 (L.R. Gorjão) has been Modified from "In preparation" to "arXiv:1907.05371".

The updated version of these chapters can be found at
https://doi.org/10.1007/978-3-030-18472-8_8
https://doi.org/10.1007/978-3-030-18472-8_13
https://doi.org/10.1007/978-3-030-18472-8_16
https://doi.org/10.1007/978-3-030-18472-8_17
https://doi.org/10.1007/978-3-030-18472-8_20

M. R Rahimi Tabar, *Analysis and Data-Based Reconstruction of Complex Nonlinear Dynamical Systems*, Understanding Complex Systems,
https://doi.org/10.1007/978-3-030-18472-8_24

Appendix A
Wilcoxon Test

To test the Markovianity of the time series, we can apply the Wilcoxon test to compare the distributions $p(x_3, t_3|x_2, t_2)$ and $p(x_3, t_3|x_2, t_2; x_1, t_1)$, with $t_3 > t_2 > t_1$ for a fixed value of x_1 [1, 2].

The null hypothesis to be tested by the Wilcoxon test is that the probability density functions $p(y)$ and $\tilde{p}(z)$ of the stochastic variables y and z are identical. In the case of Eq. 16.5, the two stochastic variables are $y = x_3(t_3)|_{x_2(t_2)}$, and $z = x_3(t_3)|_{x_2(t_2),x_1(t_1)}$. Two samples $\{y_1, \ldots, y_n\}$ and $\{z_1, \ldots, z_m\}$ of independent realizations of the variables are taken from the data. One can take values which are separated by one correlation time scale of time series to be sufficiently independent from each other for the purpose of the test, in order not to reduce the sample sizes.

Then, the number of values z_j with $y_j < z_i$ is counted for each z_i and summed over i:

$$Q = \sum_{i=1}^{m} \sum_{j=1}^{n} \alpha_{ij}, \qquad \alpha_{ij} = \begin{cases} 1 & y_j < z_i \\ 0 & y_j \geq z_i \end{cases} .$$

Under the null hypothesis, the quantity Q is Gaussian distributed with mean value $\mu_0(n, m) = nm/2$, and, for $n, m > 25$, standard deviation $\sigma_0(n, m) = \sqrt{nm(n + m + 1)/12}$. Then the quantity

$$\Delta Q = \frac{|Q - \mu_0(n, m)|}{\sigma_0(n, m)}$$

which is the absolute value of a standard normal distributed variable, has a mean value of $t_0 = \sqrt{\frac{2}{\pi}}$. To implement the test, the quantity

$$\Delta Q^* = \Delta Q / \sqrt{\frac{2}{\pi}}$$

M. R. Rahimi Tabar, *Analysis and Data-Based Reconstruction of Complex Nonlinear Dynamical Systems*, Understanding Complex Systems,
https://doi.org/10.1007/978-3-030-18472-8

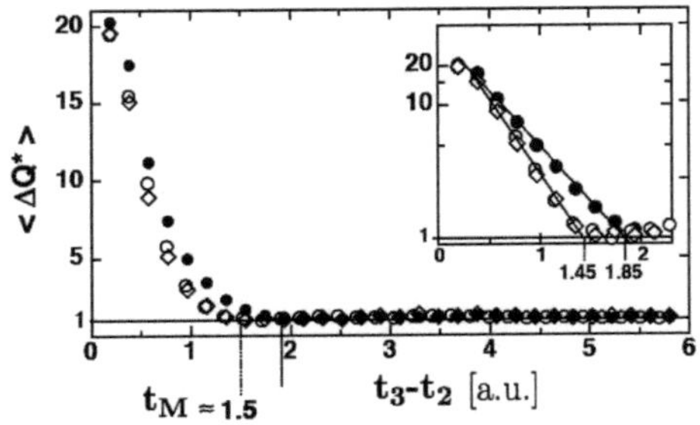

Fig. A.1 Wilcoxon Test. Typical behaviour of $\langle \Delta Q^* \rangle$ verses $t_3 - t_2$ for two time series. The Markov time scales are about $t_M \simeq 1.5$ and $t_M \simeq 1.9$

is calculated for a fixed value of x_1 for a number of for instance, 100 bins for x_2, which span the complete range of x_2. The mean value $W = \langle \Delta Q^* \rangle$, can be calculated by taking the average over the 100 values of ΔQ^*. Under the null hypothesis of the Markov property, if the relationship $p(x_3, t_3|x_2, t_2) = p(x_3, t_3|x_2, t_2; x_1, t_1)$ is satisfied, then the expected value for the expectation value of W is 1.

In the case of empirically derived time series that were sampled with a fixed sampling interval, one can apply the Wilcoxon test with varying $T = t_3 - t_2 = t_2 - t_1$ and estimate ΔQ^* with $T = 1, 2, \ldots$. By definition, the Markov-Einstein time scale t_M is that minimum value of T at which W will be 1. Typical behaviour of $\langle \Delta Q^* \rangle$ is given in Fig. A.1.

References

1 F. Wilcoxon, Biometrics **1**, 80 (1945).
2 C. Renner, J. Peinke, R. Friedrich, J. Fluid Mech. **433**, 383 (2001).

Appendix B
Kernel Density Estimator

The probability distribution function (PDF) $f_Q(x)$ is given by Eq. (2.8) as:

$$f_Q(x) = \langle \delta(x - Q) \rangle \,. \tag{B.1}$$

Let us denote $Q_1, \ldots, Q_n$ as different realizations of process Q, then, we can calculate the average of any function $g(Q)$ as:

$$\langle g(Q) \rangle = \frac{1}{n} \sum_{i=1}^{n} g(Q_i). \tag{B.2}$$

In what follows, we use the notation that the quantity without hat related to the exact values and with it introduce the estimator for corresponding object. With finite n equation Eq. (B.1), will be,

$$\hat{f}_Q(x) = \frac{1}{n} \sum_{i=1}^{n} \delta(x - Q_i) \quad . \tag{B.3}$$

Now we introduce non-negative, real-valued, integrable function K (with finite moments) with bandwidth h which satisfy the following conditions:
1. Normalisable:

$$\int_{-\infty}^{+\infty} K(x) dx = 1 \tag{B.4}$$

2. in the limit of $h \to 0$ converges to Dirac delta function:

$$\lim_{h \to 0} \frac{1}{h} K\left(\frac{x}{h}\right) = \delta(x) \tag{B.5}$$

3. and is a symmetric function:

M. R. Rahimi Tabar, *Analysis and Data-Based Reconstruction of Complex Nonlinear Dynamical Systems*, Understanding Complex Systems,
https://doi.org/10.1007/978-3-030-18472-8

$$K(x) = K(-x) \quad . \tag{B.6}$$

Density $f_Q(x)$ can be written in terms of K as:

$$\hat{f}_Q(x) = \frac{1}{n}\sum_{i=1}^{n}\frac{1}{h}K\left(\frac{x-Q_i}{h}\right) \quad . \tag{B.7}$$

This estimator converges to true PDF in the limit $n \to \infty$, $h \to 0$. The function $K(x)$ is known a kernel and Eq. B.7 define the kernel density estimator. The most frequently used kernel functions are:

$$\begin{array}{ll} Uniform & \frac{1}{2}I(|x|<1) \\ Epanechnikov & \frac{3}{4}(1-x^2)I(|x|<1) \\ Quartic & \frac{15}{16}(1-x^2)^2 I(|x|<1) \\ Gaussian & \frac{1}{\sqrt{2\pi}}e^{-\frac{x^2}{2}} \end{array} \tag{B.8}$$

where $I(|x| < 1)$ is given by:

$$I(|x|<1) = \begin{cases} 1 & |x|<1) \\ 0 & else \end{cases} \quad . \tag{B.9}$$

The estimator $\hat{f}_Q(x)$ depends on kernel function and its bandwidth. Therefore we index estimators with h.

B.1 Bias and Variance

The discussion in the remaining parts follows the references [1, 2]. To assess the quality of an estimate, one can define, the *bias* and the *variance*. These quantities will provide the optimal bandwidth to minimise the error of estimation. Be definition an estimator is said to be *unbiased* if the expectation value of the estimator equals the quantity that it is supposed to estimate. Otherwise it has a bias which is defined as the difference between the averaged values of the estimator and the quantity it is supposed to estimate. This means that the kernel density estimator is unbiased if the bandwidth is zero. For a finite bandwidth it has the finite bias. The *bias* is defined as

$$bias\left\{\hat{f}_h(x)\right\} = \left\langle \hat{f}_h(x)\right\rangle - f(x) \tag{B.10}$$

where bracket is an ensemble averaging. By substituting Eq. (B.7) in Eq. (B.10) we find:

$$bias\{\hat{f}_h(x)\} = \left\langle \frac{1}{n}\sum_{i=1}^{n}\frac{1}{h}K\left(\frac{x-Q_i}{h}\right)\right\rangle - f(x) = \left\langle \frac{1}{h}K\left(\frac{x-Q}{h}\right)\right\rangle - f(x) \quad . \tag{B.11}$$

The averaging can be written in terms of exact PDF $f(x)$ as,

$$bias\{\hat{f}_h(x)\} = \int \frac{1}{h}K\left(\frac{x-y}{h}\right) f(y)dy - f(x) \quad . \tag{B.12}$$

Now changing the variables $s = \frac{y-x}{h}$, gives

$$bias\{\hat{f}_h(x)\} = \int K(s)f(x+hs)ds - f(x) \quad . \tag{B.13}$$

By expanding f up to second order in h, one finds

$$bias\left\{\hat{f}_n(x)\right\} = f(x)\int K(s)ds$$
$$+hf'(x)\int sK(s)ds + \frac{1}{2}h^2 f''(x)\int s^2K(s)ds + \cdots - f(x) \quad . \tag{B.14}$$

The first integral with normalization condition will be unity and second one vanishes because the kernel is an even function, therefore bias will be:

$$bias\left\{\hat{f}_h(x)\right\} = \hat{f}_h(x) - f(x) = \frac{1}{2}h^2 f''(x)\mu_2(k) + \mathcal{O}(h^3) \tag{B.15}$$

where $\mu_2(k) = \int s^2K(s)ds$. It is evident that bias grows quadratically with the bandwidth and depends linearly on the curvature of the true PDF.

Another important quantity that tells us how well we approach the averaged value, is the variance of the kernel density estimator:

$$var\left\{\hat{f}_h(x)\right\} = \left\langle \hat{f}_h(x)^2\right\rangle - \left\langle \hat{f}_h(x)\right\rangle^2 \quad . \tag{B.16}$$

Variance describes the fluctuations of each ensemble, has been chosen from realizations. For small bandwidth h the leading term will be

$$var\left\{\hat{f}_h(x)\right\} = \frac{1}{nh}f(x)||k||_2^2 \tag{B.17}$$

with $||k||_2^2 = \int k^2(x)dx$. Therefore a small bandwidth leads to a large variance and vice versa.

B.2 Bandwidth Selection

We have seen that a large bandwidth leads to a low variance, but also results in a large bias and vice versa. The optimal bandwidth therefore can be found as a tradeoff between two effects. To determine the optimal bandwidth, we consider a mean square error as:

$$MSE\left[\hat{f}_h(x)\right] = \left\langle \left(\hat{f}_h(x) - f(x)\right)^2 \right\rangle \quad . \tag{B.18}$$

It will be shown that for optimization of MSE, the only needed parameter is h. It is easy to expand MSE and find

$$MSE\left[\hat{f}_h(x)\right] = var\left[\hat{f}_h(x)\right] + \left[bias\left[\hat{f}_h(x)\right]\right]^2 \tag{B.19}$$

using Eqs. B.15 and B.17 we have:

$$MSE\left[\hat{f}_h(x)\right] = \frac{1}{nh} f(x)||k||_2^2 + \frac{h^4}{4} f''(x)\mu_2^2(k) \quad . \tag{B.20}$$

Because, the MSE depends on x, we integrate over state variable and from $\int f(x)dx = 1$, we find:

$$MISE\left\{\hat{f}_h\right\} = \int MSE\left\{\hat{f}_h(x)\right\} dx = \frac{1}{nh}||k||_2^2 + \mu_2^2(k)||f''(x)||_2^2 \frac{h^4}{4} \tag{B.21}$$

where MISE is mean integrated squared error and $||f''(x)||_2^2$ is the curvature of the distribution function integrated over x. Minimizing Eq. B.21 with respect to h yields:

$$h^* = \left(\frac{||k||_2^2}{n||f''||_2^2\mu_2^2(k)}\right)^{\frac{1}{5}} \tag{B.22}$$

here h^* is the optimal bandwidth that minimize the MISE. For example, for a Gaussian PDF with variance σ^2, one obtains:

$$||f''(x)||_2^2 = \frac{3}{8\sqrt{\pi}}\sigma^{-5} \simeq 0.212\sigma^{-5} \tag{B.23}$$

and

$$\left(\frac{||k||_2^2}{\mu_2^2(k)}\right)^{\frac{1}{5}} = \left(\frac{1}{4\pi}\right)^{\frac{1}{5}} = 0.776 \tag{B.24}$$

which results in

$$h^* \simeq 1.06\hat{\sigma} n^{-\frac{1}{5}} \tag{B.25}$$

here it is assumed the underlying process has Gaussian distribution with variance $\hat{\sigma}^2$. This is sometimes called the Gaussian reference rule or the Silverman's rule of thumb (SRT) [3] bandwidth. For Epanechnikov kernel one finds $h^* \simeq 1.045\hat{\sigma}n^{-\frac{1}{5}}$ is the optimized bandwidth. The estimated standard deviation $\hat{\sigma}$ is given by

$$\hat{\sigma} = \sqrt{\frac{1}{n-1}\sum_{i=1}^{n}(\langle Q\rangle - Q_i)^2} \quad . \tag{B.26}$$

B.3 Bivariate Kernel Density

To estimate Kramers–Moyal coefficients, one needs joint $\hat{f}(x_1, x_2)$ or conditional distribution function. Here we need two kernels, one localized at x_1, and another localized at x_2, therefore,

$$\hat{f}(x_1, x_2) = \frac{1}{nh_1h_2}\sum_{i=1}^{n} K_x(\frac{x_1 - Q_{1i}}{h_1})K_y(\frac{x_2 - Q_{2i}}{h_2}) \tag{B.27}$$

where the optimum bandwidth (in d-dimensions) will given by [1]

$$h_j^* \cong \hat{\sigma_j} n^{-\frac{1}{d+4}} \tag{B.28}$$

where $\hat{\sigma_j}$ is the standard deviation estimator of x_j.

B.4 Kernel Regression: Nadaraya–Watson Estimator

The most frequently used kernel-based estimator for conditional expectation values is the Nadaraya–Watson estimator. Let us calculate the following conditional average

$$\langle Y|X = x\rangle = \int yP(y|x)dy = \frac{\int yp(x, y)dy}{p(x)} \quad . \tag{B.29}$$

We can change the variables of Eq. B.27 by Q_1 and Q_2 to X and Y,

$$p(x, y) = \frac{1}{nh_xh_y}\sum_{i=1}^{n} K_x(\frac{x - X_i}{h_x})K_y(\frac{y - Y_i}{h_y}) \tag{B.30}$$

substituting Eq. B.30 into Eq. B.29 gives

$$\int yp(x,y)dy = \frac{1}{nh_x}\sum_{i=1}^{n} K_x(\frac{x-X_i}{h_x}) \int \frac{y}{h_y} K_y(\frac{y-Y_i}{h_y})dy \quad . \tag{B.31}$$

Define variable

$$s = \frac{y-Y_i}{h_y} \tag{B.32}$$

and expand the integral of the r.h.s. of Eq. (B.31) as

$$\int \frac{y}{h_y} K_y(\frac{y-Y_i}{h_y})dy = \int (h_y s + Y_i) K_y(s)ds = h_y \int s K_y(s)ds + Y_i \int K_y(s)ds = Y_i \tag{B.33}$$

where we used normalisable and symmetric properties of kernel. Finally, we obtain Nadaraya–Watson estimator:

$$\langle Y|X=x\rangle = \frac{\sum_{i=1}^{n} Y_i K_x\left(\frac{x-X_i}{h_x}\right)}{\sum_{i=1}^{n} K_x\left(\frac{x-X_i}{h_x}\right)} \quad . \tag{B.34}$$

If we assume Y to be an increment, then can estimate Kramers–Moyal conditional moments.

References

1 M. Köhler, A. Schindler, S. Sperlich, A review and comparison of bandwidth selection methods for kernel regression. Int. Stat. Rev. **82**, 243 (2014).
2 C. Honisch, *Analysis of Complex Systems: From Stochastic Time Series to Pattern Formation in Microscopic Fluidic Films* (Diss., University of Münster (Westfalen), 2014).
3 B.W. Silverman, *Density Estimation* (Chapman and Hall, London, 1986).

MIX
Papier aus verantwortungsvollen Quellen
Paper from responsible sources
FSC® C105338

If you have any concerns about our products,
you can contact us on
ProductSafety@springernature.com

In case Publisher is established outside the EU,
the EU authorized representative is:
Springer Nature Customer Service Center GmbH
Europaplatz 3, 69115 Heidelberg, Germany

Printed by Libri Plureos GmbH
in Hamburg, Germany